Dark Tourism

BUILDING THE FUTURE OF TOURISM

Series editor: Anukrati Sharma

The world is entering the Third Millennium in which great changes are expected in all areas of human interest, life, and activity. These changes have been brought on by past and present man-made events, which have had both positive and negative consequences. The coming millennium will be marked by significant social, political, demographic, and technological changes and will definitely differ from the last century. The future will bring more leisure time, a higher standard of living, and a better quality of life for us all. This series examines recent and the most probable changes and gives a wide range of visionary insights, as well as operational takeaways.

Forthcoming Volumes

Meaningful Tourism: Strategies and Futuristic Development
Pankaj Kumar Tyagi, Vipin Nadda, and Ajit Kumar Singh

Value Proposition for Tourism "Co-opetition": Cases and tools
Adriana Fumi Chim Miki and Rui Augusto da Costa

Dark Tourism: Theory, Interpretation and Attraction

EDITED BY

ANUKRATI SHARMA
University of Kota, India

SHRUTI ARORA
University of Kota, India

AND

PARAG SHUKLA
The Maharaja Sayajirao University of Baroda, India

emerald
PUBLISHING

United Kingdom – North America – Japan – India – Malaysia – China

Emerald Publishing Limited
Emerald Publishing, Floor 5, Northspring, 21-23 Wellington Street, Leeds LS1 4DL

First edition 2024

British Library Cataloguing in Publication Data
A catalogue record for this book is available from the British Library

ISBN: 978-1-83797-337-8 (Print)
ISBN: 978-1-83797-336-1 (Online)
ISBN: 978-1-83797-338-5 (Epub)

Printed and bound by CPI Group (UK) Ltd, Croydon, CR0 4YY

INVESTOR IN PEOPLE

To my daughter "Vidushi Choudhary," who is my heart living outside of my body.
To my son "Samarth Mehta," whose laughter is my favorite sound.
Last but not the least to our wonderful contributors and readers.

Contents

About the Editors

Dr Anukrati Sharma is the Associate Professor of the Department of Commerce and Management at the University of Kota (A State Govt. University) in Kota, Rajasthan, India. She is the Director of the Skill Development Center of the same University. She is also the Dean (Honorary) of Tourism and Hospitality at Vishwakarma Skills University, Jaipur (A Govt. State University). She worked three years as the Head of the Department of Commerce and Management, University of Kota. Her special interest areas are tourism, tourism marketing, strategic management, and international business management. A member of seven professional bodies, she has attended a number of national and international conferences and presented 45 papers. She has been invited as keynote speaker and panel member by different countries such as Sri Lanka, Uzbekistan, South Africa, Nepal, Malaysia, Indonesia, and Turkey. She has been invited as Visiting Professor at Kazakhstan and Uzbekistan.

Dr Shruti Arora is currently working as Guest Faculty in the Department of Commerce and Management, University of Kota, Kota, Rajasthan, India. She has 16 years of experience in the education industry. Her core subjects are Marketing, General Management, International Business, and Customer Relationship Management. She has attended various international conferences, and her publication includes various chapters in edited books under Routledge, Emerald, and Springer and research papers in national and international UGC-approved and peer-reviewed journals. She has also been appointed as an external and internal examiner for viva voce for graduation and postgraduation students.

Dr Parag Shukla holds a PhD degree in the subject of Commerce and Business Management with focus on Strategic Marketing in the area of "Retailing." He has been working as an Assistant Professor in the Faculty of Commerce at the Maharaja Sayajirao University of Baroda [MSUB], Gujarat, India, since the year 2009. He has presented and published a total number of 90 research papers in different areas of retailing and marketing in various national and international refereed peer-reviewed journals and conference proceedings. He has also received accolades for his research papers from various academics.

About the Contributors

Shakil Ahmed is currently studying at Forestry and Wood Technology Discipline of Khulna University, Bangladesh, as a postgraduate student. He intends to pursue a career in wellness and tourism. His passion is to help the local community by ensuring sustainable forest management and nature-based tourism. Participatory forest management, nature-based tourism, and wildlife management are some of his research interests. He enjoys participating in voluntary work and has volunteered for a number of organizations, including Bangladesh Youth Initiative (BYI), Forestry and Wood Technology Club, and Bangladesh Tourism Board.

Mr Terrance Ancheary is a doctoral student of Tourism Management at the Indian Institute of Management, Sirmaur. He navigates the realms of academia, seamlessly blending passion and scholarly rigor. With an insatiable curiosity, he did his Master's of Science in Hospitality Administration from National Council for Hotel Management, Noida. Terrance envisions a future where his insights shape the trajectory of hospitality and tourism, leaving an indelible mark on the industry's sustainable and transformative growth.

Ajoy K. Bhattacharya is a retired Indian Forest Service (IFS) officer of the 1983 batch from Madhya Pradesh cadre has a penchant for forestry, environment, ecotourism, and bamboo. He was the founder and CEO of the Madhya Pradesh Ecotourism Development Mission and Board (first in India). Through his pioneering initiative, MP Forest Department set benchmarks in ecotourism, promoted public private partnership model, and institutionalized bamboo-based entrepreneurship. He is a consultant with TERI, IIFM, and other specialized organizations. Armed with a research degree and a postdoctoral, worked (December 2015 to February 2019) with the Ministry of Road Transport and Highways, Government of India as the Country Head for the Green Highways Program; which ensured greening of the highways across the country. Currently, he is on "Mission Giving Back" and working for special projects including mobilizing youth for "Mission Environment Movement" and supporting many organizations for the cause of environment, entrepreneurship, and green solutions.

Aslı Bendenay Çapa graduated from Anadolu University, Faculty of Tourism, Department of Tour Guiding in 2019. Since then, she has been working as a Professional Tourist Guide under the Ministry of Culture and Tourism in Turkey.

Between 2019 and 2022, she completed his Master's degree in the field of Tourism Management at Anadolu University Social Sciences Institute with a study on enriching the narratives of tourist guides with geo-literary reading. Since September 2022, she has been continuing her doctorate education at Anadolu University, Institution of Graduate Schools. Her main areas of interest are tourism geography, geo-literary reading, and topics related to tourist guidance.

Michael Clancy is a Professor of Politics at the University of Hartford, Connecticut, USA. His research revolves around the political economy of global tourism. He is the author of two books and the editor of a third. In addition, he has published numerous journal articles and book chapters on tourism and development, international sex tourism, cruise ship tourism, tourism and national identity, slow tourism, and over-tourism.

Ankit Dhiraj, currently working as a Senior Research Fellow at the School of Hotel Management and Tourism, Lovely Professional University, Punjab, India. He has completed his regular three-year Bachelor's degree in Business Management from Magadh University Bodhgaya, Master's in Tourism Management from Alagappa University Tamilnadu, Master's in Political Science in Swami Vivekanand Subharti University, Meerut and pursuing PhD in Tourism Management at the School of Hotel Management and Tourism, Lovely Professional University, Punjab, India. He qualified JRF in Tourism Management and NET in Political Science. His area of research interest is Buddhist tourism, wellness tourism, sustainable practices, and green economy. He has experience working with different universities in India as a Business Development Associate. Ankit Dhiraj has also participated in various national and international conferences, seminars, and research workshops. He has published more than 10 research papers in Scopus-indexed, UGC-approved, and peer-reviewed journals.

Freeda Maria Swarna M. has over 15 years of experience in the sustainability, conservation science, environment, communities, and tourism domains. Her strength lies in combining Applied Sciences and the integration of sound research with a robust implementation. As Director, Dharthi NGO was nominated to the Karnataka Ecotourism Development Board as a member of the General Body, where during her three-year tenure, she was instrumental in conducting cumulative impact assessment (CIA) studies, having consultations with stakeholders for a robust ecotourism policy for the State of Karnataka. Freeda is in the process of authoring two books, one on ecotourism, and has published articles and papers to her credit. With a penchant for writing, she is in the process of putting up her blogsite called https://highwaymybeckoning.com which will contain impactful stories of her travels, people, and places. She also runs Dharthi Green Project which focuses on the dynamics of environment and ecology and works with consulting firms that interface the corporates of India. Besides two Master's degrees in Tourism, Environment and Ecology, she also has a research degree.

Dr Ruchi Garg is an Assistant Professor in the Faculty of Commerce and Management at Nirwan University, Jaipur. She holds an MPhil and PhD in Accountancy and Business Statistics from the Department of ABST, University of Rajasthan, Jaipur. Her academic journey is marked by consistent excellence, maintaining a track record of first-class distinctions. Dr Ruchi is an accomplished Research Scholar with significant contributions to esteemed journals through her published papers and articles.

Iti Gaur is an academician, an author, and a public speaker with 12 years of teaching and 2 years of industrial experience. She has written three books and multiple research papers so far, which touch different areas related to commerce and management. Currently, she is working with Manipal University, Jaipur, India, and apart from her academic journey, she has also anchored multiple talk shows on All India Radio. She tries to find out those academic areas which are not explored by many and therefore would definitely bring a lot of untouched content for her learners in future.

Semra Günay works as a full-time Academician and full-time Professor in the Department of Tourism Management at Anadolu University Faculty of Tourism. She graduated from the Geography Department of Ankara University. She teaches and conducts research on topics such as tourism planning, geographical information systems, gastronomy and tourism, sustainable tourism, nature-based tourism activities, tourism geography, food geography, and urban tourism. She has worked and published in many national and international projects related to tourism. She worked as a Researcher in the Eskişehir Province Language Atlas project supported by TÜBİTAK. She has undertaken the management of many projects such as "Spatial Analysis and Marketing Strategy of Tourism Resources of Eskişehir Province" and "Tourist Guides Designing Tour to Phrygia Region (route design with CBS)" supported by BEBKA (Bursa Eskişehir Bilecik Development Agency).

Max Hart, an Early Career Researcher and Lecturer in Criminology at Birmingham City University, where he has taught since 2021 on a range of topics such as crime and the media, cybercrime, and investigative techniques. Max's research interests are concerned with violence, media, and the impact of technological advancements within capitalist societies.

Rami K. Isaac, born in Palestine. Rami Isaac did his undergraduate studies in the Netherlands, graduate studies in the United Kingdom, and earned his PhD from the University of Groningen, in Spatial Sciences, in the Netherlands. He is currently an Associate Professor in tourism teaching at the undergraduate as well as postgraduate levels at the Academy for Tourism at Breda University of Applied Sciences in the Netherlands. In addition, he is An Associate Professor at the Faculty of Tourism and Hotel Management at Bethlehem University, Palestine. Currently, he is the Vice President of the Research Committee 50 on International Tourism, International Sociologist Association ISA (2018–2024). His research interests are in the area of tourism development and management, critical theory, and political aspects of tourism. He published numerous articles and book chapters on tourism and political (in)stability, occupation, tourism and war, dark/heritage tourism, violence, and transformational tourism.

Md. Wasiul Islam is a Professor of Forestry and Wood Technology Discipline of Khulna University, Bangladesh. He has completed his PhD from the Business School (Tourism Cluster) of the University of Queensland, Australia. He completed his MSc in Master of Science in Forest and Nature Conservation (Minor in Leisure, Tourism, and Environment) from the Wageningen University and Research Center, the Netherlands. He also worked in Bangladesh Forest Department where he was involved in several research projects on nature-based tourism and wildlife management. His research interests include participatory protected area management, governance, social learning, nature conservation, and nature-based tourism.

Zhang Jieyao is a Senior Lecturer in the School of Management and Economics, Chuxiong Normal University, China. Her doctorate is from the Taylor's University in Hospitality and Tourism with research on Pro-environmental Behavioral Intention toward Ecotourism in China. She has two postgraduate degrees: Master's in International Hospitality Management from the Taylor's University and Hospitality Management from University of Toulouse, France. Her research areas include natural tourism, ecotourism, and behavioral studies.

Craig Kelly, Lecturer in Criminology at Birmingham City University, where he has taught since 2018, covering topics such as dark leisure and research methods. He has published research in the areas of violence and organized crime. In recent years, he has turned his attention toward the harms of consumerism, editing books on the topics of video games and dark tourism.

Shaheed Khan is academically qualified and armed with postgraduate degrees in History, Tourism, Business Administration, Law, and Research degrees. His focus areas have been Tourism, Mergers and Acquisitions, and Human Resources. He also has a micro-specialization in Strategy Management from the Indian Institute of Management Kozhikode and a Corporate Program in Leadership Management from the Harvard School of Publishing. He brings to the table multiple facets of functioning that supports any area of academics and practice in the corporate world. Having worked in the government, private, and social sectors, Shaheed has a broad understanding of Tourism, Travel, and Hospitality (TTH). Shaheed is the Head, Research and Training, with Dharthi NGO, Bangalore. He is also an Adjunct Faculty with Leading Central and State Universities and Deemed to be Universities and engages in in-class sessions for corporates of India in the domain of Strategy, M&A, Tourism, Design Thinking, and Human Resources. He also engages in training the faculty members of higher education institutions, Malaviya Mission Teacher Training Center (MMTTC) (earlier knowns as UGC-HRDC).

Burhan Kılıç, completed his undergraduate education at Gazi University Faculty of Commerce and Tourism Education, his Master's degree at Gazi University Institute of Social Sciences, Department of Tourism Management Education, and his doctorate degree at Gazi University Institute of Educational Sciences, Department of Tourism Management Education. In 2023, he became a Professor in the field of tourism. He teaches courses such as tourism marketing, consumer behavior in tourism, food and beverage management, scientific research methods, and has many national and international articles, papers, and books on basic and subcomponents of these topics such as loyalty in tourism, service quality, social media marketing, leisure tourism, and consumer attitudes and behaviors.

Janitha Kularajasingam is a Senior Lecturer in Taylor's University under the School of Management and Marketing. She has nine years of teaching experience. Furthermore, she has industrial experience in large IT organizations in Malaysia. Recently, she obtained Microsoft Innovative Educator Expert 2021–2022. She is a Reviewer in the *Journal of Applied Research in Higher Education* (Web of Science) (ESCI). Janitha has a postgraduate degree in Management and has published articles in the *Journal of Education for Business* in Scopus and other high-indexed journals. She has regularly presented several papers at national conferences. She is an active committee member of the ICIE Conference 2019–2023. She was also a conference moderator for the parallel sessions. In addition, she was also actively involved in the Enterprise Project team (Entrepreneurship Eco-System Development and Project Enterprise) at the university. She is a co-supervisor for PhD students and supervises Master's-level students.

Jeetesh Kumar is the Head (a) of Research at the Faculty of Social Sciences and Leisure Management, a Senior Lecturer at the School of Hospitality, Tourism and Events, an Associate Director for Information Management and Documentation at the Centre for Research and Innovation in Tourism (CRiT), and a Hub Leader of the "Responsible Tourism for Inclusive Economic Growth" Sustainable Tourism Impact Lab at Taylor's University, Malaysia. His doctorate is from Taylor's University in Hospitality and Tourism. He has two postgraduate degrees: a Professional Master's in Hospitality Management and International Tourism from the University of Toulouse, France, and the other in Business Administration (MBA – Marketing) from Hamdard University, Pakistan. His research areas include economic impacts, economic modelling, sustainable tourism development, SMART tourism and behavioural studies. He has worked on consultancies and research projects at the national/international level and authored 80+ publications, including indexed research articles and book chapters. 12 postgraduate students (Master's and PhDs) have graduated under his supervision, and currently, he is supervising 7 PhD scholars. He is also active in several national and international associations, conferences, and journals. He is also an Associate Editor for the *Asia-Pacific Journal of Innovation in Hospitality and Tourism* (APJIHT) and Regional Editor (Asia) for the *Event Management Journal*. Jeetesh has coedited four books under CAB International, IGI, and Emerald and is currently editing five books with Emerald, IGI, and Springer. He is an active speaker and has been invited by several institutions in the Philippines, India, Japan, Indonesia, China, Australia, and Pakistan.

Sanjeev Kumar is working as a Professor in the School of Hotel Management and Tourism at Lovely Professional University, Punjab, India. He obtained his PhD in Hospitality from Amity University Rajasthan (2015–2018). He started his academic career as a Lecturer in Hotel Management at the Institute of Hotel Hospitality Industry Training Center in Hamirpur, Himachal Pradesh. Before teaching, he worked in India and abroad with various renowned hotels. He has published more than 25 research papers, articles, and chapters in Scopus-indexed, UGC-approved, and peer-reviewed journals and books. Dr Sanjeev Kumar participated and acted as a resource person in various national and international conferences, seminars, research workshops, and industry talks. His area of research interest is F&B service, alcoholic beverages, event management, and sustainable practices.

Hande Akyurt Kurnaz, born in 1984 in Edirne. In 2005, she graduated from Ankara University Beypazarı Vocational High School Tourism Guidance associate degree program; in 2009, she graduated from Gazi University Faculty of Commerce and Tourism Education – Travel Management and Tourism Guidance Teaching. She completed his Master's degree at Muğla Sıtkı Koçman University Institute of Social Sciences, Department of Tourism Management in 2013 and her doctorate at Adnan Menderes University, Department of Tourism Management in 2018. She worked as a Research Assistant at Muğla Sıtkı Koçman University Institute of Social Sciences between 2010 and 2012 and as a Lecturer at Muğla Sıtkı Koçman University Marmaris Tourism Vocational School Tourism Guidance program between 2012 and 2019. She is currently working as an Associate Professor at Bolu Abant İzzet Baysal University, Faculty of Tourism, Department of Tourism Guidance.

Adam Lynes, an Associate Professor in Criminology at Birmingham City University, where he has taught since 2012, covering topics such as criminological theory, homicide, and transnational organized and corporate crime. He has published research on such areas as serial murder, family annihilation, organized crime, video games, and crime and dark tourism. He is currently in the process of completing a book on the crimes of the powerful for Policy Press.

Mr Puneet Mehta is a Research Scholar at Central University Kerala who explores the intricate dynamics of tourist behavior and technology. Puneet's academic focus centers on the intersection of psychology and tourist behavior, examining technology's impact on travel experiences. Puneet aspires to contribute significantly to scholarly comprehension of tourist behavior within the evolving technological landscape, heralding the start of a promising academic trajectory.

Neelima Modi is a Guest Faculty in the Department of Commerce and Management at the University of Kota, Kota, India. She has done BTech, MBA, and PhD in Business Administration. She was a gold medalist and college topper in her graduation and postgraduation. She has published and presented several research papers at national and international conferences. She was awarded best paper presentation at the international conference on the latest advancements and future trends in commerce and management, by Inspira Research Association, (2020) and Best Paper Presentation at International Conference on Prospects and Challenges Faced by Industries in 2020, Poornima University (2020).

Ravi Kant Modi is currently working as a Professor and Dean, Nirwan University Jaipur. He is a distinguished faculty member bringing with him an extensive 13 years of teaching experience. His contribution in research solidifies his position as a highly accomplished Researcher and Scholar. Dr Modi is an active member of several professional bodies, staying connected with the latest developments in his field. He has organized numerous online and offline conferences, seminars, FDP, workshops as Conference Secretary, showcasing his commitment to advancing knowledge and fostering intellectual exchange. As an author, Dr Modi has contributed to the academic community through his textbooks and edited books. Moreover, he has published over 40 papers in renowned national and international journals, including notable Scopus journals. His expertise in research is further exemplified by his position as an editorial board member for reputable journals. Dr Modi is an avid participant in academic conferences and seminars, having attended more than 50 events where he has presented his researches. His comprehensive knowledge and rich academic background make him an invaluable and a dedicated mentor to the students.

Mr Anish Mondal is a dedicated and passionate individual who is currently pursuing a Master's from NCHM IH Noida and a Master's in Tourism and Travel Management (MTTM) from IGNOU simultaneously. He completed his graduation from IHM Mumbai. He has a strong understanding of the industry's nuances and a keen interest in exploring and contributing to its ever-evolving landscape.

Dr Noëlle O'Connor is a Senior University Lecturer, Educational Developer and Researcher at the Technological University of the Shannon (Ireland). She has lectured for many years in other Irish higher education institutes including Technological University Dublin. Noëlle is on the Editorial Review Board for many of the leading tourism journals. Her research focuses on travel and tourism management, education for sustainable development, and gender equality within higher education. Noëlle has presented her research at many global conferences and in many referred journals.

Sinead O'Leary is a Lecturer in Tourism and Event Management at the Technological University of the Shannon (TUS) (Ireland). She has worked for many years as a Lecturer and a Researcher in the Irish Higher Education sector, currently in the Department of Hospitality, Tourism and Wellness in TUS. Sinéad has supervised student research at undergraduate and postgraduate levels and is actively involved with industry partners across the travel, tourism, event, and beauty industries.

Hanh Thi Pham is working as a Lecturer in Vietnam Women's Academy and Head of Tourism Department. At the same time, she's an Invited Lecturer at the University of Economics and Business, Vietnam National University, Hanoi; field Economics and Business Administration. She has a number of publications in reputed journals. Her research interest areas are tourism, dark tourism, and ecotourism.

Md. Mahfuz Ur Rahman is currently working as an intern in the USAID Ecotourism Activity supported by Bangladesh Ecotourism and Conservation

Alliance (an initiative of the USAID-Bangladesh). He did his BSc (Hon's) in Forestry from Khulna University, Bangladesh and Master's in Agroforestry from Bangladesh Agricultural University, Bangladesh. Very recently, he has been admitted at University of Padova, Italy, to continue his higher studies in Forest Science (MSc). His preferred field of research interests is the human dimensions of natural resource management, tourism, and protected area management. He prefers working with both qualitative and quantitative data.

Suduweli Kondage Lathisha Jayangi Ramanayaka I am a PhD scholar specializing in the English Language and an enthusiastic Lecturer-in Charge of the Language Unit/Assistant Lecturer in English at Sri Lanka Institute of Tourism & Hotel Management (SLITHM), driven by a deep passion for creating dynamic and interactive learning environments within the lecture hall. My commitment to inspiring and motivating students is evidenced by consistent year-on-year improvements in academic scores. I possess strong competencies in managing research projects and consistently meeting stipulated timelines and am recognized as a proficient academic writer with a track record of publishing multiple research articles annually. As an ardent multidisciplinary researcher, I have already produced 23 remarkable research papers (03 Award-Winning Research Projects on English Language (01), Tourism and Women Empowerment (01), and Tourism Entrepreneurship (01); a research paper on Rural Tourism was presented in person at the International Research Forum, 14th Asia Tourism Forum in Indonesia on 07.05.2024, 08 research papers on the English Language, English Language Teaching, and Linguistics; 02 research papers on Education; 08 research papers on Tourism and Hospitality; 01 research paper on Buddhist Jurisprudence: Community Restorative Justice and currently have 11 ongoing research projects on diverse diciplinaries. My expertise extends to setting rigorous academic standards, crafting innovative curricula, and ensuring quality assurance at a professional level. Furthermore, I am a reviewer of research articles. Additionally, I hold the prestigious title of Distinguished Toastmaster and maintain active membership in 10 hybrid US Toastmasters Clubs. In the Toastmasters community, I serve as the District 62 Area C13 Director, and my versatile roles include that of a Motivational Speaker, Panelist, Moderator, and Compere. I take pride in being an Award-Winning Speaker at Toastmasters Speech Contests and am deeply committed to mentoring and coaching individuals in the art of public speaking. I also serve as a Visiting English Lecturer and embrace the role of a Servant Leader, driven by a genuine love for serving others above all else. I am an enthusiastic social worker who provides services at elder care homes, children's homes, and for animals, including stray dogs. Above all, I am a devoted follower of the omnipotent, omnipresent, and omniscient Mother Goddess Kali.

Divya Rani, currently working as PGT in UUMV Bhairwa, Madhubani, Bihar, India. She has completed a Bachelor of Economics and Master of Economics from Magadh University Bodhgaya, Bihar, India. She has completed a Bachelor of Education from GNDU in Punjab, India. Her area of research interest is tourism impact, industry, wellness tourism, and gender inequality. Divya Rani has also participated in various national and international conferences, seminars, and research workshops. She has published more than five research papers in Scopus Indexed, UGC Approved, and peer-reviewed Journals.

Samik Ray, National Award Winner Guide (IITG since 1984), Govt. of India, received a Certificate of Excellence for Overseas Adventure Travel, South ASIA (USA). Registered trainer of skill India, guide training courses, and Tourism skill development training programs of Govt. of India and West Bengal, taught tourism and folklore at universities. Ray is a Visiting Faculty in the Department of Tourism Management (Ramkrishna Mission Vidyamandira and Swami Vivekananda Institute of Modern Science), advisor to the Board of Studies (School of Hospitality and Culinary Art, The Neotia University, West Bengal), and delivered lecture at seminars; webinars, workshops, chaired sessions at national and international seminars, and webinars on tourism, and edited the journal *Tourism: Theory and Practice* and published articles and chapters in internationally acclaimed academic journals and edited volumes.

Birendra Kishore Roy is a dedicated hotelier and academic with extensive experience in the hospitality industry and academia. He currently serves as the Training Manager in the Learning and Development Department at Hotel Suryagarh Palace in Jaisalmer, Rajasthan. With over 11 years of employment, including 4.5 years in the industry and 7 years in academic, teaching, training, and mentoring roles, Mr Roy has mentored over 500 students now employed by prestigious brands. He holds a Master's in Tourism and Travel Management from IGNOU and an MBA in Human Resources Management from Suresh Gyan Vihar University, alongside a BSc in Hospitality and Hotel Administration from IGNOU and NCHMCT. An enthusiastic researcher in tourism, hospitality, hotel management, and human resource management, Mr Roy has published four journal articles, and seven book chapters, and presented four research papers at conferences. He has completed 14 FDPs, published two patents, and holds eight prestigious certifications in Learning and Development. His work is recognized in both national and international peer-reviewed publications.

Nidhi Sharma is a Marketing and General Management Professional. Currently, she is working as an Assistant Professor at the Institute of Management Studies, Ghaziabad, and undertaking papers/subjects in the area entailing marketing, organization behavior, and business communication and advertising management. Along with this, she is also pursuing a PhD in Marketing from Sharda University, Greater Noida. She is the Faculty Coordinator of the BIG IDEAS Club. In terms of her corporate experience, she has completed her previous job assignments with Accord International, an Event management company, in New Delhi. Having experience of Academic more than 8 years. She has published various research papers in international and national journals of high repute and also presented her research work at various conferences across the globe. She has acquired a plethora of certifications in the area of marketing and research from prestigious institutions. She has a successful track record professionally and is adjudged for accomplishing multifold milestones in her carrier.

Dr Shruti Sharma is currently an Assistant Professor in the Faculty of School of Management, IMS Ghaziabad (University Courses Campus), India. Dr Sharma has worked as an Internal Trainer and Teacher in the management arena. Her doctorate degree is in commerce and management from the University of Kota, Kota Rajasthan, and she completed her dissertation research on the topic of "Managing Stress to Improve Efficiency & Motivation: A Case Study on an Academic Performance of Coaching Faculties in Kota City." She has two post-graduate degree specialties – one in Business Administration (Master's of Commerce) and another is MBA in Marketing and HR. With over 8 years of academic experience under her belt, she brings a wealth of knowledge and expertise to the table. She has authored numerous research papers in highly respected international and national journals, showcasing her expertise in the field. Additionally, she has had the privilege of sharing her research findings at prestigious conferences worldwide. Her commitment to advancing knowledge is evident through her acquisition of numerous certifications in marketing and research from esteemed institutions. Professionally, she has consistently achieved remarkable milestones, demonstrating her exceptional track record and garnering recognition for her accomplishments.

Shruti Sharma, PhD, Assistant Professor with a demonstrated history of working in education management. Skilled in lecturing, counseling, administration, mentoring, and management. Strong education professional with a PhD in Management focused on stress management and coping skills from the University of Kota, Kota.

Surbhi Sharma, PhD, is a Research Scholar at Banasthali Vidyapith, Jaipur, India. Her research interests include women empowerment, mass media, television, the portrayal of women in media, the formation of progressive female position in society, and the impact of media on the daily life of rural women.

Priya Sodani (BCom, MBA(IB), PhD). Dr Sodani has done her graduation from Mumbai University, Mumbai, and MBA(IB) and PhD from University of Kota, Kota. She was the gold medalist in Master's in International Business in 2009, University of Kota, Kota. Currently, she is working as a Guest Faculty, Department of Commerce and Management, University of Kota, Kota, and also working as an Assistant Professor in Modi Institute of Management and Technology, Kota, and also worked as a Guest Faculty of Business Administration in J.D.B. Govt. Commerce Girls College, Kota. She has work experience in the corporate field for 3 years and education field for 11 years. She is a member of the Indian Accounting Association, Kota Branch. She has contributed 11 research articles to various national and international repute journals. She has presented eight papers in national and international conferences and seminars. Dr Sodani is the author of "Indian women in Economic World," published in the year 2018. She has participated in many national and international webinars. She is an active member of social groups – Bharat Vikas Parishad, Rotary Padmini, Upasana Welfare society.

Seda Sökmen earned her academic degree from Akdeniz University and currently holds the position of Assistant Professor in the Department of Tourism Management at Anadolu University. Her academic journey includes a Master's thesis focusing on the analysis of visitors to postmodern museums. Currently pursuing a PhD, her research endeavors delve into the intricate examination of the influence of social identity, host–guest encounters, and interactions among tourists on travel experiences, approached through a social psychological lens. Her primary areas of scholarly interest encompass social psychology and museology, with a specific emphasis on understanding tourist experiences.

Neharshi Srivastava is currently working as an Assistant Professor, Amity University, Rajasthan. She is a gold medalist of Lucknow University and RCI licensed Rehabilitation Psychologist. Dr Srivastava has 10 years of counseling, research and industry experience. She is a trained CBT therapist from London. Her areas of expertise are mindfulness and positive and yogic psychology based intervention. She has more than 34 publications in international journals of Scopus/Web of Science/UGC recognized/UGC Care Listed/peer review and presented more than 32 papers in national and international conferences, contributed nine chapters in edited books and authored five books on different domains of psychology. She received "Best Young Faculty Award," "Dhruv Ratan Award," and "Jan Shakti Sema Samman" for outstanding contribution in the field of psychology. She has been invited as a Guest Speaker and Resource Person in various reputed national/international universities, news channels, and organizations like Vigyan Prasar, Department of Science and Technology, NISD, Narcotics Control Bureau, Government of India. She has supervised more than 25 Master's and Bachelor's psychology students and currently supervises four PhD scholars doing their PhD under her supervision. She is also a core committee member of Indian Academy of Health Psychology. Apart from her academics, she has around 14 years' experience in the field of electronic media. She is a Radio Jockey of AIR FM Rainbow 100.7 Mhz. Lucknow.

Beth Storey has worked for many years in the hospitality and tourism sector and has recently graduated from the Technological University of the Shannon (TUS) (Ireland) BA(Hons) Business Studies with Travel and Tourism Management. She has previously worked as an Event Coordinator for heritage events, such as the "Ború 2014" celebrations in Killaloe, Co Clare, and is currently a Tour Guide at the Irish Workhouse Centre in Portumna, Co Galway, where she is part of a team responsible for exhibition design and for creating educational resources. She is currently a PhD student at the Technological University of the Shannon.

Preface

Over the past 10 years, dark tourism has become more and more popular as individuals look for deeper travel experiences and decide to genuinely deal with our more difficult heritage instead of just going on escapist holidays. A commemorative light on the epidemic that has engulfed society can be achieved through dark tourism. An umbrella concept for travel to sites that are associated for visits to locations like prisons, slave forts, and battlefields that are connected to agony and death. The Dark Tourism edited volume is relevant to students, scholars, and academicians associated with the fields of memorialization and morality death studies, history, cultural studies, psychology, business management, museology, and heritage tourism studies, as well as religious studies. The world over, dark tourism has grown in popularity in the 21st century. Travelers' evolving tastes and preferences, together with the emergence of the notion of special interest tourism, have compelled tourism regulators globally to include this market in their mission and vision statements. The people have experienced a number of terrible catastrophes, ranging from widespread natural disasters to genocide and war. The locations of these catastrophes are now popular travel destinations for dark tourism worldwide. The reason for the surge in gloomy tourism is that there are more tragedies worldwide. The use of social media has also played a significant role in raising awareness of this kind of travel. This book will focus on war tourism, disaster tourism, banditry tourism, macabre and supernatural tourism, and the abuses and ethical problems surrounding dark tourism as well.

Creating this edited book was primarily aimed at spreading knowledge about the "Dark Tourism: Theory, Interpretation and Attraction."

This book is the conclusion of the meticulous efforts of many minds from India, Sri Lanka, Turkey, Malaysia, Ireland, USA, Netherland, China, Vietnam, England, and Bangladesh. We wish to thank all the contributors for their hard work.

Dr Anukrati Sharma
Dr Shruti Arora
Dr Parag Shukla

Chapter 1

Unveiling the Significance of Irish Workhouses as Captivating Tourism Destinations

Beth Storey, Sinead O'Leary and Noëlle O'Connor

Technological University of the Shannon, Ireland

Abstract

Dark tourism has grown as an area of study since the 1990s with a great deal of research carried out into visitor motivations in recent years. Visitor centres within the dark tourism sector have a balancing act between the education and entertainment elements of tourism and sensitivity to the difficult stories they tell as part of their remit. This study examines the importance of tourism to the survival of Ireland's remaining workhouse buildings and the role the workhouse museums play in the local community. Research found that the visitor demographics and responses correlated with other studies, with a similar spread of male and female respondents to the survey, similar reasons for visiting and many of the same emotional responses.

The workhouses are very large buildings spread over several acres of land and therefore can be difficult and expensive to maintain. Community is very important to the management at Donaghmore and Portumna, and both management teams are open to the use of the buildings for events and education. They are also aware of their place in the local communities, supporting other businesses such as accommodation and restaurants.

This chapter shows that not only are Ireland's workhouses an important part of the history of the country, but they also have a role in modern society as places to preserve and teach that history. Tourism plays an important role in the preservation of the remaining workhouses and in educating the public about a difficult time in Ireland's history.

Keywords: Irish workhouse museums; dark tourism; Ireland; visitor motivations; community engagement

Dark Tourism, 1–12

Published under exclusive licence by Emerald Publishing Limited
doi:10.1108/978-1-83797-336-120241001

Introduction

This study investigates the importance of tourism to the survival of the Irish workhouses. Through primary and secondary research, it investigates the place of tourism in the preservation of the buildings and its place in telling visitors about the history of the workhouses and making the buildings available for modern use. The aim of the research is to investigate and understand the place of workhouses as visitor destinations and the role of tourism in preserving the built heritage of the workhouses. To undertake this aim, the following steps were used:

- Provide an overview of the importance of remembering the history of the Irish workhouse system.
- Investigate the topic in relation to current trends in heritage tourism and dark tourism.
- Identify the key issues facing Irish workhouses and in using workhouse buildings for tourism.
- Undertake primary research to investigate the issues of using workhouse buildings for tourism purposes.
- Undertake primary research to investigate visitor motivations and responses to visiting the workhouse.
- Carry out data analysis of primary research findings.
- Draw conclusions from the collected data.

To carry out the above points, the topic was investigated using five main themes:

(1) Theme 1 investigated the main issues with using the workhouse buildings for tourism.
(2) Theme 2 considered the alternative uses the buildings had been put to during their lifespan and the effects of those uses on the structure.
(3) Theme 3 examined whether the need to be financially viable clashes with the other considerations of using the workhouse buildings as a tourist destination.
(4) Theme 4 investigated the role of the workhouse museums in the community.
(5) Theme 5 discussed the visitors from the viewpoint of the museum management.

The study investigates the role of tourism in the preservation of the workhouse buildings and how they have been adapted to accommodate visitors, but an important part of the study will examine visitor motivations and responses to the workhouse. It is the author's belief that without the interest of tourists, this aspect of Irish history will be lost outside of academic history circles, and that it is vital that this dark period of Irish history is remembered. Workhouse museums, where visitors can see for themselves the living conditions and hear the stories of the inhabitants from knowledgeable tour guides, have an important place in teaching Irish history, heritage, and the experiences that helped form Ireland today. There is a growing awareness of workhouses and their impact on Irish society but little in the area of tourism. This chapter aims to fill a gap in the research, providing insights into how tourism can support the preservation of workhouses and their history, and the place workhouse museums hold in the wider community.

The workhouse system was designed to industrialise the relief of poverty in Britain and Ireland. To obtain any help, destitute and starving men, women, and children were forced to leave their homes and enter a system that would strip them of their dignity and force them to work long hours for the bare minimum of food and covering. Families were split up, men, women, girls, and boys were housed in separate buildings and were only allowed to meet with permission from the staff (Higginbotham, 2012; O'Connor, 1995). The system lasted in Ireland from 1838 to 1921. Many of the buildings associated with this aspect of Irish history have been demolished or repurposed, and others have fallen into dereliction.

In 2011, the Irish Workhouse Centre in Portumna, Co Galway, opened to the public. It is dedicated to telling the story of workhouses in Ireland in a respectful and sensitive manner. Donaghmore Famine Workhouse Museum in Co Laois tells the story of the workhouse in the context of the Famine and is also an agricultural museum, and Carrickmacross Workhouse in Monaghan tells the story of Carrickmacross Workhouse during the famine. Workhouse museums, where visitors can see for themselves the living conditions and hear the stories of the inhabitants from knowledgeable tour guides, have an important place in teaching Irish history, heritage, and the experiences that helped form Ireland today.

Donaghmore Workhouse opened in 1853, one of the last workhouses to be built in Ireland, and it closed as a workhouse in 1886. In 1927, the buildings became the site of the Donaghmore Co-operative which established a creamery to assist local farmers produce and sell butter. By the following year, Donaghmore was producing 400,000 gallons of milk at 2s 5d per gallon. In 1968, Donaghmore Co-op amalgamated with other creameries to form Avonmore, and more recently, Avonmore amalgamated with Waterford Creamery to form Glanbia PLC. Glanbia is still active on the site (Donaghmore Famine Museum, 2023).

In 1988, a voluntary committee formed to renovate and interpret the two front buildings, which are the most complete on the site. The former girls' building houses a large collection of agricultural and farmhouse implements, and the former boys building has a very well-preserved boys' dormitory as well as the Master's House, where the workhouse Master lived and had his office. The museum opened in 1992 and is still run by a voluntary committee with staff from Community Employment, Tús, and Rural Social Schemes. The Donaghmore Workhouse building is now owned by Laois County Council.

Portumna Workhouse opened in 1852 and closed in 1922. The infirmary continued to function for a few years longer before finally closing. After the workhouse closed, the buildings were put to a variety of uses, including a knitting factory, vegetable cooperative, potato packing company, metal body factory and offices for Inland Waterways, the Office of Public Works, as well as a storage yard for Galway County Council roads department. The County Show was held in the grounds in the 1950s, but the buildings gradually fell into disuse from the 1970s onward. By 1992, the buildings were derelict and covered in ivy. Work began in 1999 to conserve the buildings, and they officially opened in 2011. Today, the ground floor of the girls' building, two floors of the women's building, and the dining hall are accessible to the public. There is a museum showing items from workhouses around the country and several permanent exhibitions, including work by sculptor Kieran Tuohy, and an exhibition about Polish humanitarian and explorer Pawel Strzelecki. Temporary exhibitions have included "1845 Memento Mori" by Paula Stokes, and an exhibition about Midwifery in the Workhouse. The Irish Workhouse Centre hosts regular conferences and seminars; the upstairs rooms of the girls' building are used for classes and courses. In 2022, the Irish Workhouse Centre hosted a play by *The Old Tuam Society* that was attended by the President and Mrs Higgins.

Literature Review

Interest in places of suffering and death dates back centuries (Lewis et al., 2021) but, although not new as a phenomenon, the study of dark tourism as a concept is more recent (Kunwar & Karki, 2019). The term, dark tourism, was coined in the 1990s, with the publication of a research paper on the topic by John Lennon and Malcolm Foley (Foley & Lennon, 1996). It was not the only attempt to understand the draw of death and suffering as a tourist activity. The same year, Seaton introduced the term thanatourism in relationship to visitors to battlefield sites (Seaton, 1996) and other researchers coined their own, including "morbid tourism" (Blom, 2000). Stone defined dark tourism as "visiting real or recreated places related with death, suffering, disgrace, or the macabre" (Stone, 2006). Lennon and Foley defined dark tourism as "phenomena which encompass the presentation of real and commodified death and disaster sites' (Foley and Lennon, 1996, p. 198). It is an umbrella term for classifying and understanding the places associated with death and dark emotions, as well as being used as an analytical instrument in conceptual research, rather than a clearly defined mode of tourism (Zerva, 2021). By focussing on death, Seaton's interpretation is much narrower than Lennon and Foley's definition (Light, 2017) making it more restrictive and concise while dark tourism can be symbolic and incorporate macabre feelings (Zerva, 2021). The simplest definition is that of a journalist in the Irish Times who states that they are "sites where man's inhumanity to man was perpetrated" (Boland, 2020). This definition shows the breadth of dark tourism in a way that many others don't, incorporating all aspects of the subject rather than a narrow focus on death and death sites. Boland writes of dark tourism as branding or marketing, suggesting that it is our interest that makes the site dark.

In tourism, it is the stories which are interesting to the visitor (Lussetyowati, 2015). Sharpley (2005) gives four types of dark tourism consumption – experience, where the consumption of the tourism product elicits feelings, thoughts, and emotions; play, the sharing of an experience where the site is a product to be used;

integration, where the experience is integrated into the sense of self; and classification. More recent research has shown that visitors to dark tourism sites are more often seeking knowledge and understanding of the events (Mangwane et al., 2019) which is no different to other types of culturally motivated tourist using their leisure time to gather new information and experiences, raising their cultural level, and satisfying cultural needs (Edwards, 2013). Sharpley questions whether there is a "dark" tourism sector at all, arguing that tourism motivated by fascination with death is rare and that the supply and demand of dark tourism sites is driven by other factors, suggesting that the many forms of dark tourism are alternative experiences (Sharpley, 2005). Tourism has a responsibility to engage visitors in powerful and transformative learning experiences (Packer & Ballantyne, 2016), and dark tourism certainly offers this.

A large part of tourism in Ireland is based around heritage that would be considered "dark," such as Kilmainham Gaol, Spike Island, Strokestown House Famine Museum, and the Titanic Museum in Cobh, to name some of the best known. Despite this, there is currently little research into dark tourism in Ireland, although the area is growing. Some recent studies include narrative construction at Kilmainham Gaol (Ogliari, 2022), a historical study of Glasnevin Cemetery (Callan, 2019) and an investigation into the revitalization of the Harland and Wolff shipyard in Belfast (Hodson, 2019).

Author Gillian O'Brien, in an Irish Times article, suggests that the term "dark tourism" implies a certain exploitation of people (Boland, 2020), and it is this view of dark tourism being exploitative that could make scholars in Ireland reluctant to study it. Within dark tourism, it is necessary to be wary of exploitation, and ethics must be the priority; such sites need to be unbiased and to ensure images and artifacts are appropriately explained.

One study of dark tourism in Ireland investigates it as a tourism product for the Wild Atlantic Way (McLoughlin & Reis, 2019). They discuss how authenticity is important to the modern tourist, and that this is something that purpose-built museums lack. The authenticity of a site, such as a prison, is regarded as an essential element of the visit, engendering empathy and sympathy (McLoughlin & Reis, 2019). Research into dark tourism in Northern Ireland found that there were generational differences in how the use of areas affected by the troubles were viewed, with younger generations believing that use should be made of the history and the older generations believing that they were too recent (Simone-Charteris et al., 2018).

It is impossible to discuss heritage as a tourist product without considering the tourists. A very simple division is between "general" tourists who consume culture as part of a holiday and "specific" tourists who travel purposefully to engage in cultural tourism (Galí-Espelt, 2012; Richards, 1996). By understanding tourist motivations, providers will be able to offer a product that meets the needs of all stakeholders. This can revitalize the host community, giving heritage buildings a new economic value (Kordej-De Villa and Šulc, 2021). Cultural visitor satisfaction is delivered through authenticity (Park et al., 2019), through sense of place (Ashworth, 2009), and through a combination of visitor-related and context-related factors (Falk, 2011 cited by Richards, 2018). In a 2019 study of the Irish Workhouse Centre, it was found that visitor's main motivation was to learn about the past, rather than to visit a place of suffering (Cunniffe, 2019), and that their reasons were very similar to visitors at other types of heritage sites.

Methodology

The study sought to understand the role of tourism in preserving the heritage, and particularly built heritage, of workhouses and the benefits of using the workhouses for tourism. The questions differed depending on whether the respondents were visitors to the workhouse or staff and were based on the research objectives outlined previously. Question 1 was investigated using an exit survey offered to visitors to the workhouse; questions 2–4 formed the basis of in-depth interviews with the workhouse management.

- *Research Objective 1:* What are the motivations of visitors to the workhouse, and what is their response to the visit?
- *Research Objective 2:* What are the main issues with using heritage buildings such as the workhouse for tourism?
- *Research Objective 3:* Does the need to be financially viable clash with other considerations (if any) of using the workhouse buildings as a tourist destination?
- *Research Objective 4:* What role, if any, do the workhouse museums play in the local community?

The research method used both a questionnaire and in-depth interviews. The intention was to gather both qualitative and quantitative data to allow a full understanding of the topic. Through the quantitative research, the intention was to understand emotional responses to visiting the workhouse and whether that differed with regards to demographics and visit intention. The qualitative research would enable in-depth understanding of how tourism affects the workhouse, as built heritage and as a dark heritage site. For this study, the reasons for choosing the mixed method approach were (i) triangulation, to find convergence and corroboration from the different methods, (ii) complementarity, seeking to enhance and elaborate upon findings, and (iii) to expand the enquiry and to access the breadth of knowledge held by professionals within the industry as well as the data from visitor surveys.

A questionnaire was offered to visitors to the Irish Workhouse Centre, to understand their motivations and responses to their visit. This was left at the reception desk, and visitors were asked to complete it at the end of their visit. They were made aware that the questionnaire was completely optional, and that it would only take a few minutes to complete. The research was carried out between October and early December 2022 during an extended season at the workhouse, meaning that there were fewer visitors than normal, and therefore, fewer respondents to the survey than would have been available had the research been carried out at high season. Most of the questions asked were closed, with boxes to tick, but there was also an opportunity for the respondent to add a comment about their visit.

In-depth, semi-structured interviews were carried out with management at Donaghmore Famine Museum and the Irish Workhouse Centre to gather qualitative data. The goals of these interviews are to understand the experiences and perspectives of stakeholders responsible for the preservation and promotion of these heritage sites and how tourism has affected the buildings, whether tourism is the best use for the sites and what they think the future could be for the buildings.

Results

The purpose of this research was to investigate the importance of tourism to the Irish workhouse museums. The intention of the primary research was to answer a combination of research objectives and themes and to place the research in context with the secondary research carried out previously. There were four main themes to the research: visitor motivation; the main issues with using the buildings, and how they are used; financial viability; and the role of the museums in the community. It was found that the visitor information from the limited surveys corresponded with findings from other studies, and the interviewees were in agreement that the workhouse museums play an important role in educating visitors about the history of the workhouse, preserving the buildings for future use and supporting the local communities.

Research Objective 1: Visitor Motivation

The first objective was to explore the motivations of visitors and how they responded to their visit. This was done through visitor surveys and was also investigated from the provider's perspective. It was found that most visitors already had some knowledge of, or some level of interest in, workhouses and their visits were intentional rather than spontaneous, although the research did find that two of the three visitors who only entered because they saw the sign were also interested in history, and one ticked the interested in the famine box too. There has been a great deal of research recently into visitor motivations at dark tourism sites, with Cunniffe (2019), in a master's study at the Irish Workhouse Centre, finding that the interest in visiting the workhouse is very similar to that at other heritage sites. The interview respondents were in agreement that the visitors are vital to the long-term survival of the workhouse buildings, with a consensus that, without the tourist, there would be no need to keep the buildings in repair. Many workhouses were converted into schools, hospitals, and county homes, but others, such as Kells, have been demolished entirely. In telling the story of the workhouse, in authentic surroundings, it is possible to keep the history alive and to teach visitors about this dark period in authentic surroundings that make the telling rawer and more visceral.

Research Objective 2: The Main Issues Using the Buildings

This investigated the main issues using the workhouse buildings as a tourism destination. The interview respondents identified several main areas of interest. The predominant issue was with the repair and maintenance of the buildings themselves. The workhouses were built to house hundreds of people and therefore are very large, with dormitories and work rooms spread across numerous buildings. Over the years, there have been alterations to suit industrial uses as well as the decay resulting from abandonment, and this has left many of the buildings dangerous to access and with the need for expensive repairs. The museums rely on grant aid for the work, but both museums agree that authenticity is vital, and that repairs and rebuilding is undertaken using modern best practice. This means that, in areas that are important to telling the story of the workhouse, the spaces are left raw and authentic, but it also creates the opportunity to use more modern materials where appropriate, such as the glass doors in the Irish Workhouse Centre Museum. It also allows for necessary modifications for modern use to be made without compromising the integrity of the buildings, such as toilet facilities and kitchens. These interventions are essential to the long-term viability of the spaces.

Research Objective 3: Financial Viability

While the main purpose of the museums is to educate visitors about the history of the workhouses, both museums rely on public funds, such as Pobal, or schemes such as the Community Employment Scheme to do so. This can mean the long-term survival of the museums is precarious, and therefore, management at both venues is conscious of the need to expand the usage of the museums. Both management teams are open to community use of the museums as venues for music, plays, and art exhibitions, while the Irish Workhouse Centre has built a reputation for seminars, conferences, and training courses. Both teams also discussed the possibility of improving the buildings to create a more modern event space. Interestingly, both identified the same area, the former dining hall, as the best place for this.

Research Objective 4: Role in Community

Both management teams are aware of their role in the local community, providing facilities for the use of the community, for example, as an arts venue for Portumna Shorelines Arts Festival. They are also aware of the role they play drawing visitors to the area where they also spend in local businesses. One of the managers stated in the interview that he prefers to take phone calls rather than online bookings as then he can give suggestions for other destinations in the area as well as make recommendations for accommodation and dining. Management at both venues indicated that they are interested in expanding their role in the community.

Conclusions

The rise of dark tourism as a field of study has spurred extensive research on visitor motivations in recent years. Within the dark tourism sector, visitor centres face the challenge of balancing education, entertainment, and sensitivity to the difficult stories they convey. This study focuses on the importance of tourism for the survival of Ireland's remaining workhouse buildings and the role played by workhouse museums within the local community. The research findings align with other studies, revealing similar visitor demographics, motivations, and emotional responses.

Workhouses are sprawling structures spread across many acres of land, making their maintenance difficult and costly. The management of workhouse museums recognise the significance of preserving these buildings as sites of historical interest while also ensuring their availability for local community use. Striking a balance between authenticity and the necessary modifications for modern purposes is crucial, and both museums studied diligently follow best practices, working with the existing spaces to the best of their abilities. The museums at Donaghmore and Portumna prioritise their connection with the community, willingly offering the use of the buildings for events and education. Additionally, they understand their role in supporting other local businesses, such as accommodations and restaurants.

Although this study was limited in scope, it highlights that Ireland's workhouses not only hold immense historical importance but also serve a modern societal role as preservers and educators of that history. Tourism plays a vital role in preserving the remaining workhouses and educating the public about a challenging period in Ireland's past.

Recommendations

This was a very small and limited study conducted off-season. One recommendation would be to allow more time for interviews and surveys and to expand the pool of interviewees. There are only three workhouse museums in Ireland, Portumna, Donaghmore, and Carrickmacross. The manager of Carrickmacross was not interviewed for this study, but it would have allowed a more complete study had there been time to do so. It might also be beneficial to interview the heritage officers for Counties Galway, Laois, and Monaghan. It would also give another dimension to the study as it could be argued that managers and directors of the museums are biased toward the use of workhouses as tourist destinations.

The survey was created before much secondary research had been done, and there was no time to pilot the study properly. This was done so that it was possible to access the last visitors to the workhouse, during an extended season. It led to some of the questions offering confusing answers that were difficult to interpret. If the questions were developed after the secondary research had been carried out, they could have been based on research and therefore better tailored to the study. Carrying out the survey during the season would give a larger respondent base, and the information could be extrapolated with greater certainty. Possible future areas to research could be to compare the approach to the workhouse museum in Ireland to that of the United Kingdom where they appear to be better established and supported, another could be to investigate the ethics of using workhouses as visitor attractions.

References

Ashworth, G. J. (2009). Heritage and economy. In R. Kitchin & N. Thrift (Eds.), *Encyclopedia of human geography* (pp. 104–108). Elsevier Science. Accessed on March 27, 2023.

Blom, T. (2000). Morbid tourism - a postmodern market niche with an example from Althorp. *Norsk Geografisk Tidsskrift - Norwegian Journal of Geography, 54*(1), 29–36. https://doi.org/10.1080/002919500423564.

Boland, R. (2020). Shining a light on Ireland's places of famine, death and rebellion. *Irish Times*. https://www.irishtimes.com/life-and-style/travel/shining-a-light-on-ireland-s-places-of-famine-death-and-rebellion-1.4387180. Accessed on November 13, 2022.

Callan, P. (2019). "This cemetery is a treacherous place". The appropriation of political, cultural and class ownership of Glasnevin Cemetery, 1832 to 1909. *Studi Irlandesi. A Journal of Irish Studies, 9*(9), 251–270. https://doi.org/10.13128/SIJIS-2239-3978-25516. Accessed on November 13, 2022.

Cunniffe, R. (2019). *Management of dark tourism sites: Examining visitor experiences to the Irish Workhouse Centre*. Masters Chapter. Athlone Institute of Technology.

Donaghmore Famine Museum. (2023). Home page. http://donaghmoremuseum.ie/index.html. Accessed on December 13, 2022.

Edwards, J. (2013). Managing heritage and cultural tourism resources. In R. Raj, K. Griffin, & N. Morpeth (Eds), *Cultural tourism CABI* (p. 13). https://doi.org/10.1079/9781845939236.0013. Accessed on March 29, 2023.

Foley, M., & Lennon, J. J. (1996). JFK and dark tourism: A fascination with assassination. *International Journal of Heritage Studies 2*(4), 198–211. https://doi.org/10.1080/13527259608722175.

Galí-Espelt, N. (2012). Identifying cultural tourism: A theoretical methodological proposal. *Journal of Heritage Tourism, 7*(1), 45–58.

Higginbotham, P. (2012). *The workhouse encyclopedia*. The History Press Ltd.

Hodson, P. (2019). Titanic struggle: Memory, heritage and shipyard deindustrialization in Belfast. *History Workshop Journal, 87*, 224–249. https://doi.org/10.1093/hwj/dbz003. Accessed on November 14, 2022.

Kordej-De Villa, Ž., & Šulc, I. (2021). Cultural heritage, tourism and the UN sustainable development goals: The case of Croatia. In M. B. Andreucci, A. Marvuglia, M. Baltov, & P. Hansen (Eds.), *Rethinking sustainability towards a regenerative economy, future city* (Vol. 15). Springer. https://doi.org/10.1007/978-3-030-71819-0_19. Accessed on October 17, 2022.

Kunwar, R., & Karki, N. (2019). Dark tourism: Understanding the concept and recognizing the values. *Journal of APF Command and Staff College, 2*(1), 42–59. https://doi.org/10.3126/japfcsc.v2i1.26731. Accessed on March 23, 2023.

Lewis, H., Schrier, T., & Xu, S. (2021). Dark tourism: Motivations and visit intentions of tourists. *International Hospitality Review*, *36*(1), 107–123. https://doi.org/10.1108/ihr-01-2021-0004. Accessed on November 26, 2022.

Light, D. (2017). Progress in dark tourism and thanatourism research: An uneasy relationship with heritage tourism. *Tourism Management 61*, 275–301.

Lussetyowati, T. (2015). Preservation and conservation through cultural heritage tourism. Case study: Musi riverside Palembang. *Procedia - Social and Behavioral Sciences*, *184*, 401–406. https://doi.org/10.1016/j.sbspro.2015.05.109. Accessed on March 23, 2023.

Mangwane, J., Hermann, U., & Lenhard, A. (2019). Who visits the apartheid museum and why? An exploratory study of the motivations to visit a dark tourism site in South Africa. *International Journal of Culture, Tourism and Hospitality Research*, *13*(3), 273–287. Accessed on March 29, 2023.

McLoughlin, E., & Reis, J. (2019). The potential for dark tourism as a Niche tourism product along Ireland's Wild Atlantic Way (WAW). In *Conference: 10th International Conference on 'Sustainable Niche Tourism'*. https://www.researchgate.net/publication/333602579_The_Potential_for_Dark_Tourism_as_a_Niche_Tourism_Product_along_Ireland's_Wild_Atlantic_Way_WAW. Accessed on November 08, 2022.

O'Connor, J., (1995). The workhouses of Ireland: The fate of Ireland's poor. Anvil Books.

Ogliari, E. (2022). Reframing "the most holy spot in Ireland": An analysis of the narratives from and around Kilmainham Gaol. *Altre Modernità*, 78–97. https://riviste.unimi.it/index.php/AMonline/article/view/18687. Accessed on November 13, 2022.

Packer, J., & Ballantyne, R. (2016). Conceptualizing the visitor experience: A review of literature and development of a multifaceted model. *Visitor Studies*, *19*(2), 128–143. https://doi.org/10.1080/10645578.2016.1144023. Accessed on March 29, 2023.

Park, E., Choi, B., & Lee, T. (2019). The role and dimensions of authenticity in heritage tourism. *Tourism Management*, *74*, 99–109. https://doi.org/10.1016/j.tourman.2019.03.001. Accessed on March 29, 2023.

Richards, G. (1996). Production and consumption of European cultural tourism *Annals of Tourism Research*, *23*(2), 261–283.

Richards, G. (2018). Cultural tourism: A review of recent research and trends. *Journal of Hospitality and Tourism Management*, *36*, 12–21.

Seaton, A., (1996). Guided by the dark: From thanatopsis to thanatourism. *International Journal of Heritage Studies*, *2*(4), 234–244.

Sharpley, R. (2005). Travels to the edge of darkness: Towards a typology of "dark tourism". In C. Ryan, S. J. Page, & M. Aicken (Eds.), *Taking tourism to the limits: Issues, concepts and managerial perspectives* (pp. 217–228). Elsevier. https://www.researchgate.net/publication/284256980_Travels_to_the_Edge_of_Darkness_Towards_a_Typology_of_Dark_Tourism. Accessed on November 13, 2022.

Simone-Charteris, M. T., Kirkpatrick, J., & McLaughlin, C. (2018). An investigation of the differences that exist between generations in relation to supporting dark tourism in Northern Ireland. *DBS Business Review*, *2*. https://doi.org/10.22375/dbr.v2i0.26. Accessed on November 14, 2022.

Stone, P. (2006). A dark tourism spectrum: Towards a typology of death and macabre related tourist sites, attractions and exhibitions. *Tourism: An Interdisciplinary International Journal*, *54*(2), 145–160. https://whc.unesco.org/en/about/. Accessed on October 17, 2022.

Zerva, K. (2021). Dark tourism on Netflix: From place to person-dependent. *Tourism Management Perspectives*, *38*, 100823.

Chapter 2

Dark Tourism, Memory, and the Reproduction of National Identity: The National September 11th Museum

Michael Clancy

University of Hartford, USA

Abstract

Many dark tourism sites, in particular battlefields, memorials, and museums, are instrumental in constructing and reinforcing narratives of national identity. They serve as sites of secular pilgrimage and are central in denoting self and other. They also serve to identify key national moments and involve themes of sacrifice on behalf of the nation. This chapter examines the US National September 11 Museum in New York, which opened in 2014 from this standpoint. Scholars have argued the events of 9/11 were so profound and shocking that they created a "void of meaning" for Americans. This void has been filled in several ways, but important among them has been the creation of memorials and museums and the specific messages of US national identity included within them. The national September 11th Museum plays a particular leading role here in reinscribing dominant popular ideas of American national identity.

Keywords: National identity; pilgrimage; manicheanism; September 11th; innocence; national greatness; affect

Introduction

Dark tourism takes many forms and serves many purposes. An often overlooked role is that of national identity formation and maintenance. Nationalism requires frequent reimagining and the signs, symbols, and myths that do this work can be found everywhere. Many attract tourists. While the content of national identity varies considerably, one common theme is that of national tragedy and sacrifice.

Dark Tourism, 13–25

Published under exclusive licence by Emerald Publishing Limited
doi:10.1108/978-1-83797-336-120241002

Monuments, markers, and museums serve as tangible sites of collective memory that act as foundational narratives of the nation. Tourists come to consume these narratives in acts of learning and secular pilgrimage. For many scholars of nationalism, this is a crucial and necessary practice. Devotion to the nation must be recreated constantly in order to maintain consent. The National 9/11 Memorial Museum, which opened to the public in May 2014, constitutes a useful contemporary case study. Messages embodied within that museum reinforce and reinvigorate a particular version of national identity – one of US muscular nationalism in response to aggrieved innocence.

The remainder of this chapter first discusses popular sources of national identity formation and specifically the role of tourism and then turns to the museum itself. Here its creation and role in constructing particular notions of United States national identity are outlined.

Nationalism, National Identity, and Tourism

Nationalism and national identity have long been pressing topics for scholars (Anderson, 1991; Billig, 1995; Connor, 1993; Gellner, 1983; Hobsbawm, 1992; Smith, 1991, 2009, 2013). Nationalism undergirds the formation and maintenance of the modern nation, and is central to consent to government authority. To the extent that national identity is constructed through social processes, what are the relevant sources? The phenomenon is multifaceted, fluid, and therefore slippery to get a handle on. Smith (1991, p. 14) argues that national identity involves an identifiable people "sharing a historic territory, common myths and historical memories, a mass, public, culture, a common economy and common legal rights and duties for all members." Within foreign policy literature, much of the focus is on elites. Jackson's (2005) critical treatment of the US response to September 11 relies almost exclusively on public statements made by top officials in the Bush Administration in the days and months immediately after the attack, including speeches by President Bush himself. Similarly, Ó Tuathail and Agnew (1992) refer to "intellectuals of statecraft," among them the President, whom they describe as the "chief *bricoleur* of American political life, a combination of storyteller and tribal shaman" (p. 195). It is these leaders and intellectuals who provide myths and meanings, or what Ó Tuathail and Agnew call "geopolitical reasoning" (1992, p. 194). Although useful, such treatments are incomplete. They tacitly posit a one-way transmission belt between elite mythmaking and resulting public opinion and attitudes. In other words, they assume elite attitudes and preferences become swallowed whole by the nation at large. Instead, we also must pay attention to how ordinary elites consume narratives of national identity.

If we accept that there are many authors of the nation, where should we look? Certainly education is central (Benei, 2007), as are official monuments, holidays and commemorations. Public speeches from elites elicit important historical narratives and symbols as well. Popular culture is also important, from television shows and films to books, comics, newspaper articles, and posters (Dittmer, 2010; Edensor, 2002; Nieguth, 2015). Daily practices, including food preparation and consumption also contribute (Ronta & Ichijo, 2022). Billig's (1995) conception of banal nationalism also focuses on the everyday, including acts of performing fealty to the nation. Similarly, Fox and Miller-Idriss (2008) suggest that people are consumers of all the entities that make up the nation and create their own everyday meanings.

Although tourism is primarily treated as a leisure activity and form of consumptive practice, scholars have long recognized it also as an important source of culture making (Greenwood, 1977; Macleod & Carrier, 2010; Sharma, 1999; Shepherd, 2002, 2003). As tourism has become an important economic engine, governments around the world have increasingly promoted the activity and taken on a greater role in shaping it. Most national governments have a tourism department or ministry that in many cases possesses significant political clout. Government tourism agencies have also taken on a growing role in what was once called destination of place marketing but increasingly spills into nation branding. Nation branding introduces the tools of public relations and marketing to world politics. Governments increasingly hire global marketing and branding firms to create and promote fundamental messages about the nation. (Aronczyk, 2013; Olins, 1999).

Outside of official government branding efforts, tourism connects to national identity formation and maintenance more broadly (Bhandari, 2022; Frew & White, 2011; Huang, 2022; Pitchford, 2008; Pretes, 2003). In the US case, many sites may not attempt to represent America but nevertheless contain elements of national values and ideology embodied within them. Disney World, one of the single most popular tourist destinations in the United States, comes to mind, as does Las Vegas. In contrast, as Frew and White (2011) remind us, specific tourism sites, in particular those focusing on culture and heritage, in effect stage the nation to the extent that they identify peoples, shared history and other national myths in a way that helps shape a common national identity among an otherwise diverse population. Their audience is made up of not just foreigners but also the domestic citizenry. Official and quasi-official sites are particularly powerful in this sense. Pretes suggests monuments are powerful tourism sites relating to national identity maintenance as they help to "popularize a hegemonic nationalist message of inclusion" (2003, p. 127). Huang and Santos (2011) argue capital cities represent state-sanctioned showpieces of the nation. They argue Washington D.C. represents three main themes central to American national identity: (1) glorifying the nation by highlighting its wealth, power, and superiority; (2) reflecting American ideals such as liberty, justice, diversity, and individualism; (3) arousing feelings of patriotism through continued references to sacrifice and wars.

Similarly, Sather-Wagstaff (2011) places sites of death and tragedy – so called dark tourism – as central in this process, referring to this segment of tourism as "a significant part of the cultural terrain through which tragic events, past and present, are apprehended, comprehended, and commemorated" (p. 127). Other tourism scholars also point to the importance of war, citing battlefield tourism as a source of creating national cohesion by marking shared sacrifice (Butler & Suntikul, 2013; Ryan, 2007; Winter, 2011). Sather-Wagstaff (2011) reminds us that dark tourism, which Cole (1999, p. 114) defines as "gazing at someone else's tragedy" has a long history, dating back to religious pilgrimages going to shrines containing human remains of saints (p. 69). Early Thomas Cook tours included the site of the Battle of Waterloo and the Grand Tour often included sites of death and destruction as well. Among today's dark tourism sites, many center on war memorials, battlefields, and the resting places of military leaders (Budreau, 2011; Butler & Suntikul, 2013; Ryan, 2007). For Sather-Wagstaff, "the educational, cultural and historical dimensions of tourism and tourism destinations are overwhelmingly centered on the dead" (2011, p. 70). In most cases the dead in question sacrificed their lives for the nation.

Similarly, Pretes (2003) suggests monuments in particular represent something supposedly shared by all citizens, helping to "popularize a hegemonic nationalist message of inclusion" (p. 127). Young describes monuments as being of little intrinsic value, yet central to the story of a nation:

> The matrix of a nation's monuments emplots the story of ennobling events, of triumphs over barbarism, and recalls the martyrdom of those who gave their lives in the struggle for national existent – who, in the martyrological refrain, died so that a country might live. (1993, p. 2)

Sturken's (2007) work takes up this theme more systematically. She is interested in how American culture encourages what she calls a touristic relationship to history. Examining monuments to the terrorist attack in Oklahoma City and Ground Zero in New York, she argues that consumerism through a form of tourism that can be called "nationally sanctioned remembrance" (p. 13) enables the production of a particular type of nationalist myth that depoliticizes the political subject, thereby offering consent for government actions of all sorts. Similar to monuments, museums collect and chronicle official and quasi-official accounts of history and help shape public memory. In short, tourism informs national identity and monuments and museums such as the one at the WTC site, constitute a rich source for investigation.

The National 9/11 Museum

Much has been written about the National 9/11 Memorial and the larger site itself but much less attention has been paid to the museum. The National 9/11 Museum opened on May 15, 2014, some three years after the neighboring *Reflecting Absence* Memorial opened. The five-level, 110,000-square foot, primarily underground museum was the product of work by two separate architectural firms, Norwegian firm Snøhetta, which designed the pavilion, and Davis Brody Bond (DBB), responsible for the museum itself. Together the memorial and museum costs have been estimated to be as high as $1 billion, though organizers dispute that figure. The museum contains to acquire items each year and by 2022 held more than 74,000 items, including 2,600 oral histories, 21,500 artifacts, 700 video and audio files, and 46,000 digital and print images. Operating costs are reportedly more than $80 million per year. Due to the underground nature of most of its space, security, engineering, and maintenance amount to roughly one-third of its operating budget (Small, 2021). The COVID-19epidemic was devastating for the museum and in response upwards of 60% of museum staff were furloughed or laid off by 2021. Daily Admission to the museum is as much as $33 in 2023.

Then-President and CEO of the National 9/11 Memorial and Museum Joe Daniels writes that the museum has two primary roles: Commemoration and education:

And it is through the 9/11 Memorial Museum that we fulfill a sacred obligation to preserve this crucial history for future generations. The Museum is the global focal point for telling the story of what happened on 9/11 through first person accounts, artifacts, and digital documentation (Daniels, n.d.).

This statement raises the question of how well a museum can simultaneously serve as a source of education, a place to bear witness and honor victims, and publicly process emotion effectively. Museum planners engaged with leading exhibition design firm Thinc, which in turn hired trauma specialists while completing the design. Undoubtedly this involved recognizing, if not encouraging, emotional responses from visitors. The primary stories told at the museum are of personal tragedy and experience, mainly during the moments of the September morning in question. Thinc strategically placed tissue boxes and silent spaces throughout the exhibits. Psychologist Billie Pivnick, who worked with Thinc, suggested the goal was to "to keep it feeling alive and present without making it so alive and present that it's unbearable" (Peltz, 2014). Pivnick, an expert in trauma and loss, suggested one primary goal was for visitors to experience and re-experience troubling emotions in a "safe-space" (Miller, 2014).

The museum space occupies the footprint of the original towers and much of it is located directly underneath the memorial pools. As with other memorial sites located exactly where historical events took place (Pearl Harbor, Auschwitz) placement is meant to be cathartic. Carl Krebs of primary museum architectural firm DBB said, "You are entering somewhat sacred ground within an very important and even sacred site as a whole" (Davis et al., 2014). Structurally, two main exhibits occupy the main space 70 feet below street level: a memorial exhibit honoring the lives of the victims that occupies the footprint of the South Tower, and a historical exhibit that takes up the footprint of the entire North Tower. Here the events of that day are chronicled minute by minute. In addition, "interstitial areas" surrounding the main exhibits contain artifacts such as building remnants and artifacts damaged by the attacks and are meant as connected spaces that "function as a complete, thought-provoking experience" (Core77 Design Awards, n.d.).

Early critics largely panned the museum. *Los Angeles Times* critic Christopher Hawthorne (2014) described it as a "relentless" "overstuffed" reaction to the neighboring minimalist 9/11 Memorial. Writing in the Washington Post, Philip Kennicott argued the museum "serves as an oversized pit of self-pity and voyeurism, where visitors can feel personally touched by the deaths of people they didn't know" (2014). What is clear is that the museum targets emotion over understanding. Holland Cotter of the *New York Times*, argued, "[t]he first thing to say about it, and maybe the last, is that it's emotionally overwhelming" (Cotter, 2014).

The museum and memorial together are examples of the recent proliferation of what Senie (2016) calls "victim memorials." Originating in the United States with Maya Lin's Vietnam War Memorial and also including the US Holocaust Memorial Museum, the museum to the Oklahoma City bombings and the Columbine shooting memorial, such installations focus on individual death and the triumph of hope while simultaneously creating "a community of mourners, individuals who are suddenly linked by grievous loss" (Senie, 2016, p. 5). Ultimately, this is done at the National 9/11 museum at the expense of knowledge and deeper understanding. Kennicott identifies a crucial danger of such entities: "memorial museums remain a problematic hybrid form, with an unstable balance between education and emotion. Often they are activated not just by haunting reminders of the victim, but continuing anger at the perpetrator" (2014).

Although much more may be said about the entire museum, three underappreciated aspects of it are particularly important with respect to the discussion of narratives of meaning and national identity. The first is affect. Nonrepresentational theory tells us that the biological realm has an important effect on how we understand the world around us. Connolly's (2002) research on biological processes, specifically chemical producing glands and neuro-responses demonstrate that many of these emerge from stimuli such as surroundings and operate at the precognitive level. From the moment visitors enter the National 9/11 Museum premises, they are met first by surprise and then overstimulation that leads to a sense of personal disjointedness. The first experience visitors have is not nominally part of the museum at all; it is virtually the same security theater they go through at airports. Upon entry patrons immediately meet security personnel, are told to take off coats and jackets, place bags and purses, contents of pockets on security belts, and walk through high-tech imaging machines. The important point here is that, as with airports, people receive the clear message they are entering a space of potential danger. This is not just any museum; rather it is an active, particularly vulnerable site in the forever global war on terror (GWOT). Photography in and around the screening area is prohibited "for security reasons," and this is the one place in the museum where visitors are treated rather gruffly. In short, for museum patrons, their very first experience is a biopolitical encounter. Not only must they themselves be discounted as a threat; the effect is intended and clear, and they are reminded to be afraid.

The security screening also possesses particular power because it taps into the dominant historical national discourse that simultaneously distinguishes between the safe homeland and the danger outside of it, part of what Ó Tuathail and Agnew refer to as the "production of geographical knowledge" (1992). This theme goes back to the founding fathers and has been repeated throughout US history. In 1988, for instance, then President Ronald Reagan stated, "My friends, I have often expressed my belief that the Almighty had a reason for placing this great and good land, the 'New World,' here between two vast oceans. Protected by the seas, we have enjoyed the blessings of peace" (cited in Ó Tuathail & Agnew, 1992, pp. 196–197). Yet it simultaneously questions that very binary in a way that conjures up the Cold War and the Red Scare danger from within. Although in many past conflicts we could clearly distinguish the safety of the domestic from foreign danger, now – as with the Cold War – we must simultaneously worry about the danger from within.

Past the security screening, most museum visitors make a long descent downstairs or escalators from the airy, glass-dominated pavilion, past huge steel tridents, the first of many remains of the towers and surrounding buildings, to the muted light of the main lobby below where the tours of the museum begin. The *New York Times* describes this initial entrance into the museum as "carefully choreographed" and Cotter reminds us of the purposeful religious overtones:

> Its nearest equivalent I can think of is the dynamic of religious pilgrimage sites, whether Christian churches, Buddhist temples or Sufi shrines. There, the mortal remains of saints, and objects sanctified by their touch, are the focus of attention. Here, you also walk a long, sanctified route, stopping at the equivalent of side chapels and altars, contemplating icons, talismans and embodied miracles (2014).

Again the focus here is on affect. We are entering a somber place, a holy place where a profound tragedy took place. Emphasis is on reverence and honoring the dead, some 14,000 unidentified bodily remains of which were moved away from, and then back to the World Trade Centre (WTC) site just prior to the museum opening.

Second is victimization and innocence. Visitors are told there is no set itinerary or path through the museum, but the architecture creates a logic of its own here, taking patrons first through "the ramp," the introductory exhibition. Those visitors seeking a clear historical understanding of what happened that day and why, instead face the first of many confusing and contradictory narratives. One of the most basic questions one might ask, for instance, has to do with the subject of the attacks. Who were the targets and who were the victims? Throughout we are given incomplete and contradictory answers.

The introductory exhibition contains first an oversize photograph of lower Manhattan taken from Brooklyn on the morning of September 11. The day has often been described as a "perfect morning," suggesting not just the ordinary, but also the idyllic, an Eden of sorts. Behind this is the "We Remember" exhibit, an overlapped image of a world map over vertical panels that contain multimedia testimonials of people from all over the world regarding where they were and how they experienced the attacks. The museum has collected more than 7,000 stories from all over the world in more than 15 languages. The exhibit, created by Thinc and the architectural and digital media firm Local Projects, contains varied soundscapes, and projections of the words from the oral histories presented in a manner that constructs the world's continents. As many as two billion people worldwide witnessed the events of September 11, 2001, either in person or through television imagery. The words are those of the observers directly experiencing trauma themselves through the act of observation. In essence we are told they, and we, were all victims that day.

Elsewhere, however, the dominant message focuses on the occupants of the towers that morning, those on the hijacked planes, and especially the first responders. Sturken (2007) and Langewiesche (2003) remind us that in the aftermath of the attacks a sort of hierarchy of victimhood emerged. Although the remains of the occupants, the first responders and the hijackers were all interspersed within the rubble, firefighters and police urged construction crews to treat the remains of first responders with particular reverence during the recovery and reclamation efforts:

> These tensions flared especially over the differing treatment of human remains – on the one extreme, the elaborate flag-draped ceremonials that the firemen accorded their own dead, and on the other, the jaded "bag 'em and tag 'em" approach that they took to civilians. Langewiesche (2003, p. 70, cited in Sturken, 2007, p. 181)

This raises the larger issue of the often-contradictory themes associated with the exhibits and museum. In a museum that is almost wholly devoted to those who died that day, victims are portrayed at once as both innocent and heroic, themes ultimately tied to deep notions of American national identity. This construction of innocence is done through various means, most notably through the display of the ordinary within artifacts. Wallets, purses, shoes, hotel keycards, watches, and teddy bears – all are prominently displayed and labeled as meaningful. Sturken emphasizes everyday photos of missing victims that are reproduced or collected and displayed throughout the museum as "powerfully evok[ing] a kind of prior innocence" (2007, p. 183). For both victims and witnesses, what took place that morning was so unimaginable in large part *because* of that very innocence.

National innocence has long been a central motif of American national identity. The United States being an "indispensable nation," somehow outside the community of nation-states, goes back to colonial times and part of this is due to the accompanying assertion of innocence associated with both its founding and its natural landscape. (Madsen, 1998; Philpott & Mutimer, 2009). These claims of innocence have been asserted throughout US history, in particular during times of armed conflict. To the extent that innocence is so deeply engrained within American national identity and the attacks primarily targeted the innocent, the prevailing message is that the attacks were, ultimately, attacks on America.

This portrayal of innocence acts not merely an end in and of itself. Rather it confers upon the United States first the ability to speak for those who hold similar ideals; second, asserts moral and broader superiority over others; and, finally, acts as judge and jury with regard to how to respond. In addition to motivating consent for all this, the claim of prior innocence also relieves visitors from taking history seriously, especially that history which would lead a politicized opposition group to plan and carry out such an attack.

One related theme is that of the buildings themselves as innocent victims. Images of the towers are everywhere at the museum. As Sturken points out, their photographs and assorted relics, parts, and ruins are displayed as artifacts much like those of the people who died that day. There is anthropomorphism at work here. Museum patrons are encouraged to honor and mourn the loss of the buildings themselves much like they mourn the dead. But equally important, the towers – which were once largely dismissed by critics as uninspired examples of banal modernism (Sturken, 2007, p. 201), and the result of advances that "made megalomania compatible with economics" (Huxtable, 1973) – have been rehabilitated after their destruction. This is the source of the third theme: American power and ingenuity. At the museum the buildings represent among other things, man's domination over nature through technology. One of the largest "exhibits" in the entire museum is known as "the bathtub," the outside slurry wall originally built to contain the Hudson River. The wall of the bathtub survived the attacks and was reinforced during museum construction, becoming an attraction in and of itself.

The original architectural scale model of the WTC site, including the towers, makes up part of the current historical exhibition, as does an exhibit relating to the building of the Twin Towers. In a museum supposedly devoted to honoring and memorializing the victims of 9/11, it is peculiar that so much attention is devoted to the original creation of the buildings. Yet doing so suggests the existence of a uniquely American ingenuity, perhaps even superiority – not unlike the ancient Egyptians or Mayans building their cities and pyramids. The message is these are clearly an advanced people. Only Americans, the displays seem to suggest, could create not one, but two tallest buildings in the world simultaneously. The final display of the intact buildings, again, oddly placed near the *end* of the dimly lit historical exhibition after patrons have watched video after video of them crashing to the earth, is a giant photograph of the towers. Adjacent stands a quote from Austin J. Tobin, executive director of the Port Authority in 1966 when construction began: "In spirit, the trade center is a United Nations of Commerce." By referring to the United Nations, Tobin was suggesting WTC to not only be international in scope but also, like the United Nations, a force for peace and cooperation. This echoed a similar sentiment from twin towers architect Minoru Yamasaki's, that the WTC "would become a physical expression of the universal effort of men to seek and achieve world peace" (Yamasaki, 1979, p. 112, quoted in Morton, 2006, p. 20). The Tobin quote and the larger presentation of the twin towers assert both America's unique scientific and technical prowess and its naïve innocence.

Interestingly, the area that has drawn the most attention and criticism within the museum is the seven-minute film "The Rise of al-Qaeda" downstairs that comes almost as an afterthought to the museum experience. Visitors – only after passing through the opening exhibition, the long ramp that leads them past the vast overview of the Foundation Hall, the survivors' stairs that took WTC building occupants to safety on Vesey Street, the oversized Virgil quotation on memory, the remnants of elevator motors and a badly damaged NYFD ladder company 3 fire truck, Memorial Hall with its remembrance of each individual victim, and finally the vast historical exhibition, chock full of artifacts and a minute-by-minute timeline – only then finally reach the al-Qaeda film. Most patrons are by then spent and exhausted. Finally, at that point, they are given some historical context and outline of the grievances held by the militant group. This comes literally at the end of the logical tour of the museum. By then it feels much too late. Emerging visitors ascend a long escalator back to ground level to mournful military music. They are, in effect, leaving a national funeral.

Conclusion

Museums and memorials play a unique role in framing national identity. They tell the foundational myths of the nation. To the extent that governments officially or quasi-officially create them, the narratives they contain possess special powers. The fact that the monuments and museums exist at all signals key moments or points of profound national change and meaning. Because of the nature of nationalism itself it is hardly surprising that most such sanctioned museums revolve around violence and war. Nations are born of war and subsequent conflicts legitimize the primary national narrative. In this sense, the National 9/11 Museum is necessary. It documents a significant historical moment in the American experience. Such museums combine evoke emotion, shared history and sense of collective suffering, contributing to and reinforcing a dominant national identity narrative.

All of this also dovetails with tourists and tourism. Tourism – especially dark tourism – often presents some essence of the nation for consumption. It combines reverence and a quest for education, similar to antecedents of modern tourism, the religious pilgrimage and the Grand Tour. Graburn (1977) refers to travel and tourism as modern manifestations of "the sacred journey." Today the nation constitutes a primary source of sacralization. Sturken (2007, p. 199) argues that sacralization has historically been secular in the United States, not religious, and is tied more to patriotism. Yet religious references are often everywhere. The claiming of the WTC site as "hallowed ground" is just one example. For the National 9/11 Museum, presenting narratives of the nation are not open-ended, but must hit on earlier frames of reference to be effective. Ultimately it does just that. It constructs the United States as exceptionally innocent, equates that identity with the unique embodiment higher civilizational virtues, and creates a simplistic dualism that contrasts the United States with the rest of the world.

References

Anderson, B. (1991). *Imagined communities: Reflections on the origin and spread of nationalism*. Verso.

Aronczyk, M. (2013). *Branding the nation: The global business of national identity*. Oxford University Press.

Benei, V. (2007). *Manufacturing citizenship: Education and nationalism in Europe, South Asia and China*. Routledge.

Bhandari, K. (2022). Tourism and commercial nationalism. *Annals of Tourism Research*, *95*(6). https://doi.org/10.1016/j.annals.2022.103443

Billig, M. (1995). *Banal nationalism*. SAGE.

Budreau, L. M. (2011). *Bodies of war: World War I and the politics of commemoration in America, 1919–1933*. NYU Press.

Butler, R., & Suntikul, W. (2013). *Tourism and war*. Routledge.

Cole, T. (1999). *Selling the Holocaust: From Auschwitz to Schindler; how history is bought, packaged and sold*. Routledge.

Connolly, W. E. (2002). *Neuropolitics: Thinking, culture, speed*. University of Minnesota Press.

Connor, W. (1993). *Ethnonationalism*. Princeton University Press.

Core77 Design Awards. http://designawards.core77.com/home/award_permalink?id=30765&category_id=4. Accessed on March 14, 2016.

Cotter, H. (2014, May 14). Sept. 11 memorial museum at ground zero prepares for opening. *The New York Times*. http://www.nytimes.com/2014/05/14/arts/design/sept-11-memorial-museum-at-ground-zero-prepares-for-opening.html

Daniels, J. (n.d.). President's message. 9/11 Memorial and Museum website. https://www.911memorial.org/presidents-message. Accessed on March 14, 2016.

Davis, L., Desantis, A., Roberts, G., & Ruby, M. (2014, May 14). A New story told at ground Zero. *The New York Times*. http://www.nytimes.com/interactive/2014/05/14/arts/design/September-11-Memorial-Museum.html

Dittmer, J. (2010). *Popular culture, geopolitics, and identity*. Rowman & Littlefield Publishers.

Edensor, T. (2002). *National identity, popular culture and everyday life*. Bloomsbury Academic.

Fox, J. E., & Miller-Idriss, C. (2008). Everyday nationhood. *Ethnicities*, *8*(4), 536–563.

Frew, E., & White, L. (2011). *Tourism and national identity: An international perspective*. Taylor & Francis.

Gellner, E. (1983). *Nations and nationalism*. Cornell University Press.

Graburn, N. H. H. (1977). Tourism: The sacred journey. In V. L. Smith (Ed.), *Host and guests: The anthropology of tourism* (pp. 17–32). University of Pennsylvania Press.

Greenwood, D. (1977). Culture by the pound: An anthropological perspective on tourism as cultural commoditization. In V. L. Smith (Ed.), *Hosts and guests: The anthropology of tourism* (pp. 129–138). University of Pennsylvania Press.

Hawthorne, C. (2014, May 26). Architecture review: At 9/11 Memorial Museum, a relentless literalism. *Los Angeles Times*. http://www.latimes.com/entertainment/arts/la-et-cm-911-museum-review-20140526-column.html

Hobsbawm, E. J. (1992). *Nations and nationalism since 1780: Programme, myth, reality*. Cambridge University Press.

Huang, Y. (2022). Consuming geopolitics and feeling maritime territoriality: The case of China's patriotic tourism in the South China Sea. *Political Geography*, *98*(8). https://doi.org/10.1016/j.polgeo.2022.102669

Huang, W.-J., & Santos, C. A. (2011). Tourism and national identity in the United States: The case of Washington, DC. In *Tourism and National Identities* (pp. 13–25). Routledge.

Huxtable, A. L. (1973, April 5). Big but not so bold: Trade center towers are tallest, but architecture is smaller scale. *New York Times*. http://timesmachine.nytimes.com/timesmachine/1973/04/05/103218617.html?pageNumber=34

Jackson, R. (2005). *Writing the war on terrorism: Language, politics and counter-terrorism*. Manchester University Press.

Kennicott, P. (2014, June 5). The 9/11 Memorial Museum doesn't just display artifacts, it ritualizes grief on a loop. *The Washington Post*. https://www.washingtonpost.com/entertainment/museums/the-911-memorial-museum-doesnt-just-display-artifacts-it-ritualizes-grief-on-a-loop/2014/06/05/66bd88e8-ea8b-11e3-9f5c-9075d5508f0a_story.html

Langewiesche, W. (2003). *American ground: Unbuilding the World Trade Center* (Reprint edn). North Point Press.

Macleod, D. V. L., & Carrier, J. G. (2010). *Tourism, power and culture: Anthropological insights*. Channel View Publications.

Madsen, D. L. (1998). *American exceptionalism*. Edinburgh University Press.

Miller, A. (2014). Memorializing a tragedy. *APA Monitor*, *45*(8), 28. http://www.apa.org/monitor/2014/09/tragedy.aspx

Morton, P. A. (2006). Document of civilization and document of barbarism: The World Trade Center near and far. In D. J. Sherman & T. Nardin (Eds.), *Terror, culture, politics: Rethinking 9/11* (pp. 15–32). University of Indiana Press.

Nieguth, T. (2015). *The politics of popular culture: Negotiating power, identity, and place*. McGill-Queen's Press - MQUP.

Ó Tuathail, G., & Agnew, J. (1992). Geopolitics and discourse: Practical geopolitical reasoning in American foreign policy. *Political Geography*, *11*(2), 190–204. https://doi.org/10.1016/0962-6298(92)90048-X

Olins, W. (1999). *Trading identities: Why countries and companies are taking on each others' roles*. The Foreign Policy Centre.

Peltz, J. (2014, May 21). Tissues, counselors help ease pain at 9/11 Museum. *CNS News*. http://www.cnsnews.com/news/article/tissues-counselors-help-ease-pain-911-museum

Philpott, S., & Mutimer, D. (2009). The United States of Amnesia: U.S. Foreign policy and the recurrence of innocence. *Cambridge Journal of International Affairs*, *22*(2), 301–317.

Pitchford, S. (2008). *Identity tourism: Imaging and imagining the nation*. Emerald Publishing Limited.

Pretes, M. (2003). Tourism and nationalism. *Annals of Tourism Research*, *30*(1), 125–142. https://doi.org/10.1016/S0160-7383(02)00035-X

Ronta, R., & Ichijo, A. (2022). *Food, national identity and nationalism: From everyday to global politics*. Springer Nature.

Ryan, C. (2007). *Battlefield tourism: History, place and interpretation*. Routledge.

Sather-Wagstaff, J. (2011). *Heritage that hurts: Tourists in the memoryscapes of September 11*. Left Coast Press.

Senie, H. F. (2016). *Memorials to shattered myths: Vietnam to 9/11*. Oxford University Press.

Sharma, K. K. (1999). *Tourism and culture*. Sarup & Sons.

Shepherd, R. (2002). Commodification, culture and tourism. *Tourist Studies*, *2*(2), 183–201. https://doi.org/10.1177/146879702761936653

Shepherd, R. (2003). Fieldwork without remorse: Travel desires in a tourist world. *Consumption Markets & Culture*, *6*(2), 133–144. https://doi.org/10.1080/1025386032000109596

Small, Z. (2021, July 12). 9/11 Museum's 20th-anniversary exhibitions become victims of cuts. *The New York Times*. https://www.nytimes.com/2021/07/12/arts/design/9-11-museum-budget-cuts.html

Smith, A. D. (1991). *National identity*. University of Nevada Press.

Smith, A. D. (2009). *Ethno-symbolism and nationalism: A cultural approach*. Routledge.

Smith, A. D. (2013). *Nations and nationalism in a global era*. John Wiley & Sons.

Sturken, M. (2007). *Tourists of history: Memory, kitsch, and consumerism from Oklahoma City to ground zero*. Duke University Press.

Winter, C. (2011). Battlefield tourism and Australian national identity: Gallipoli and the Western Front. In E. Frew & L. White (Eds.), *Tourism and national identities: An international perspective* (pp. 176–189). Routledge.

Yamasaki, M. (1979). *A life in architecture*. Weatherhill.

Young, J. E. (1993). *The texture of memory: Holocaust memorials and meaning*. Yale University Press.

Chapter 3

Dark Tourism: Educational or Exploitative Side of Tourism?

Iti Gaur

Manipal University Jaipur, India

Abstract

The concept of tourism is often a rosy experience for many, where one would even associate the term with amazing sight-seeing or some romantic experiences. In the recent past a different side of tourism has been gaining popularity. What is that? This side of tourism is based mainly on natural or man-made disasters. This chapter would be talking about Dark Tourism. People around the world are more inclined to know about tragedies, deaths, and brutal killing of people in masses. This inclination is making people visit such sites which further makes them understand, imagine (the reality of past), and know in depth about such sad occurrences.

This chapter would then give you a tour around more of similar type of tourisms namely disaster tourism, ghost tourism, and more. Tourists lately are proud of being known as "Dark Tourists" and they prefer to pay visit to places like the Auschwitz's Holocaust concentration camps, Catacombs, Hiroshima's Peace Memorial Park, Bhangarh, and many more. This section would highlight the questionable sides of this tourism.

In the last segment of this chapter, the conclusion, would be drawn toward identifying whether Dark tourism is educational, like the other sides of tourism, or it is an exploitative side of the tourism that is merely trying to make profits from the tragedies happened across the globe.

Keywords: Dark tourism; tragedy; death; atrocities; Auschwitz; catacombs; disaster

Dark Tourism, 27–40

Published under exclusive licence by Emerald Publishing Limited
doi:10.1108/978-1-83797-336-120241003

Introduction

Imagine few decades back, if 10 persons would be asked, whether they would like to travel to a haunted place as a part of tourism, a good number of them would be in utter shock that what is the need of traveling to such places when the earth is full of beautiful landscapes to see. Apparently if the same question would be asked today, a good number of them now would be excited to travel such places.

What is it? Or what actually has changed that lead to the change of decisions of the people over the span of few decades. Two renowned academicians from Scotland named Malcolm Foley and J. John Lennon in their book titled "Dark Tourism: The attraction to Death & Disaster," came with the term "Dark Tourism." It is not like people never used to travel to the tragic destinations before 1996. Public hangings in the historic days, used to have hundreds of people as audience (Cramer, 2022). Though the concept came up more in light after the term "Dark Tourism" was coined by the said authors. Death, disasters, and atrocities gain tourists' attraction who wishes to gaze upon real and recreated death (Stone, 2006).

Dark tourism is a phenomenon that has gained considerable attention in recent years. The immense popularity of TV show like Chernobyl and a popular web series on an OTT Platform-Dark Tourist happens to be another good reason, why dark tourism is becoming popular. This form of tourism is characterized by a growing interest in destinations such as former battlefields, disaster zones, and sites of atrocity, where visitors seek to engage with the darker aspects of human history. However, as this industry is growing, there are debates surrounding the good and the evil side of this tourism.

This chapter delves into the complex dichotomy between the educational merits and exploitative nature of dark tourism. It further aims to provide a comprehensive analysis of this controversial travel trend, considering its impact on both visitors and the destinations themselves. This chapter also throws light upon the educational benefits of and the inherited potential exploitation of dark tourism.

Through an exploration of case studies, theoretical frameworks, and research, this chapter will offer insights into the motivations that drive individuals to engage in dark tourism, as well as the ethical dilemmas that arise from the commodification of tragedy and suffering. It will investigate how very sensitive events that took place in the past have the ability to affect the perceptions of the visitors – related to these tragic sites and at the same time how all the stakeholders of the this tourism are responsible for being respectful and responsible for such places.

Ultimately, this chapter aims to stimulate critical discourse and encourage a nuanced understanding of the moral complexities surrounding dark tourism. By addressing the ethical considerations and educational potential of visiting sites of tragedy and trauma, it aspires to contribute to the ongoing dialogue about the role of tourism in shaping collective memory, fostering empathy, and preserving historical consciousness.

What Is Dark Tourism?

The term tourism is most commonly, connected to all the good thoughts of a person. This term often makes us imagine, the lovely range of mountains or even some magical beaches with crystal clear water touching the feet. In contrast to this, in the recent past, tourism is not only restricted to the rosy experiences or spectacular views but has moved beyond this.

A different side of tourism has emerged in the last few decades, where people do not want to experience the leisure but want to visit the sites of tragedy and atrocities. A good bunch of people want to participate in the tours which take them to the places which are connected to death, tragedy, torture, atrocities, and other forms of disaster. The people who want to visit these sites, call themselves "Dark Tourist." The entire human race that lives on this planet, is not the same, in terms of their respective nature and behavior, thereby we can understand that why some wants to visit the beautiful sites, while the others want to experience the darker places – those which are connected to any sort of tragedy.

Death is a paradox of holidays. This is not something to be celebrated or enjoyed. Perhaps it is the case, but some argue that isn't it necessary for people to know about the history, if it is then how can Dark Tourism be bad. People around the world want to have a firsthand experience of these places and wants to know more, beyond what is available in print form.

Various events take place across the countries, which are able to engage a big part of the population, where people could participate and learn about the tragedies. Moreover, there are Dark exhibitions, museums, and trips being organized, to give a taste of dark tourism to people, which further adds to their desire of getting deeper into the events.

What Are the Other Forms of Dark Tourism?

The intent of the dark tourists is something that brings along the differences in the dark tourism itself. Based on the sites and the intention of the visitors, the dark tourism could further be fragmented as Disaster Tourism, War/Battlefield Tourism, Ghost Tourism, Prison Tourism, Cemetery Tourism, and also Holocaust Tourism (Forenseca et al., 2016). Dorina Maria Buda, a famous academician and writer, expressed the fact that the intentions of the tourist matters a lot in the sphere of dark tourism (TEDx Talk, 2017).

- *Disaster Tourism:* Any tragic event, that has taken place at a certain location, may it be man-made or natural, drives the dark audiences move toward the same. These set of tourists are the "disaster tourists," who want to dig deep into these locations. The mishappenings like the explosion at the Chernobyl nuclear power plant, the earthquake at Kathmandu, or a hurricane in New Orleans.
- *War/Battlefield Tourism:* Dark tourists also wish to visit the spots where war or any battle took place. The remains from the war or the urge to know more about these historical events make them persuade toward these spots. Places of World War II in Berlin witnesses several tourists round the year.

- *Ghost Tourism:* "Ghost" a term that often repels a good number of people but at the same time attracts a ton. The need to see ghost or experience the paranormal activities is the reason why Ghost tourism is gaining popularity day by day. Many hotels across multiple locations are said to be having a past where someone committed suicide or some tragedy occurred. People rather than staying away from such places are ready to pay for staying inside these haunted buildings. Places like Bhangarh, located in India, and the French Quarter, New Orleans, are a few of the examples of the ghost tourists' attraction.
- *Prison Tourism:* There are several prisons present on this planet and the history has experienced a few of them to be the most horrible form of prisons. Some of these prisons have seen the stay of freedom fighters and the others have been home for many culprits and gangsters. No matter for which reasons, people were behind the bars, but the reason why some prisons are popular among the others are may be their location, the tortures done on those places, or some of them are even giving you an option to stay there as a prisoner for an entire day. *Kala Pani*, Cellular Jail Located in Andaman & Nicobar Islands, *Alcatraz*, also known as "The Rock," in its heyday was said to be the most secure prison around the globe. These and other few prisons are the major tourist attractions for the prison tourists. Srinivasa Rao Apparasu, reporter of Hindustan Times, stayed in one of the Indian prisons, located at Telangana, India after paying 500 INR back then. During his stay, he was serving as a labor at the prison and also was served the same lunch as the prisoners (Hindustan Times, 2016). Earlier sites of punishment and prisons gain a lot of tourist attention (Strange & Kempa, 2003).
- *Cemetery Tourism:* The urge of the tourists, to get the firsthand experience of the famous graveyards or cemetery is making the Cemetery tourism touch boom. These tourists are also known as "Tombstone tourists." Few of the famous cemeteries receives as much as 3.5 billion tourists per year. The cemeteries are not only reflection of deaths but also of the societal and cultural backgrounds of the places (Saunders, 2017). Some of the famous cemeteries are Maqbaratoshoara, Tabriz, Iran and Mount of Olives Jewish Cemetery, Jerusalem, Israel.
- *Holocaust Tourism:* Post World War II, holocaust is the considered to have propounded the term "genocide," as a conductor of the crime that the Nazis did to the Jews and others in Germany. Even after so many years of the Holocaust, so many layers are still being uncovered and probably some are yet to be known. For understanding the past, knowing about how people now respond to the genocides and other atrocities, getting in depth details of the particular historical event these trips are organized, and Auschwitz Concentration Camp, tops the chart in this category.

Popular Dark Spots Around the World

Any place that has experienced mass killings or any other type of natural or man-made disaster attracts the dark tourists toward itself. Around the globe, there are numerous such places, which compels the dark tourists toward itself, though some of them top the chart. The following part of the chapter will discuss about the atrocities that occurred at a few renowned places across the globe:

- *Auschwitz Concentration Camp:* This cannot be possible that we are talking about the dark sites and Auschwitz will not be counted. The said concentration camp is a reflection that an urge of a tyrant, to become the most powerful person, could do such harm to humanity. The visitors visit this place because they want to imagine what sort of torture approximately 1.3 million people went through.
  Gas chambers where almost thousands of people were killed, the execution wall, and the train tracks through which the innocent people were brought to their place of death, all these are the most popular spots among the dark tourists, at the Auschwitz Concentration Camp. Lucy (an Auschwitz's survivor) in her interview (Santenello, 2023) stated that her mommy and sister were taken to the gas chambers right in front of her eyes. The prisoners like her were made bald and were given wooden shoes. To support her words, the visitors can find the big pile of hairs of the jews' and also their shoes. These experiences make the visitors imagine what torture the prisoners must have had gone through. Infact, those who brought this torture to reality were some really normal looking people (criminals and crime fighters, 2023).
- *Catacombs of Paris, France:* A famine that occurred in paris led to the death of millions of people and for the traditional way of their burial, sufficient places were not available. Therefore a decision was made to keep the dead bodies in the underground tunnels, which were located right beneath the city. Gilles Thomas, Author of The Catacombs of Paris (History, 2021), stated that the bodies were stopped to be buried in 1780 and toward 1785 the bones were transported to the catacombs, during the night hours, when the city was sleeping. He further added that the catacombs were officially opened for the public (Initially for a few top officials and later for the public) in 1809 and since then it has attracted billions of tourists. In the words of Jonathan Young, Founding Curator, Joseph Campbell Archives (History, 2021), a whole network of tunnels is filled with the human bones. The entrance of the catacombs has a written warning in French, which when translated to English says "Stop! This is the empire of the dead." The structure holds the bones of approximately 6 million people.
  These catacombs represent the final resting place for so many people but the same resting place of countless souls is a picnic spot for many. Many people have been found enjoying their picnics, Wine-tasting ceremonies, parties organized by drug addicts, and more in these tunnels, which is surely not at all a respectable way to treat the departed souls.

- *Aokigahara Forest, Japan:* Situated near Mount Fuji in Japan, there is a dense forest, more popularly known as the "Suicide Forest" and is one such another dark spot. The said forest is spread into a vast area and has witnessed a very big number of suicidal cases. The spot is also currently an attraction spot for the dark tourists.
  Somewhere down the lane in the forest, one can easily find strings lying on the ground of the forest. The local people say that, those who enter the forest with an intention to commit suicide carry these strings with them, just in case there is a change of heart inside the heart of the forest. These strings would help them come out of the densely spread forest. It is also said that those who went inside, with such intention, never come out.
  Logan Paul, a famous YouTuber, went inside the forest and recorded a video. He even found a dead body in there but to the utter shock, he treated it in the most disrespectful manner and had to witness a huge backlash (Duff, 2018).
- *Chernobyl Nuclear Plant, Ukraine*: It was the night of April 26, 1986, when the nuclear reactor number 4 was facing some issues in its operation and soon after that, the night witnessed the world's biggest nuclear disaster (Rathee, 2022). The location was Soviet Union. The dangerous radioactive material started to come out with the explosion and soon started to spread all over. The ill effects of the radioactive material were not only confined to Ukraine but reached Spain, Sweden, and other neighboring places too. The instance, even lead to radioactive-based rain in some parts of the United Kingdom (Rathee, 2022). The official name of the plant was *Vladimir Lenin Nuclear Power Plant*. Though the government initially shut down the entire plant but still the other units remain active and continue to look after the energy needs of Ukraine.
  The Exclusion zone, 30 km of radius of the plant, has been declared by the government as a no entry zone. Apparently, the same zone is the tourist attraction today. The dark tourists, find it fascinating to enter into that zone and they try to connect with the past there. The visitors try to know that exactly how the events would have played out. The place is still full or almost has a sea of masks being worn by people there and depression surrounds the air. Despite all these facts, the place tops the chart for dark tourists.
- *Pompeii, Italy*: August 24, 79 AD, was the day to witness the world's most terrifying volcanic eruptions till date. It was Mount Vesuvius, which after giving a few signals through a bit of shaking the earth, took its most horrifying form. People back then were reliving their lives peacefully, people taking care of their infants, shopping, enjoying in their gardens, and enjoying the luxury that the city of Pompeii used to experience.
  All of a sudden, the volcanic eruptions started to come from the sky, in the form of ash. People began to run to save their lives, some running to escape from the place, while others trying to hide within their homes. But soon the entire area was under 21 feet of volcanic eruption.
  Today after years of archeological discoveries, the place has become one of the most popular tourist attractions in Italy. People want to visit the place to see what kind of civilization the place was, back then. At the same time, the

time frozen bodies of the population are also a popular attraction for the visitors. The city is still referred to as "The frozen city of Pompeii."

All of the aforementioned spots gain tourist popularity for numerous reasons. Some of them find themselves, as though they can understand more, behind what would have happened at that place. Some of them are for the sake of capturing pictures, making reels and brag in their inner circle about the places they have been to. These are not the only places around the globe that hold attention of the dark tourists, infact there is a long list of such places but the said chapter would not be sufficient to do justice with the chapter.

What Drives the Tourists to Visit the Dark Spots?

Dorina Maria Buda, a famous academician and writer (TEDx Talk, 2017), stated that emotions are the ultimate drivers of tourism. The emotions that drive a person toward beaches and sunshine must not be the same as to, what drives people toward the dark locations. What type of emotions, actually govern this side of tourism that is to be talked about. The research indicates that the "Dark Tourists" are compelled by the below mentioned reasons to visit these places:

- *How Exactly the Event Would have Played Out:* People when are exposed to some event, may be through newspapers, articles, stories, media and even movies, create a lot of questions in their minds about the event. Sometimes, they even want to dig deep, to understand that how exactly the event would have occurred. To satiate their curiosity, the idea about traveling to these places, originates in their minds.
- *Fascination of Death*: Though this might seem to be a bit non-convincing but in the race of understanding the dark reality, the researchers do come across the fact that, the tourists find "death" fascinating. Public hangings, assassinations, tragedy and any apocalypse are the things, which pose to be an attractive feature, to the dark tourist. People want to see and yes, some of them enjoy, going through the entire journey of someone's pain.
  Lucy, an Auschwitz survivor (Santenello, 2023) said that they were given only one meal a day, that too was made with rotten potatoes and other rotten items. A tourist visiting the Auschwitz Concentration Camp, who had never skipped a meal in his life, might not understand the pain that starvation brings.
- *The Places Turning Abandon:* The streets or areas, which once saw people living their normal lives, are now abandoned. The dark tourists, wish to experience with their naked eyes, that what the places, which once witnessed people roaming around and the other day death and disasters, looks like now. Like in the case of Pompeii, one can easily find the depictions in books, over internet, and more sources that what it used to be like at the time of Mount Vesuvius explosion. People there (as per the depictions) were living really normal life and within a day, the entire city was buried under hot ashes.

  While touring the Chernobyl Nuclear Plant, one may find the pictures that reflect the kind of life the workers were living before the blast took place

(BuzzFeed News, 2022). The dark tourists want to go back to the point, in their minds and compare it with the after effects of the tragedies. After that they connect themselves, with the pain that must have had occurred.

- *Urge to Get into a Forbidden Area:* Like it has been mentioned before, all the tourists around the world are not the same; their intent, their motives, and their ideologies differ. For these reasons, some of the people find it really amusing, to get inside the areas, which are forbidden to enter.
  Reason could be anything, such as bragging among the circle about their adventurous journeys, doing something that most of the people are afraid of doing, or may be just to put the pictures taken on those spots, over social media handles.
- *Consider this an Adventure*: People who visit the dark places, are often found not to be respectable toward the tragedy that had happened at that spot. They instead, consider the whole trip to be an adventurous one. A good number of articles and videos are the proofs that people go beyond the boundaries of respect at those places. One must not click a smiling selfie, at a place which had heard cries of millions of people, who were burnt alive, and within few hours were turned into ashes. This is clearly not an example of an adventure.
- *Unique*: While people in general take the advices of their friends and family, as to which place they may travel to. Some find themselves to be doing the acts which are different than the others. For the said reason, they want to travel at those "UNIQUE" spots, which have been experienced only by a hand full of people.
- *Genuine Desire to Gain Knowledge*: It is not only like the dark tourists are a bunch of people who wish to cause harm to the places or want to show off about the places they have traveled to. Instead, there are also the people who participate in this type of tourism, who wish to know about the history of those places and bring out the voices, the depictions of the occurrences, and also the harsh realities of the said places.
- *Search of Authenticity*: Its not that the dark tourists only visit the dark places because they hold dark motives. A segment of them also carries the curiosity of gaining firsthand knowledge about those places, without having any intermediaries in between. This is the reason they turn into dark tourists.
- *Brag on Social Media*: Without a doubt, life today has somehow become centered on the opinion of others. This is so because, whatever people are doing in general today is, may be, to put on social media, to brag about, to showcase one's status in the society, or may be the urge for becoming talk of the town. They visit these tragic spots to gain the attention of the netizens.

This behavior sometimes becomes ruthless and therefore people in order to satisfy these kinds of needs go beyond the boundaries of respect and responsibility and participate in the things which may or may not be of their interest.

Which Side of Dark Tourism Is Getting More Power Educational or Exploitative Side?

Indeed, a million dollar question it is. Is it right to visit the sites where something terrible has happened? Will it be respectful to take selfies at a concentration camp? There are several debates that might arise if we are going to talk about whether it is respectful or disrespectful to visit the places that once witness major disasters or atrocities.

That being said, due to the media coverage, opinions of public at large, and rapidly increasing tourism in this sphere has also led to development of both of these sides. On one side where the educational aspect of dark tourism is helping the world to better understand the history related to the events, to the other side the exploitative part is more inclined toward gaining more profits out of the atrocities and the tragedies which in turn adversely affect the sentiments of those who have suffered due to those events taking place.

To assess and to understand that which side is gaining more power in contemporary times, it becomes crucial that we dig deep into the media coverage, trends, and behavior of public at these tragic spots. For getting a deeper understanding analysis of public opinions and the initiatives taken by authorities to regulate and promote responsible tourism practices is required.

Due to the given sensitivity of this subject, it becomes clear that the stakeholders must continuously assess and monitor the effects that the dark tourism industry is having on both the educational and exploitative fronts. This would further help to lay the foundation of policies and practices that the visitors must adhere to so that the said industry would not be causing any direct or indirect harm to the places or the affected people.

The Educational Side

The educational side of dark tourism has gained increasing recognition and power in recent years. While the phenomenon remains controversial, it has demonstrated its potential to serve as a powerful tool for educating visitors about historical events, tragedies, and the complexities of the human experience. Several aspects contribute to the growing emphasis on the educational aspect of dark tourism:

- *Historical Awareness*: Dark tourism provides an opportunity for individuals to engage with the past on a more profound and empathetic level. By visiting sites of historical significance, tourists can gain a firsthand understanding of the events that shaped the course of history and appreciate the narratives and experiences of those involved.
- *Preservation of Memory*: Visiting sites associated with tragedy and trauma can contribute to the preservation of collective memory. Through guided tours, exhibitions, and interpretive materials, these sites can serve as living memorials, ensuring that the stories of the affected communities are not forgotten and their experiences are honored.

- *Cultural Understanding*: Dark tourism encourages cultural understanding and empathy by providing insight into the lived experiences of communities affected by past events. By immersing themselves in the historical and cultural context, tourists can develop a more nuanced understanding of the complexities and challenges faced by individuals and societies during difficult times.
- *Promoting Dialogue and Reflection*: Responsible dark tourism initiatives often include opportunities for reflective engagement and meaningful dialogue. Through guided discussions, workshops, and interactive activities, tourists can engage in critical reflection, fostering a deeper understanding of the social, political, and ethical dimensions of historical events. Holding trips to the dark spots like Auschwitz Concentration Camp can provide an opportunity to combat any kind of increasing anti-muslim or antisemitism sentiment in Germany (DW Documentary, 2011).
- *Academic Research and Scholarship*: Dark tourism has also spurred academic research and scholarship, leading to a more comprehensive understanding of the impacts of tragedy and trauma on societies. This scholarly engagement has contributed to the development of multidisciplinary fields, such as memory studies, heritage tourism, and cultural preservation, fostering a deeper appreciation for the educational value of visiting these sites.

By acknowledging the educational potential of dark tourism, stakeholders in the tourism industry can develop responsible and ethical practices that prioritize historical accuracy, cultural sensitivity, and community engagement. Through these efforts, dark tourism can continue to serve as a means of promoting historical awareness, fostering empathy, and preserving the memory of past events for future generations.

The Exploitative Side

The exploitative side of dark tourism has been a subject of concern within the tourism industry. While dark tourism can provide educational value and historical awareness, its exploitative aspects can overshadow the ethical considerations, potentially leading to various negative impacts. Some of the key elements characterizing the exploitative side of dark tourism include:

- *Sensationalization and Commercialization*: The commodification of tragedy and suffering for commercial gain is one of the primary concerns associated with dark tourism. In some cases, the focus on sensationalizing or dramatizing historical events may lead to the trivialization of human suffering, disrespecting the dignity of those affected. People must not click pictures of those who underwent a tragic event and are in their moments of pain and grief. Just to generate revenue from selling such pictures is really a bad idea.

- *Disrespectful Behavior*: Tourists engaging in insensitive behavior, such as taking inappropriate photographs, displaying a lack of empathy, or engaging in disrespectful activities, can contribute to the exploitation of the sites and communities. This behavior can exacerbate the pain and trauma experienced by individuals and communities connected to the historical events. Let us recall the incidence that happened with Logan Paul, a YouTuber; he while making a Vlog in Aokigahara forest, encountered a dead body. Instead of being respectable toward someone who lost his life or showing grief or sorrow, he made fun of the deceased person and also used bad words. This kind of behavior is completely unacceptable at such places.
- *Insensitive Marketing Practices*: The promotion of dark tourism sites using insensitive or sensationalist marketing strategies can contribute to the exploitation of tragic events. Emphasizing shock value or promoting sites solely for their sensational appeal can disregard the historical significance and the dignity of the people involved.
- *Lack of Community Involvement*: When dark tourism sites are developed and managed without the active participation of local communities, there is a risk of marginalizing the voices and perspectives of those directly affected by the events. This exclusion can perpetuate a sense of exploitation and disregard for the communities' needs and concerns. Moreover talking about the community, it is also important that the visitors really need to be cautious about the time of their visits to the tragic spots. The sufferers are often in need of time to heal from the disasters or tragedies. Visiting these spots, when the incidence is still fresh is not acceptable.
- *Preservation Issues*: Over-commercialization and excessive tourism can lead to environmental degradation, physical damage to historical sites, and the erosion of the authenticity and integrity of these locations. Without adequate preservation efforts and responsible management, the sites can lose their historical value and significance, leading to further exploitation of the tragic events they represent. Take Taj Mahal, one of the seven wonders of the world and a spot which is also termed as a dark spot by many, has names of people written all over the walls of the structure. People have tried to take out the expensive carvings from the said structure and eventually ended up ruining some of the most beautiful art works.

To address the exploitative side of dark tourism, it is essential for stakeholders, including tour operators, local communities, and policymakers, to prioritize ethical and responsible practices. This includes promoting sensitivity, cultural understanding, and community engagement while implementing guidelines that respect the dignity of the sites and the individuals connected to them. By adopting a responsible and respectful approach, the tourism industry can mitigate the negative impacts of exploitation and strive to uphold the integrity of these sensitive historical sites.

Research Box

In the recent past, "Dark Tourism" has gained immense popularity and thereby lays a lot of scope for further research in the said industry:

- *Travel Intentions of the Visitors:* People who travel or wish to travel the dark spots often have different intentions that drive them toward the spot. In some cases the motives are surrounding the educational areas; in contrast sometimes the motives are more inclined toward casual visit or showcasing the pictures over social media. Hence, the future endeavors could be streamlined toward getting more grasp as to understanding the intent of the visitors to the dark spots.
- *Development of Dark Tourism from Economic Perspective:* This is undeniable that though a lot of disrespectful acts have been noticed on or around the dark spots, dark tourism poses to bring in a lot of revenue to the dark spot sites and is healthy for the economic situation of especially those places that have faced some disasters. The said revenue helps the affected areas to reconstruct itself. Therefore, the relation of dark tourism and economic development could be further studied.
- *Impact over the Communities:* The communities which have got affected due to the major disasters or the tragic events, often witness gazing eyes over them, in their moments of grief. The people around might be trying to overcome the incidents or may be in need of some people to talk about the event. Looking at what the impact could be, further work of research could be built upon understanding the impact that dark tourism or dark tourists have on the affected communities.

Conclusion

Haley Stainton (TEDx Talk, 2023) stated that travel is the best form of education. People learn from tourism. Since the ancient days, when people used to travel around the world for the trade purposes, they not only earned money but also learned about different cultures, different cuisines, different languages, and more. All of this happened because of trade. In a nutshell it could be said that traveling and learning go hand in hand but sometimes people do both learn and unlearn, while traveling.

The urge to learn about the dark spots, make them visit such places but while being there in order to learn the past, they often forget about the basic respectful ethics that they require to follow there. Rather than paying respect, the students were found walking and enjoying on the railway tracks of Auschwitz, which once carried tons of people towards their deaths. Some others were found capturing smiling selfies inside the concentration camps (Sampson, 2019). This kind of behavior never runs parallel to learning about the events rather goes the other way.

It cannot be denied that the trips organized for dark tourism, brings in money, growth of the economy, means of support, facilitation infrastructural development, repair the host community, income of tour operators and airlines or other modes of transportation. But at the same time bad behavior of tourists, disrespectful, smiling, or laughing in someone's hardships, treating people as if they are exhibits of a museum, inappropriate remarks, wearing disrespectful clothes, signs or words of disrespect, reflects that many of us have become sore at heart.

Dark tourism, with its intricate interplay of educational potential and exploitative tendencies, remains a subject of profound ethical and moral contemplation within the broader tourism industry. Throughout this chapter, we have critically analyzed the multifaceted dimensions of this phenomenon, grappling with the complexities of its impact on both visitors and the destinations themselves.

It has become clear that dark tourism, when approached with sensitivity and respect, can serve as a powerful educational tool, fostering empathy, historical awareness, and a deeper understanding of the human experience. It has the capacity to memorialize the past and honor the memories of those who have suffered, thereby preserving the narratives that are integral to shaping collective memory and societal consciousness.

However, our exploration has also illuminated the potential pitfalls of this form of tourism, emphasizing the need for responsible management and ethical engagement. The risk of exploitation, sensationalism, and the commodification of human tragedy underscores the necessity for a conscientious approach that prioritizes the dignity and integrity of the sites, the communities, and the individuals affected by these historical events.

Moving forward, it is imperative for stakeholders in the tourism industry to adopt a framework that balances the educational benefits of dark tourism with a profound respect for the sensitivity of these sites. This requires the implementation of guidelines and practices that prioritize ethical considerations, community involvement, and sustainable preservation efforts, ensuring that the narratives of these sites are portrayed with integrity and compassion.

In conclusion, while the educational value of dark tourism is undeniable, it is crucial to navigate this terrain with a deep sense of responsibility, empathy, and ethical awareness. By acknowledging the complexities and sensitivities inherent in these visits, we can aspire to foster a form of tourism that not only educates but also cultivates a genuine respect for the past and a profound appreciation for the resilience of the human spirit. Only through such conscientious efforts can dark tourism truly contribute to a more empathetic and enlightened global society.

Discussion Questions

- How dark tourism differs from the other forms of tourism?
- What impact dark tourism has on the people, who were the victims of the incidences?
- Is it ethical to make Vlogs at the tragic spots?
- What are the core intentions of the dark tourists?

References

BuzzFeed News. (2022, January 26). *Meet a Chernobyl tour guide*. YouTube. https://www.youtube.com/watch?v=fIZJjTpwnjg&t=2s

Cramer, M. (2022, October 8). Beaches? Cruises? 'Dark' tourists prefer the gloomy and macabre. *The New York Times*. https://www.nytimes.com/2022/10/28/travel/dark-tourism-travel.html#:~:text=The%20term%20%E2%80%9Cdark%20tourism%E2%80%9D%20was,Attraction%20to%20Death%20and%20Disaster.%E2%80%9D

criminals and crime fighters. (2023, August 25). *Revealing the horrors of the holocaust*. YouTube. https://www.youtube.com/watch?v=6SYN82l6Ves&t=231s

Duff, H. (2018, January 11). *Logan Paul finding a dead body 18+*. YouTube. https://www.youtube.com/watch?v=8NX4KOZ5nvI&rco=1

DW Documentary. (2011, October 11). *Young Germans visit Auschwitz*. YouTube. https://www.youtube.com/watch?v=VQJqZCQj3SQ&t=503s

Forenseca, A. P., Seabra, C., & Silva, C. (2016). Dark tourism: Concepts, typologies and sites. *Journal of Tourism Research & Hospitality*. https://doi.org/10.4172/2324-8807.S2-002

Hindustan Times. (2016, September 6). *Prison tourism: How it feels to spend a day in this 220-year-old Telangana jail*. YouTube. https://www.youtube.com/watch?v=2nZ6h9nfH6Y

History. (2021, December 15). *The UnXplained: Darkness lurks in the Paris Catacombs*. YouTube. https://www.youtube.com/watch?v=4CfazQ2P8D8

Rathee, D. (2022, July 28). *Chernobyl nuclear disaster*. YouTube. https://www.youtube.com/watch?v=t_cSk9tQQkk

Sampson, H. (2019, November 13). Dark tourism explained. *Washington Post*. https://www.washingtonpost.com/graphics/2019/travel/dark-tourism-explainer/?tid=wa_hardregwall_auth&=undefined

Santenello, P. (2023, July 1). *The last witness: 94 – Year – Old holocaust survivor us*. YouTube. https://www.youtube.com/watch?v=b90CcDgeSxo&t=1544s

Saunders, T. (2017, February 11). *The world's most fascinating cemeteries to visit before you die*. Thrillist. https://www.thrillist.com/travel/nation/famous-cemeteries-to-visit

Stone, P. (2006). A dark tourism spectrum: Towards a typology of death and macabre related tourist sites, attractions and exhibitions. *Tourism: An International Interdisciplinary Journal*, *52*, 145–160.

Strange, C., & Kempa, M. (2003). Shades of dark tourism: Alcatroz and Robben Island. *Annals of Tourism Research*, *30*, 386–405.

TEDx Talk. (2017, October 5). *(Dark) tourism and emotions in places of turmoil*. YouTube. https://www.youtube.com/watch?v=lTsEE7TCSuM

TEDx Talk. (2023, September 15). *Tourism and death- the rise of the dark tourism industry*. YouTube. https://www.youtube.com/watch?v=QMCVoL1DfAM

Chapter 4

Prison Tourism: The Emotional Responses to Robben Island, South Africa

Rami K. Isaac

Bethlehem University, Palestine & Breda University of Applied Sciences, The Netherlands

Abstract

The purpose of this chapter is to investigate the motivations and emotions of Western visitors to Robben Island. The methodology used is a qualitative research approach, with 25 participants conducting semi-structured interviews to acquire a better understanding of the visitors' emotions and motives. Visitors' motivations are "historical interest," "earning and understanding," "education," "must-see," "part of itinerary," and "remembrance," according to the findings. Furthermore, this study found that "sadness" is the most frequently encountered emotion by visitors to Robben Island. These insights can benefit a variety of tourism industry stakeholders, including the Robben Island Museum. The data can be used by the Robben Island Museum to improve the visitor experience. By understanding the range of emotions experienced by visitors, what evokes these emotions, and how emotions influence behavior, Robben Island Museum can focus on more effectively evoking these emotions. This will enhance the customer experience and create a greater impact and lasting impression.

Keywords: Motivations; emotions; prison tourism; Robben island; sadness; must-see

Introduction

Tourism is a discretionary leisure time activity freely indulged in for the pleasure it conveys (Ashworth & Hartmann, 2005). However, besides seeking this pleasure in leisure time, other motives for tourism have been brought to light recently. This

Dark Tourism, 41–59

Published under exclusive licence by Emerald Publishing Limited
doi:10.1108/978-1-83797-336-120241004

other side of tourism, where people actively seek experiences associated with death and atrocities, is called dark tourism (Foley & Lennon, 1996).

Although the concept of dark tourism was only scientifically been introduced in 1996 (Foley & Lennon, 1996), and therefore is the youngest subset of tourism (Kunwar & Karki, 2019), much earlier examples of dark tourism can be found. Events with death and suffering at the core have existed long before 1996. According to Stone (2005), one of the earliest examples of dark tourism is seen in the patronage of the Roman gladiator games. Death and suffering were the main attractions of these events, which makes that these events can be considered one of the first dark tourism attractions. Moreover, these attractions with a strong tie to death, tragedy and suffering continued to exist and expand. According to Lennon and Foley (2000), in this past decade, the growth of the number of dark tourists has been significant. This growth in the number of dark tourists and therefore the attention given to events of death, suffering and atrocity is related to the curiosity of people toward mortality. Furthermore, it has become an increasingly popular topic among researchers and consequently, several subsets of dark tourism have presented themselves (Isaac, 2022). One of these subsets on which this research will be focused is prison tourism.

Prison tourism relates to the act of visiting penal establishments, which were used in the past to exile prisoners by separating them from society and putting them in remote locations (Ashworth & Hartmann, 2005). Prisons that are seen as past sites of punishment confinement and oppression have been turned into tourism attractions all over the world. Robben Island, located just off the coast of Cape Town, South Africa, is one of the most famous penal establishments in the world.

Currently, there is a gap in existing literature, since no research has been done regarding the motivations and emotions of Western visitors to Robben Island, South Africa. There are several case studies on Robben Island (Deacon, 2004; MacGonagle, 2016). However, only one study related to motivations and emotions at Robben Island by Strange and Kempa (2003) is found in existing literature. Strange and Kempa (2003) completed their research many years ago, so it's interesting to observe how the discipline has progressed since then. As a result, the disparity persists. There is no research on the reasons and emotions of Western tourists to Robben Island. Therefore, this chapter explores the motivations and emotions of Western tourists, visiting Robben Island. The focus is on Western Tourists of all ages, because of their significant presence in global tourism (Hall & Page, 2014) and because of the connection they feel toward the history of Apartheid. Western tourists can be defined as tourists traveling to destinations from the United States, Europe, Canada, and Australia. This research aims to better understand what kind of emotions these visitors experience during their visit to Robben Island. Furthermore, it explores their motivations, as to why these visitors are attracted to this site. The relevance of this study lies in the provision of insight into the consumption of dark tourism. The outcome will show motives that drive tourists to visit dark tourism sites, emotions evoked at dark tourism sites, and how these emotions can influence tourists' behavior. These insights can be valuable for various stakeholders in the tourism industry, specifically Robben Island Museum. The data can be used by Robben Island Museum to enhance the customer experience. By understanding the range of emotions experienced by visitors, what evokes these emotions, and how emotions influence behavior,

Robben Island Museum can focus on more effectively evoking these emotions. This will enhance the customer experience and create a greater impact and lasting impression.

Literature Review

Dark tourism was introduced in 1996 by Foley and Lennon (1996). However, long before the introduction of the term, travels with death and atrocity at its core have taken place. Furthermore, the first guided tour to places of death was recorded in England in 1838 (Stone, 2005). During this guided tour in England, tourists witnessed the hanging of two convicted murderers. Nowadays, the term dark tourism has expanded, and many places are considered dark tourism sites. However, some of the most famous dark tourism sites nowadays are John F. Kennedy's death site in Dallas, Texas, and the Ground Zero 9/11 Memorial in New York (Strange & Kempa, 2003). Additionally, ever since the introduction of the term in 1996, multiple subsets of dark tourism have been introduced in academic research such as war/battlefield tourism, disaster tourism, cemetery tourism, ghost tourism, holocaust tourism, and prison tourism (Fonseca et al., 2016).

The term dark tourism refers to the act of traveling to sites associated with death and atrocities. According to Foley and Lennon (1996), the development of this term has been possible because of the media. Therefore, dark tourism is both a product of circumstances of the late modern world and a significant influence upon those circumstances. However, politics, sociology, and current technologies also play an important role in creating these dark tourism sites (Foley & Lennon, 1996).

Prison Tourism

Prison tourism is a subset of dark tourism (Fonseca et al., 2016). Prison tourism refers to the act of visiting a prison that is no longer in use (Aslan, 2015). Prison tourism as a subset of dark tourism has not yet been explored in depth in academic papers, which means the information about the subset of prison tourism is limited (Aslan, 2015). According to Strange and Kempa (2003), the opportunities for prison tourism have only recently arisen, in the late 20th century. This is because only in the last two decades empty prisons have been transformed and put to new use. Some have been transformed into hotels, others into museums or tourism sites. Therefore, the lack of academic papers on this topic can be explained. However, it can be seen that curiosity related to the history of penal establishments and the features of punishment is increasing. Therefore, the opportunity for prison tourism to blossom is created (Strange & Kempa, 2003).

Although there are not many academic papers regarding prison tourism, as a subset of dark tourism, there are many decommissioned prisons that are transformed into touristic attractions. According to Aslan (2015), there are more than 100 prison museums worldwide. These prisons often offer an entertaining, historical, educational, and cultural experience because of the history behind these prisons. According to Isaac and Çakmak (2016), the two most well-known prisons that have opened up for the public are Alcatraz and Robben Island. The latter of the two, Robben Island, plays a central role in this research. The island served as a maximum security prison from 1961 until 1991 (Robben Island Museum, 2023). In January 1997, Robben Island officially opened to the public (Robben Island Museum, 2023). Therefore, Robben Island is seen as a prison tourism destination due to its historical significance as a former prison and its current significance as a tourism destination that is a symbol of the struggle against Apartheid.

Motivations for Dark Tourism

According to Beh and Bruyere (2007; as cited by Isaac et al., 2019), the term motivation refers to the inner force that drives and directs human behavior. There are many studies regarding the motivations for visiting places of death and atrocity; dark tourism sites. This provides different perspectives and opinions on the topic, which also causes discussions among different researchers. Therefore, there are several reasons provided about what the motivations for visiting dark tourism sites can be.

There are multiple arguments presented (Ashworth & Hartmann, 2005) as to why people are attracted to this phenomenon which is death or atrocity. The first argument is the curiosity argument. It is stated that people are naturally curious about the unusual or the unique. Therefore, they might be drawn to sites of atrocity or death. The second argument presented is the empathy argument, which relates to the identification of the visitor with the victims. However, this raises an issue because although the message might be intended to raise empathy for the victim, the tourists might interpret it differently. This leads to the possibility of visitors empathizing with the perpetrators instead of the victim. The third argument is the horror argument, which refers to the attraction to horrific occurrences because of the emotions of fear and fascination that it causes. Although these motivations seem legitimate, other researchers looked into the motivations for dark tourism, which resulted in other outcomes. According to Seaton (2009), the main motives for visiting dark tourism sites are the pleasure of seeing others' misfortune and the contemplation of your death.

However, according to Biran et al. (2011, as cited by Kunwar & Karki, 2019) dark tourism motivations can be divided into four factors: See it to believe it; Learning and understanding; Famous death tourist attractions; and Emotional heritage experience.

To summarize all motivations found in tourism literature, a list of all conceptual and empirical motivations has been created by Isaac et al. (2019) (Table 4.1). To provide a complete overview, this table has been updated with more recent findings within the literature.

Overall, it can be stated that the motivations for engaging in dark tourism are complex and can vary a great deal. This is due to the wide range of dark tourism sites, the individuals partaking in dark tourism, and their connection to the site, which results in different types of motivations. Therefore, it is interesting to analyze the underlying motivations of tourists visiting a "dark" tourism site, Robben Island.

Table 4.1. List of Conceptual and Empirical Motivations.

Authors	Empirical Motivations	Conceptual Motivations
Foley and Lennon (1996)		Education, commemoration, entertainment
Ashworth (2002; as cited by Isaac et al., 2019)		Horror, curiosity, empathy and understanding
Ashworth (2004; as cited by Isaac et al., 2019)		Pilgrimage, identity, education, horror, empathy
Ashworth and Hartmann (2005)		Curiosity, out of ordinary
Dann (1998; as cited by Isaac et al., 2019)		Desire to overcome phantom, novelty, nostalgia, celebration of crime, deviance, bloodlust, interest, and mortality
Rojek (1997; as cited by Isaac et al., 2019)		Sense of social and cultural identity, lifestyle, media attention, emotional and psychological motives
Seaton (1999; as cited by Isaac et al., 2019)		Moral witness, social experience, seeking to sacralize the site, pride, pilgrimage
Seaton and Lennon (2004; as cited by Isaac et al., 2019)		Schadenfreude (the pleasure of viewing others' misfortune)
Cole (1999; as cited by Isaac et al., 2019)		Voyeurism
Tarlow (2005)		Sense of insecurity, sense of gratitude, feelings of humility, feeling of superiority, curiosity, must do

(Continued)

Table 4.1. *(Continued)*

Authors	Empirical Motivations	Conceptual Motivations
Stone (2012; as cited by Isaac et al., 2019)		Education, entertainment, memorization, moral instruction
Isaac (2022)		Challenging own sense of mortality
Dunkley et al. (2007; as cited by Isaac et al., 2019)	Pilgrimage, remembrance, validation, special interest	
Ryan (2007; as cited by Isaac et al., 2019)	Special interest, thrill/risk seeking, validation, authenticity, self-discovery, iconic sites, convenience, morbid curiosity, pilgrimage, self-remembrance, empathy, and contemplation	
Hughes (2008; as cited by Isaac et al., 2019)	Impulse purchase, must-see	
Bigley et al. (2010)	To experience opposing political regimes, curiosity, historical interest, educational purposes, pleasure, unique experience, to feel the importance of freedom and peace	
Braithwaite and Leiper (2010; as cited by Isaac et al., 2019)	Recreational purposes, dark emotions	
Biran et al. (2011; as cited by Isaac et al., 2019)	See it to believe it, learning and understanding, famous death tourist attraction, emotional heritage experience	

Table 4.1. *(Continued)*

Authors	Empirical Motivations	Conceptual Motivations
Kang et al. (2012; as cited by Isaac et al., 2019)	Learning and obligation, social reasons and curiosity, emotional heritage experience	
Podoshen (2013; as cited by Isaac et al., 2019)	Simulation, emotional contagion, topographic comparison	
Raine (2013)	Interest in death, horror, and morbid curiosity	
Isaac and Çakmak (2014)	Self-understanding, curiosity, conscience, a must-see place, and exclusiveness	
Biran et al. (2014; as cited by Isaac et al., 2019)	To reap the benefits of contemplating life, one's mortality, education	
Korstanje and George (2015)	Fascination with death	
Isaac and Çakmak (2016)	Remembrance, worth visiting, learning and understanding, paying respect, must-visit	
Isaac et al. (2019)	Memory, gaining knowledge and awareness, exclusivity	
Mangwane et al. (2019)	Novelty, knowledge-seeking, remembrance and respect for victims, curiosity, recommendations, value for money, fun, family, museum attributes	
Lewis et al. (2021)	Satisfying curiosity, interaction with the paranormal, escape from everyday life	

Source: Author's table.

Emotions

A definition that is used by multiple researchers is that emotions are felt, short-lived responses to external stimuli (Isaac & Çakmak, 2016; Kunwar & Karki, 2019). Moreover, according to Hupka et al. (1999; as cited by Isaac & Budryte-Ausiejiene, 2015), the types of emotions that people experience are largely universal. Furthermore, it is stated by Frederickson and Losoda (2005; as cited by Isaac & Budryte-Ausiejiene, 2015) that emotions can have a big effect on how people think and behave. Lastly, emotions are believed to have long-term consequences for the quality of life (Frederickson & Losoda, 2005; as cited by Isaac & Budryte-Ausiejiene, 2015).

To understand how emotions influence behavior within the tourism industry, it is of great importance to understand the impact of emotions. However, as noted by Isaac and Budryte-Ausiejiene (2015), there is a lack of empirical studies that explore the emotional association or the meaning visitors assign to the destinations they visit.

The emotion shame is seen as the central emotion to the definition of dark tourism as a whole (Biran & Poria, 2012; as cited in Ashworth & Isaac, 2015). Moreover, according to Dann (2015, as cited in Ashworth & Isaac, 2015), psychotic disturbance is likely to occur at dark tourism sites.

Multiple researchers researched emotions at dark tourism sites. For example, Tarlow (2005) claims four basic emotions interact with the psychological state of dark tourists. These emotions are insecurity, gratitude, humility, and superiority. Moreover, it is stated (Austin, 2002; Brown, 2016; Dunkley et al., 2011; Isaac & Çakmak, 2016; Kidron, 2013; Mowatt & Chancellor, 2011; Zhang et al., 2016; as cited in Isaac, 2022) that the most commonly found emotions within dark tourism are sadness, sorrow, grief, anger, and horror. Additionally, disappointment (Podoshen, 2013; as cited by Isaac, 2022), shock and fear (Buda et al., 2014; Zheng et al., 2017; as cited by Isaac, 2022), and a sense of hope (Koleth, 2014; as cited by Isaac, 2022) are also mentioned as frequently experienced emotions at dark sites.

Therefore, it can be stated that although emotions are part of any type of tourist experience (Isaac, 2022), the emotions experienced at dark tourism sites are often darker. However, this does not rule out the possibility of people experiencing positive emotions at dark sites (Nawijn et al., 2016). Positive emotions often experienced at dark sites, include a sense of hope and stimulation of pride (Cheal & Griffin, 2013; Koleth, 2014; Sharpley, 2012; as cited by Isaac, 2022). However, according to Ashworth and Isaac (2015, p. 320), classifying negative and positive emotions in terms of the outcome is not always fair since negative emotions can also lead to positive outcomes and the other way around. This is also confirmed by Oren et al. (2021; as cited by Isaac, 2022) since they state that negative emotions can have a positive contribution which in turn contributes importantly to visitors' satisfaction. Table 4.2 lists empirical emotions and conceptual emotions described in the literature.

Table 4.2. List of Conceptual and Empirical Emotions.

Authors	Empirical Emotions	Conceptual Emotions
Austin (2002)	Anger, anguish, sorrow, sadness	
Strange and Kempa (2003)	Sadness, admiration, respect, courage, forgiveness	
Dunkley et al. (2011; as cited by Isaac, 2022)	Sadness, sorrow, grief, anger, horror	
Mowatt and Chancellor (2011)	Sadness, anger	
Podoshen (2013; as cited by Isaac, 2022)	Disappointment	
Koleth (2014)	Hope	
Brown (2016)	Emotional, moved, choked	
Isaac and Çakmak (2016)	Horrific, scary, depressing, shocking, sadness, disturbing, intense, powerful, eye-opener, unforgettable, glad I visited	
Weaver et al. (2018)	Hatred, anger, depression, sadness, shock, contemplation, pity	
Tarlow (2005)		Insecurity, gratitude, humility, superiority
Ashworth and Isaac (2015)		Curiosity, shame, empathic grief, psychotic disturbance, offense/anger, fear
Cave and Buda (2018)		Fear, fun, excitement, fascination with death, anger
Isaac (2022)		Sadness, sorrow, grief, anger, horror, disappointment, shock, fear, sense of hope, repulsion, disgust

Source: Author's table, Isaac.

Methodology

For this research on examining the emotions and motivations of Western visitors at Robben Island, a qualitative research approach is adopted. Creswell (2013) states that qualitative research is particularly well-suited for exploring the experiences that individuals have and what meaning individuals attach to those experiences. It focuses on gathering non-numerical data, making it ideal for creating a deeper understanding of subjective experiences, opinions, and attitudes (Creswell, 2013). Qualitative research can be used to gain insight through interviews, participant observations, or existing documentation (Baarda, 2020).

Semi-structured interviews have been conducted with Western tourists who just visited Robben Island. According to Adeoye-Olatunde and Olenik (2021), semi-structured interviews are particularly beneficial when wanting to be focused, but also having the opportunity to explore pertinent ideas that come up during the interview. Other researchers explored dark tourism through the use of interviews (Light & Ivanova, 2021; Strange & Kempa, 2003). Therefore, a sample size of 25 interviews has been established for this research. To ensure the correct data is gathered through the interviews, the interview is divided into different sections: motivations, emotions behavior, and intentions. Intentions refer to the intentions of the visitor for the future, so are they willing to recommend or revisit Robben Island? To get to Robben Island, visitors need to take a 30–45-minute ferry ride. This means, that after they participate in the tour of the island, they need to take the ferry back to the mainland. This presented the ideal opportunity to conduct interviews, as during this time the visitors were on the ferry with no other distractions. Therefore, they were available to participate in interviews. Moreover, the experience at Robben Island was freshly in their memories, which led to more valuable responses. In total 25 interviews, of 15–25 minutes, with Western Tourists were done. To ensure that the participants fit into the target group, Western tourists, the interviewer asked about their nationality before conducting the interview. The semi-structured interviews were conducted on March 9th and 10th 2023 at Robben Island. The interviews are transcribed and analyzed using the Grounded theory (Wagner et al., 1968), and biographical details of the interviewees are given in Table 4.3.

Findings and Discussion

Sociodemographics and Travel Planning

The study includes participants with a wide age range, from 18- to 78-years old. The average age of the participants is 49-years old, which suggests the inclusion of participants from different age groups. Additionally, the sample is diverse when it comes to gender; 60% of the participants identify as female and 40% of the participants identify as male. Moreover, Western tourists from several nationalities participated in this research. The majority of the participants come from the United States and the United Kingdom, followed by participants from the Netherlands. The study also includes participants from France, Ireland, Malta,

Table 4.3. List of Participants.

Participant	Age	Gender	Country of Residence
Participant 1	76	Male	France
Participant 2	61	Female	France
Participant 3	29	Female	The Netherlands
Participant 4	63	Male	The United States
Participant 5	40	Female	The United States
Participant 6	36	Female	The United Kingdom
Participant 7	78	Female	The United States
Participant 8	53	Male	The Netherlands
Participant 9	73	Female	The United Kingdom
Participant 10	43	Male	The United Kingdom
Participant 11	20	Female	The Netherlands
Participant 12	29	Male	The United States
Participant 13	61	Female	The United Kingdom
Participant 14	67	Male	The United Kingdom
Participant 15	66	Female	The United Kingdom
Participant 16	33	Female	The United States
Participant 17	49	Female	The Netherlands
Participant 18	50	Male	The Netherlands
Participant 19	61	Female	The United States
Participant 20	56	Female	The United States
Participant 21	63	Male	England
Participant 22	64	Male	Ireland
Participant 23	30	Male	Malta
Participant 24	18	Male	The Netherlands
Participant 25	23	Female	Sweden

Source: Author's table, Isaac.

and Sweden. However, the latter groups represent a smaller portion of the sample compared to the former three. Additionally, the research points out who the participants' travel companions are, while visiting Robben Island. Forty six percent of the participants are visiting with their family, 25% are accompanied by their partners, and 17% are visiting with friends. The remaining 12% reported visiting with other travel companions, such as a group tour or colleagues. To get a better insight into the diversity of the sample, figures have been created, and Table 4.4 shows the age composition of the interviewees.

Table 4.4. Participant's Age.

Age Range	Count of Participants
18–29	5
30–39	3
40–49	3
50–59	3
60–69	8
70–79	3
Total	25

Source: Author's table, Isaac.

Apart from sociodemographics, there are several findings related to travel planning and execution that are worth mentioning. For 76% of the participants, this was their first time visiting South Africa. Additionally, for 100% of the participants, it was their first time visiting Robben Island. Moreover, when it comes to booking their trip, the vast majority of the participants mentioned booking with a travel agency or tour operator. A smaller amount of participants suggested they booked online by themselves. Although the majority of the participants suggested they booked their trip to South Africa with a tour operator, this wasn't the case when booking their tour to Robben Island. Thirteen people booked their tour to Robben Island online, by themselves. Two participants specifically pointed out that they booked through the Website of Robben Island and three people mentioned booking with Get Your Guide. Only three people mentioned that the tour to Robben Island was prebooked by a tour operator or an agency.

Motivations

A holiday is the most mentioned motivation for visiting South Africa, with nine respondents mentioning this as their main motive for traveling there. "It is vacation, it is a trip I have been wanting to take since I was a kid and now he agreed to come along" (Participant 19). Additionally, weddings were mentioned as a motivation to travel to South Africa, four times. "Our eldest nephew is getting married here. His first man lives over here, so he thought it would be a nice idea for us all to come" (Participant 14). Next to that, two respondents stated that they traveled to South Africa for their honeymoon. This suggests that South Africa is a popular destination for weddings and/or honeymoons. Furthermore, safaris and seeing the big five are mentioned three times as a motivation. "We were all very excited to see all of the large animals in the wild and enjoy the nature" (Participant 17). Other motivations that are mentioned less frequently include business, visiting a new destination, interest in the story of South Africa and Apartheid, and recommendations from friends.

Moreover, the findings from the interviews reveal a diverse range of motivations for visiting Robben Island. The main motive, mentioned by 13 respondents, is a deep historic interest. "For me, it was historical reasons, yeah the general interest about what happened and what they went through" (Participant 21). "While you are here I think it is something you have to see, to learn more about the history of the country as well" (Participant 25). These 13 respondents mention being interested in learning about the history, and express the importance of visiting places with a history like Robben Island to create awareness and remember what happened. "I think it is a very important piece of history that needs to be told. To make sure this will never happen again" (Participant 17).

What is interesting to note is that 10 respondents mention already having some knowledge of the history of South Africa. They mention either having heard about it on the news, or reading about it because they were interested. "On TV we saw people with t-shirts and listening to songs about freeing Nelson Mandela. We remember the Apartheid movement and to have seen it then and to see it up close now. We are re-living it" (Participant 13). "I have done a lot of reading about Mandela and some of the other freedom fighters of the Apartheid era and wanted to see it in real life. It is still very shocking" (Participant 19). Seven out of ten participants who already had some knowledge of the history of South Africa are older than 60. From this, it can be stated that the generation of over 60-years old are already especially knowledgeable about the history before they visited Robben Island. Additionally, many respondents state Apartheid and Nelson Mandela as the main reasons for visiting Robben Island. Five respondents mentioned being interested in Apartheid, and nine respondents mentioned Nelson Mandela. "I always wanted to see where Nelson Mandela was. That is the main purpose" (Participant 22). Next to that, educational purposes are mentioned five times by the respondents. All of the abovementioned reasons relate to each other, since Nelson Mandela and Apartheid are a big part of South African history and people visit to educate themselves on this important part of history.

Furthermore, multiple respondents mention Robben Island is a must-see when visiting South Africa. "You just got to do it don't you? It is a must-see" (Participant 13). Another motive mentioned a few times is the fact that it is part of the itinerary or that it is simply because they are in Cape Town anyway. What is interesting to see is that the participants under 25-years old are all on holiday with their families. "My parents booked this tour because they wanted this holiday to be educative as well" (Participant 11). Moreover, other participants who mention Robben Island being part of their itinerary are above the age of 60. Therefore, it can be stated that for these age groups, it is not their own choice to visit Robben Island, but their parents' choice or the travel agencies' choice. Furthermore, less frequently mentioned motivations for visiting Robben Island are recommendations from friends, curiosity, and getting more cultural background.

All in all, the results from the interviews highlight the importance of history in motivating Western Tourists to visit Robben Island. The number of participants who mentioned these visits, and the reasons they cited, are shown in Table 4.5.

Table 4.5. Visitors' Motivations for Visiting Robben Island.

Motivation	Amount of Participants that Mentioned it
Historical interest	13
Apartheid	5
Nelson Mandela	9
Educational	5
Itinerary	5
Must-see	5
Awareness/remembrance	5
Other[a]	4

Source: Author's own table, Isaac.

[a]Responses included curiosity, recommendations from friends, and more cultural background.

Emotions

After a thorough analysis of the responses, it can be concluded that sadness and anger are the most commonly mentioned emotions by tourists visiting Robben Island. "Anger and sadness, because the people they were treated so unfairly" (Participant 8). What is interesting to note is that both of these emotions are mentioned multiple times by several participants, indicating this emotion was of higher intensity for these individuals. Anger was of high intensity since it has been mentioned four times by one participant, indicating its very high intensity for that individual. "Infuriating, enraging. Enraging is the word. This made me angry" (Participant 12). Many participants were moved by the fact that an ex-prisoner gave the tour inside the prison and were deeply affected by their stories. This can be validated by the observations as well, since from observations it became evident that visitors were deeply impacted by the stories the guide told. People showed signs of shock, and sadness, were horrified, and felt empathy. Therefore, this was identified as the most important emotional trigger. It evoked a sense of empathy among the visitors, who also felt ashamed of the unfair treatment that the prisoners had to endure. "I think especially speaking to a prison survivor. It is very difficult to envision now that people were treated that way, and kept in such environments purely based on their political beliefs" (Participant 25). The intense and unjust treatment of prisoners which the ex-prisoner tells them about during the tour is a significant emotional trigger for many participants. As a result of this, a sense of disbelief is mentioned frequently by the participants, as they struggled to understand how people could treat each other like that. "The fact that how much food or clothes you would receive was determined by race or how hard you would work. It is just so sad and unbelievable" (Participant 17).

Some people even mentioned getting emotional during the tour. "I cried while listening to the tour guide telling us about what he had experienced himself and why it is so important that all of their stories are being heard. Like he said 'so that this will never happen again'" (Participant 17). One participant became emotional during the interview again. When asked about the living circumstances of the prisoners he started crying and said "Oh it is actually, it's horrific. It is just awful, there is no need for it" (Participant 14). Furthermore, several participants mention being shocked by the stories, and several participants mention feeling scared as a response to this.

Although the majority of the emotions experienced are negative, several participants mention feeling positive emotions as well. The positive emotion that is mentioned most frequently is impressed, followed by gratitude, hope, and moved. These participants mentioned experiencing a sense of gratitude for the privileges they have in their own countries, which they realize they should not take for granted. "Hope and also thankful for how we live in Sweden and the privileges that we have. So gratitude, yes" (Participant 25). They mentioned being impressed by the stories and feeling hope and encouragement. "You feel sad. But also, encouraged I guess by how prisoners like Mandela managed to keep their spirits and stay committed and focused and driven during their whole sentence" (Participant 20). Further details are given below in Table 4.6.

Table 4.6. Visitors' Emotions Related to Their Experience at Robben Island.

Type of Emotion	**Emotions**	**Total Quantity Mentioned**	**Amount of People that Mentioned it**
Negative emotions	Sadness	19	10
	Anger	10	4
	Shame	5	4
	Disbelief	8	7
	Fear	3	3
	Emotional	11	10
	Shock	4	3
	Empathy	3	1
	Other[a]	7	7
Positive emotions	Impression	10	6
	Hope	3	1
	Gratitude	4	2
	Moved	3	3
	Other[b]	4	4

Source: Author's own table, Isaac.

[a]Responses included pity, disappointment, awe, astonishment, and confusion.

[b]Responses included fascination, a sense of wonder, encouragement, and excitement.

Discussion

The findings suggest that the most important motivation of Western tourists visiting Robben Island is "historical interest." Remarkably, existing literature on dark tourism does not explicitly mention historical interest as a motive. Although learning and understanding (Isaac & Çakmak, 2016; Biran et al., 2011; as cited by Kunwar & Karki, 2019) can be identified as similar, historical interest is not mentioned literally. Within the research by Biran et al. (2011; as cited by Kunwar & Karki, 2019), learning and understanding refers to the specific interest in being educated about the Second World War and the atrocities that took place in Auschwitz. Therefore, these historical events are what motivated tourists to visit this dark tourism site. In this particular research, historical interest refers to the interest in being educated about the history of South Africa, specifically the Apartheid era. Moreover, within this research, the connection between Auschwitz and Robben Island came to light. Several participants either compare Robben Island to Auschwitz or mention Robben Island triggering their interest in visiting other dark tourism sites in the future, Auschwitz in particular. Consequently, it can be stated that learning and understanding (Biran et al., 2011; as cited by Kunwar & Karki, 2019) can be identified as similar to historical interest, making it a usual motivation within dark tourism.

Furthermore, other important motivations arising from this research include "education," "must-see," "part of itinerary," and "remembrance." Curiosity is a common motivation for visiting dark tourism sites (Ashworth & Hartmann, 2005; Isaac & Çakmak, 2014; Mangwane et al., 2019). Additionally, learning and understanding (Isaac & Çakmak, 2016; Biran et al., 2011; as cited by Kunwar & Karki, 2019) and knowledge-seeking (Mangwane et al., 2019) are existing motivations in literature, which can be viewed as similar to educational motivations. The educational aspect of learning and understanding (Biran et al., 2011; as cited by Kunwar & Karki, 2019) is why it relates to the newly explored motivation of "education" in this research. Moreover, according to Mangwane et al. (2019), knowledge-seeking is the main motivation for visiting dark tourism sites that arises from their research. It refers to the urge to learn about history and to increase knowledge about sites of human tragedies. Therefore, it can be seen as similar to the motive arising from this research, which is to educate. Thus, "education" can be seen as a recurring motivation for visiting dark tourism sites.

According to some participants, it is one of the most well-known sites in South Africa. Therefore, the majority of the participants feel they have to visit Robben Island while they are in South Africa. This motivation aligns with findings from research by Isaac and Çakmak (2016), which reveals must-see as an important motivation within dark tourism. In this research by Isaac and Çakmak (2016), must-see specifically relates to visitors considering the site as a must-see attraction. Thus, these motivations mentioned in both the research reports are the same, making "must-see" a common motivation within dark tourism.

Moreover, "remembrance" has been documented in the literature as well (Isaac & Çakmak, 2016; Mangwane et al., 2019), referring to the urgency people feel, not to forget what happened in the past. Finally, "Part of itinerary," is a new finding, not yet identified in previous studies. Less frequently mentioned motivations that are in line with existing literature include "recommendations from friends" (Mangwane et al., 2019), "curiosity" (Ashworth & Hartmann, 2005; Isaac & Çakmak, 2016; Mangwane et al., 2019), and "budget-friendly" (Mangwane et al., 2019).

This research points out that the most frequently experienced emotion of visitors at Robben Island is "sadness." This can be supported by Isaac (2022), since according to him one of the most found emotions at dark tourism is sadness. Furthermore, other emotions most present according to the findings are "emotional," "disbelief," "impressed," "anger," and "shock." Within the literature review of this report, it becomes apparent that anger (Ashworth & Isaac, 2015; Isaac, 2022) and shock (Isaac, 2022) have been discovered by other researchers. According to Isaac (2022), anger is, along with sadness, one of the most found emotions within dark tourism literature. Therefore, it can be stated that anger and sadness are both frequently experienced emotions by tourists visiting dark tourist sites.

Conclusion

This chapter explores the motivations and emotions of Western tourists, visiting Robben Island. It can be stated that the findings from this research are partially in line with existing literature on the topic. Therefore, this research contributes to creating a deeper and new understanding of the motivations and emotions of Western tourists visiting Robben Island. This research has practical implications. First, adding more visuals throughout the entire tour. Based on the collected data, it becomes evident that a notable number of visitors express a desire for enhanced visual elements throughout the tour. These individuals emphasize the importance of incorporating pictures to create a more vivid understanding of prison life during the Apartheid era. Second, implementing an effective online marketing strategy. The main motivations for visiting Robben Island are historical interest, must-see, educational purposes, itinerary, and remembrance. From the findings, it became apparent that the vast majority of the visitors book their tour to Robben Island online.

Finally, there are several suggestions for future research that could provide valuable insights. The first suggestion would be to include interviews with professionals managing the site, to provide a broader and different angle. Furthermore, instead of using semi-structured interviews to collect data, a survey could be used. Surveys do not require presence at the site, and no direct contact with the participant. Therefore spreading surveys could lead to a bigger sample size.

References

Adeoye-Olatunde, O. A., & Olenik, N. L. (2021). Research and scholarly methods: Semi-structured interviews. *Journal of the American College of Clinical Pharmacy*, *4*(10), 1358–1367.

Ashworth, G. J., & Hartmann, R. (2005). *Horror and human tragedy revisited: The management of sites of atrocities for tourism*. Cognizant.

Ashworth, G., & Isaac, R. K. (2015). Have we illuminated the dark? Shifting perspectives on 'dark' tourism. *Tourism Recreation Research*, *40*(3), 316–325.

Aslan, S. (2015). Prison tourism is a form of dark tourism. *The Journal of International Social Research*, *8*(40), 600–608.

Austin, N. K. (2002). Managing heritage attractions: Marketing challenges at sensitive historical sites. *International Journal of Tourism Research*, *4*(6), 447–457.

Baarda, B. (2020). *Research: This is it* (3rd ed.). Noordhoff.

Bigley, J. D., Lee, C., Chon, J., & Yoon, Y. (2010). Motivations for war-related tourism: A case of DMZ visitors in Korea. *Tourism Geographies*, *12*(3), 371–394.

Brown, L. (2016). Tourism and pilgrimage: Paying homage to literary heroes. *International Journal of Tourism Research*, *18*(2), 167–175.

Cave, J., & Buda, D. (2018). Souvenirs in dark tourism: Emotions and symbols. In *The palgrave handbook of dark tourism studies* (pp. 707–726). Palgrave Macmillan.

Creswell, J. W. (2013). *Qualitative inquiry and research design: Choosing among five approaches* (3rd ed.). SAGE.

Deacon, H. (2004). Intangible heritage in conservation management planning: The case of Robben Island1. *International Journal of Heritage Studies*, *10*(3), 309–319.

Foley, M., & Lennon, J. J. (1996). JFK and dark tourism: A fascination with assassination. *International Journal of Heritage Studies*, *2*(4), 198–211.

Fonseca, A. P., Seabra, C., & Silva, C. (2016). Dark tourism: Concepts, typologies and sites. *Journal of Tourism Research and Hospitality*, *0*(2), 1–6.

Hall, C. M., & Page, S. J. (2014). *The geography of tourism and recreation: Environment, place and space* (4th ed.). Routledge.

Isaac, R. K. (2022). *Dark tourism studies* (1st ed.). Routledge.

Isaac, R. K., & Budryte-Ausiejiene, L. (2015). Interpreting the emotions of visitors: A study of visitor comment books at the Grūtas Park Museum, Lithuania. *Scandinavian Journal of Hospitality and Tourism*, *15*(4), 400–424.

Isaac, R. K., & Çakmak, E. (2014). Understanding visitor's motivation at sites of death and disaster: The case of former transit camp Westerbork, The Netherlands. *Current Issues in Tourism*, *17*(2), 164–179.

Isaac, R. K., & Çakmak, E. (2016). Understanding the motivations and emotions of visitors at Tuol Sleng Genocide Prison Museum (S-21) in Phnom Penh, Cambodia. *International Journal of Tourism Cities*, *2*(3), 232–247.

Isaac, R. K., Nawijn, J., Van Liempt, A., & Gridnevskiy, K. (2019). Understanding Dutch visitors' motivations to concentration camp memorials. *Current Issues in Tourism*, *22*(7), 747–762.

Koleth, M. (2014). Hope in the dark: Geographies of volunteer and dark tourism in Cambodia. *Cultural Geographies*, *21*(4), 681–694.

Korstanje, M. E., & George, B. (2015). Dark tourism: Revisiting some philosophical issues. *E-review of Tourism Research*, *12*, 127–136.

Kunwar, R. R., & Karki, N. (2019). Dark tourism: Understanding the concept and recognizing the values. *Armed Police Force Command and Staff College Journal*, *2*(1), 42–59.

Lennon, J. J., & Foley, M. (2000). *Dark tourism the attraction of death and disaster* (3rd ed.). Thomson Learning.

Lewis, H., Schrier, T., & Xu, S. (2021). Dark tourism: Motivations and visit intentions of tourists. *International Hospitality Review*, *36*(1), 107–123.

Light, D., & Ivanova, P. (2021). Thanatopsis and mortality mediation within "lightest" dark tourism. *Tourism Review*, *77*(2), 622–635.

MacGonagle, E. (2016). *History and memory in an African context* (1st ed.). Cornell University Press.

Mangwane, J., Hermann, U. P., & Lenhard, A. I. (2019). Who visits the apartheid museum and why? An exploratory study of the motivations to visit a dark tourism site in South Africa. *International Journal of Culture, Tourism and Hospitality Research*, *13*(3), 273–287.

Mowatt, R. A., & Chancellor, C. (2011). Visiting death and life: Dark tourism and slave castles. *Annals of Tourism Research*, *38*(4), 1410–1434.

Nawijn, J., Isaac, R. K., Van Liempt, A., & Gridnevskiy, K. (2016). Emotion clusters for concentration camp memorials. *Annals of Tourism Research*, *61*, 244–247.

Raine, R. (2013). A dark tourist spectrum. *International Journal of Culture, Tourism and Hospitality Research*, *7*(3), 242–256.

Robben Island Museum. (2023). *Robben island museum tour*. Tour.

Seaton, T. (2009). Chapter 5. Purposeful otherness: Approaches to the management of thanatourism. In *Multilingual matters eBooks* (pp. 75–108). Bristol, Channel View.

Stone, P. R. (2005). Dark tourism – An old concept in a new world. *Tourism Magazine by the Tourism Society*, *125*. https://works.bepress.com/philip_stone/26/

Strange, C., & Kempa, M. (2003). Shades of dark tourism. *Annals of Tourism Research*, *30*(2), 386–405.

Tarlow, P. (2005). *Dark tourism: The appealing 'dark' side of tourism and more* (pp. 47–58). Elsevier eBooks.

Wagner, H., Glaser, B. G., & Strauss, A. L. (1968). The discovery of grounded theory: Strategies for qualitative research. *Social Forces*, *46*(4), 555.

Weaver, D., Tang, C., Shi, F., Huang, M., Burns, K. J., & Sheng, A. (2018). Dark tourism, emotions, and postexperience visitor effects in a sensitive geopolitical context: A Chinese case study. *Journal of Travel Research*, *57*(6), 824–838.

Chapter 5

Dark Attraction Sites: Understanding Motivations, Intentions, and Behaviors of Tourists

Neelima Modi[a] *and Priya Sodani*[b]

[a]University of Kota, India
[b]Modi Institute of Management and Technology, India

Abstract

Dark Tourism is a newly recognized niche tourism type that is first coined in 1996 by Lennon and Foley. This new term has attracted academicians and researchers globally. Till mid of 20th century, major work on dark tourism was limited to understanding its concept and giving it various names such as death tourism, black tourism, horror tourism, thanatourism, morbid tourism, and many more. After the mid-20th century, the focus of researchers was shifted to tourists' psychology specifically on tourists' motivation. Researchers suggested various motivational factors such as 'curiosity of unusual,' 'education and learning,' 'historic interest,' 'inner purification,' and 'interested in death-related sites' that influence tourists to seek dark attraction places. Recent research identified impact of dark places on visitors' emotions and experiences. Some has argued that after visit tourists' experience negative emotions such as depression or horror. While some suggested that dark tourism have positive emotion experience. This chapter is based on a secondary database, descriptive in nature, and aims to depict the motivation, purposes, and behavior of tourists toward dark attraction places that are linked to places of dark history.

Keywords: Black tourism; dark tourism; motivation; psychology; dark attraction

Dark Tourism, 61–73

doi:10.1108/978-1-83797-336-120241005

Introduction

Tourism is the phenomenon of traveling to places to fulfill some purpose and motive. People started to move from one place to another in search of knowledge and recreation. Earlier 'Pleasure' and 'leisure' was not the major purpose of travel as in the modern world. Romans used to visit pilgrimage sites, shrines, and temples. After World War II, the trend of tourism increased both within countries and internationally. Countries have recognized the contribution of tourism to the economy and established programs to promote international travel.

The motive of tourists to visit a place is changing with time. Earlier it was to meet friends, family, and relatives. Now it is to earn new learning experiences, understand different cultures, rituals, the history of the place, relax, entertain, and be involved in recreational activities. Numerous emerging trends of tourism are taking place to fulfill different purposes of tourists. Dark tourism is a new bond of tourism that is performed by tourists' having varying motives and purposes. The motive behind the visit to a place is led by beliefs, thoughts, ideas, emotions, culture, and environmental factors. For the growth of tourism business enterprises and the long-term benefits of the tourism sector, it is vital to understand the tourist's psychology and demand. This chapter aims to understand tourists' motives behind their visit to unusual places of dark attraction.

Objective of the Chapter

The objective of this chapter is to understand the concept of the unusual travel phenomenon of dark tourism and the motivation behind such visitation in a comprehensive way. Understanding tourists' psychology is foremost important for tourism stakeholders and institutes. This study determines all conscious and unconscious factors of tourists' minds that influence them to visit places of suffering, sorrow, pain, and death. Understanding tourists' minds and behavior is important for the successful growth of tourism businesses and operations linked to dark tourism.

Methodology

This study is descriptive type, adopted the conceptual method of understanding tourist behavior and motivation in special reference to dark attraction places related to suffering, pain, atrocity, and death. For the study, previous research papers, literature, journals, articles, and books are appropriately taken and exercised.

Dark Tourism: An Introduction

The success of Netflix's 2018 series 'Dark Tourists' has captured great attention and proved that people are interested in visiting places of dark attractions. The notorious camp of Auschwitz, Poland is the world's most infamous Holocaust concentration camp of World War II that is now collecting a large number of tourists (2 million 320 thousand in the year 2019) every year (Museum Reports/Museum/Auschwitz-Birkenau, n.d.). The Hiroshima Peace Memorial Park and Museum (affected bomb blast area) also attracts approximately one million tourists every year (Main Building of Hiroshima Peace Memorial Museum Reopens with New Exhibits That Vividly Convey Day of A-Bombing, n.d.). The victims of the 9/11 terrorist attack 'The Ground Zero 9/11 Memorial' of New York City registered millions of tourists coming each year to pay honor to the dead people (9/11 Memorial Museum, n.d.). This unusual visitation is not new (Stone, 2011). Places of sorrow, tragedy, disaster, atrocity, and death are always been attractions for tourists. Romans used to visit gladiator games to watch hangings in England and visit sites of pilgrimages from the medieval ages. Such travels are some examples of dark tourism which was first understood and enlightened in the year 1996. Lennon and Foley (1996) and Foley and Lennon (1997) have introduced and defined 'dark tourism' as visiting places that are directly or indirectly associated with death (mass or individual) and disaster.

'Dark' refers to a past that is full of suffering, tragedy, and pain. It includes cemeteries, graveyards, battlefield sites, genocide places, holocaust sites, mass cremation places, war sites, memorial sites, prisons, suicide points, haunted places, natural disaster sites, etc., and has been historically significant in some way by societies. This unique bond of the tourism industry has caught the attention of academicians, researchers, and the media. Dark tourism, where 'dark' refers to death/disaster/sufferings and 'tourism' refers to entertainment, are two conflicting terms making it difficult to accept as tourism product. Some countries like Japan and the United States have fully accepted and developed dark tourism and getting benefits in terms of socioeconomic growth. Not only this, dark tourism holds important lessons of human life and augment us with bitter truth of death. In the article "Cemetery tourism: coming to term with death?" (2010), by the director of the Centre for Thanatology at Radboud University, Eric Venbrux has emphasized on the importance of dark tourism travel to see burial places, death rites, and cemeteries as such visitation teach us to deal with bitter truth of life.

At the beginning of the 20th century, a large number of researchers focused on the conceptual framework, defining the spectrum of dark tourism and theoretical construction of the dark attraction travel phenomenon. Academicians gave different names like 'Thanatourism' (Seaton, 1996), 'Black spot tourism' (Rojek, 1993), 'Atrocity heritage tourism' (Ashworth, 2004), 'morbid tourism' (Blom, 2000), 'penal/prison tourism' (Strange & Kempa, 2003), 'grief tourism' (Dunkley et al., 2007), 'disaster tourism' (Robbie, 2008), 'gothic tourism' (McEvoy, 2016), and many more. 'Dark tourism' is considered an umbrella term covering all such visitation of places that are explicitly or implicitly associated with places of suffering, atrocity, pain, disaster, and death. It is noted that these places must have significant dark history. Many times, dark tourism visits are considered as heritage tourism, but visiting dark sites as typical sites of historical places have quite differences. According to Light (2017), dark tourism and thanatourism are distinct form of tourism and different from heritage tourism. Further theoretical studies on the degree of darkness, the supply side of dark tourism, types of dark tourism, etc., were done. Some places like Holocaust sites, concentration camps, battle field sites, and genocide places where tragedy happened are taken as darkest tourism sites, while some commercialized places such as ghost houses and fun factories purposely developed for entertainment are considered as lightest dark sites (Stone & Sharpley, 2008).

After the mid-20s, the focus of researchers and academicians shifted from theoretical construction to exploratory research. However, fewer empirical-based studies analyzed the motives of tourists, their experiences, and other impacts of dark place visitation. This chapter has systematically taken previous studies with special emphasis on tourists' motivation and experiences at dark places for understanding tourist psychology.

Dark Tourism Motivation

Understanding tourists' motivation is quite a complex process. Motivational factors for dark sites depend on tourists' intentions, individual perceptions, sensibility, and what is "dark" connected with the site. Many researchers have hypothesized and analyzed the driving and motivational factors behind traveling to dark places. Lennon and Foley (1996, 1999, 2000) and Foley and Lennon (1997) have analyzed the practice of traveling to places of death and disaster and have hypothesized that such visit is due to the emotional connection of tourists with the place, for learning purpose, or entertainment purpose, or to pay honor to victims, and it can be an incidental visit involved in the itinerary of an organized travel package. Visiting places of World War I was for gathering historical information and/or collecting knowledge and/or to know the truth behind the story (Lennon & Foley, 1999). Travel to places of war or battlefield may also be the expression of a hobby or interest (Lennon and Foley, 2000). Lennon and Foley (1999, 2000) differentiated two types of dark tourists: one having a special curiosity about a specific site/event and others having no connection with the site but travel for other reasons.

According to Seaton (1996) degree of personal interest in 'death' turns on the travel motive to the dark place. While Dann (1998) has argued that possible motivation components are 'to pursuit for distinction or uniqueness,' 'to overcome the fear of phantoms,' 'elemental bloodlust,' 'to feel nostalgia,' 'for the sense of mortality,' 'to celebrate crime/deviance,' 'dicing with death or challenging journey,' and 'Inner purification' was also considered an influencing factor to visit dark sites (Blom, 2000).

Lubbe (2003) determined that the 'need' is the origin of tourists' behavior and leads to generating motivation, which leads to performing the actual travel behavior. Yuill (2003) has determined the motivation factors behind visiting museums and found that influencing factors are getting informed, collecting images of historical events, education, and confirmation of one's identity. The major motivations are to fulfill the purpose of history learning, education, and memorial purposes. He argued that survivors' guilt motivates them to visit that destination to awaken the value of one's life in the visitors, even when the site is attractive or not. The purpose of 'remembrance,' 'interested in history,' and 'personal connection with the place' was supported by Poria (2004). Wilson (2004) gave reasons behind travels to dark destinations are 'morbid' and 'entertainment.'

Ashworth and Hartmann (2005) explained the three grounds for visiting sites of brutality are 'curiosity about the unusual,' 'horror attraction,' and 'compassion with the victims.' They argued that travel to places of tragedy is the outcome of 'curiosity.' As per Poria et al. (2006), when tourists take in the dark place as their heritage, they express more attention and more curiosity in learning and feel more connected and involved. In consequence, people visit dark sites to visit places of their heritage connection.

According to the research paper "A Dark Tourism Spectrum" by Stone (2006), the motivation and attraction of visitors to places of dark attraction can be differentiated by during and after-visit motivations. He has formulated a 'spectrum of supply' of dark tourism places, pacing them in ranges from 'darkest' to the 'lightest' based on several factors like geographical factors, location, political factors, size of the place, historical importance, ideological factors, and temporal factors. He comparatively analyzed two dark sites Alcatraz and Robben Island and found that Alcatraz is less dark as commercialization and development of the place faded its darkness, while Robben Island is less commercialized and so holds originality, seriousness, and darkness. The more the darkness of the place, the more the attraction of tourists. The originality of the place, development, seriousness, etc., impact the degree of darkness and so degree of attraction of visitors.

Sharpley (2006) argued that the majority of tourists do not know their motivation and drivers behind travel to dark attractions. He proposed a 'shades of dark site' framework to define dark places in four shades of attraction based on various factors including the degree of interest of tourists in death and disaster.

Wight and Lennon (2007) suggested that 'knowing about the country' is the reason behind a visit to dark sites. According to Stone and Sharpley (2008), dark tourism motivation and after-visit experiences of tourists are impacted by their state of society and sociocultural environment. According to Bowman and Pezzullo (2009), death rites are performed usually as formal ceremonies for giving empathy to the family, society, and associated people of the dead person rather than marking the passing of the deceased. Thurnell-Read (2009) studied on analyzing the experiences of young visitors who went to see the site of Holocaust 'the concentration camp Auschwitz-Birkenau' in Poland. It has found that native came to the visit Holocaust, not by their intention or attraction but recommended by their friends.

As suggested by Dunkley et al. (2011), graveyards and cemeteries are places of attraction for the intention of achieving religious, spiritual, retrospective, and contemplation purposes. Visiting sites of disaster or crime is for affirmation and recognition of events that happened. Additional influencing factors can be learning, thrill and risk-seeking, desire to visit sites of immortalized historic events, acknowledging the truth of a place or person, convenience for visitation, and curiosity to visit the recent disaster-affected area. Robinson et al. (2011) has interestingly explained that the motivation for seeking dark attraction places is to closely examine the place of death and gather proof of experiences.

According to Bittner (2011), the phenomenon of traveling to places of suffering and death is not a novel practice but the demand for such dark sites is scarce. He studied through the research paper "Dark Tourism – Evaluation of Visitors Experience after Visiting Thanatological Tourist Attractions," the historical and cultural aspects of thanatourism sites and the extent of the attractiveness of sites with the potential for economic contribution. He analyzed the motivation of visitors to revisit the same or similar thanatourism sites. Results investigated that tourists are involved more in places of unique themes such as the Holocaust that portray mass suffering. The majority were motivated to visit such places by cultural needs and a sense of knowing more is generated during the visit to a thanatological destination. 'Personal interest' provokes visitors to revisit the same or similar thantological place.

Stone (2011) has highlighted the role of dark places to remind people of their mortality and the importance of their lives. Keyes (2012) also listed drivers of dark travel including education, pilgrimage, remembrance, and entertainment. 'Personal connection' and 'considered as own heritage' is again supported by Mowatt and Chancellor (2011).

White and Frew (2013) supported the factor 'curiosity' and 'fascination with dark sites for reflecting mortality' are influencers of dark sites. On a similar concept, Raine (2013) stated that 'curiosity in the past or death' is the major factor behind visits to dark places. Thrill seekers and morbidly curious pay visits to funeral places to experience and witness death. According to Farmaki (2013), commemorative purposes are the driving factor toward dark tourism with the purpose of remembrance, respect, and education. Dalton (2015) explained that tourists pay visits to cemeteries for a 'sense of duty' or 'moral obligation' while doing visits to crime-related dark sites to capture memories of suffering, trauma, and death.

Pearce in Lew et al. (2014) enlightened that tourists' motivation is the sum of biological and sociocultural forces. Dark visitation is led for the purpose of memorialization, commemorative, or education purposes. According to Issac and Cakmak (2014), 'curiosity' is the motivating factor for tourists traveling to dark sites. Tinson et al. (2015) found that educated people visit places of death or tragedy for self-identity, for a better understanding of historical events, and for learning. 'Interest in history or death' is the motivation factor according to Yankholmes and Mckercher (2015). As per Yan et al. (2016), tourists pay visits to places of dark historical significance to understand and learn about past happenings at the site.

Tourists' Behavior, Experiences, and Emotions at Dark Attraction Places

Experiencing unusual place has impact on emotions of tourists. To understand impact of dark place experience on tourist emotions, some previous empirical studies has taken in this study. It is to understand that emotions of a human are a complex parameter to understand and measure. Observing behavior and self-assessment, and adequately designing questionnaires can be utilized to measure the emotions. According to Izard and Malatesta (1987), physical conditions and external environmental factors impact an individual's emotions.

Miles (2002) has also comparatively studied two dark sites of similar attributes and events. One site is the original place of the event and the other is a museum dedicated to victims. The study concluded that the intensity of tourists' experiences is high for a place of the original event (Auschwitz concentration camp taken in the study) as compared to the site of simulation of the event (the Washington Holocaust Museum taken in the study). According to Austin (2002) and Miles (2002), majorly tourists experience negative emotions like fear, horror, depression, sadness, vengeance, empathy, etc., after paying a visit to a dark attraction place. While Biran et al. (2014), Chang (2017), Thurnell-Read (2009), Nawjin and Fricke (2015), Best (2007), Lisle (2004), and many others supported the fact that death-related sites generate positive emotional experiences of love, fascination, interest, pride, hope, gratitude, and so on. According to Tarlow (2005), four basic emotions insecurity, gratitude, humility, and superiority interact with the tourists visiting dark sites.

Kang (2010) has utilized a framework based on the benefits and enduring involvement of tourists to find out the motive of travel, experiences, and benefits that tourists gain. It has been found that the major motive is a 'sense of obligation' and 'personal duty' behind the visit of tourists. Experiencing dark places contributed to gaining positivity and emotional experience. Degree of involvement of the tourists impacts recognition of dark event. Tourists those are less involved acquire less information and knowledge about the place. Such benefits and experiences of tourists resulted by the study also depends on previous knowledge of tourists and friendliness with the dark event.

According to Bittner (2011), 'attraction of the place' and 'information collected prior to the visits' impacts visitor's experiences. Anna Farmaki (2013) has studied the demand and supply side of motivation by exploratory investigating at dark heritage sites in Cyprus. He has analyzed that the demand for dark tourism is 'supply-driven.' The attraction of the site, projection of dark sites, personal factors of tourists, and time passed by the event impact the demand for dark tourism as well as the interpretation of the site by tourists.

Li-Hui Chang (2017) analyzed the emotional impact of dark places on tourists through his research paper 'Perception of dark tourism and how visits impact their emotional experience.' The researcher has considered the war museums of Kinmen Island and collected data through a survey. It has been found that there is a constructive association between dark places and their emotions and strong geopolitical disposition. A tourist's perception and emotional experiences are positively associated. Awareness about the history of the place strongly bonds the dark place and optimistic emotional experiences of visitors as compared to those who were unaware. Tourists with past experiences with dark places and familiar with attributes of present dark places have found lesser positive emotional experiences. Past experiences with darker places also lessen the excitement of visitors and positive impact on emotions. A factor of tourists' education level also reflects their emotions and experiences of dark tourism. higher level of education and awareness about the place motivate visitors to seek truth and leads to higher positive emotional experiences.

According to Handayani et al. (2017), not just academicians but policymakers have also admitted the significance of dark places as an instrument to benefit suffered community and site development. In his research paper "Smart Tourism for dark sites: The Sacred Site of the Dead, Trunyan Cemetery," the pulling factor of dark sites with smart tourism is focused. The study determined that the attraction of dark sites and emotional aspects of the place can be enriched through the application of information and communications technology (ICT) developments on death sites. The ICT application would resolve ethical concerns in dark places and online communities could experience places through social media.

Kusumawardhana et al. (2020) has utilized strengths, weaknesses, opportunities, and threats (SWOT) model and threats, opportunities, weaknesses and strengths (TOWS) matrix to investigate capabilities, opportunities, threats, and lacunas of sites of dark attraction places and space operations. The research paper "Dark Tourism's Products Development: The Cases of Fatahillah Dungeon and Taman Prasasti Museum" provided a framework for value product development for dark tourism to stand out from other tourism products. The value creation framework explains that dark tourism sites need the development of initiatives with product description, spectrum distinction, and assessment of the SWOT model and TOWS matrix. This will strategically create value for the places and enhance tourist experiences at dark attraction places.

The empirical study of Schneider et al. (2021) on tourists' motivation to dark sites of Romania shows that 72% of respondents has prior experience of dark destinations while approx. 28% are interested to visit such unusual places in Romania. The Theory of Planned Behavior (TPB) model with five constructs (dark experience, unique learning experience, engaging entertainment, and casual interest) of dark tourism was utilized by the researcher and analyzed that experiencing a dark place and unique learning experience has a significant positive effect on the attitude of visitors. It has also concluded that places with paranormal activities are highly interesting and visitation to such places is based on personal beliefs and personal choices rather than external influences. The researcher also emphasized policy formulations for effectively developing and understanding the right sense of dark tourism and rightly promoting places by providing tour packages.

Conclusion

Dark tourism is a unique form of tourism practice. Usually, tourism is taken as leisure activity but dark tourism is not at all associated with leisure. Phenomenon of paying visits to places of 'death' and 'disaster' is quite complex. So, this unusual tourism is not for everyone. But these are places of inherent information of past significant to know at present.

Tourism to places associated with death or pain is not a new phenomenon. Long ago, people were attracted to places associated with death, suffering, violence, or disaster, such as watching ancient Roman gladiatorial performances, medieval pilgrimages, visit to cemeteries and graveyards, etc. Such travels are influenced by various factors and tourist psychology. This chapter emphasized on conceptual understanding of dark tourism, consumption, tourists' behavior, motivation, and experiences of tourists at places of dark attraction.

It has found that people are interested in places related to death and disaster involving several factors ranging from the curiosity in the history of the place (Ashworth & Hartmann, 2005; Dalton, 2015; Dunkley et al., 2011; Farmaki, 2013; Issac & Cakmak, 2014; Poria et al., 2004; Raine, 2013; Tinson et al., 2015; White & Frew, 2013; Yankholmes & Mckercher, 2015), paying honor and respect (Ashworth & Hartmann, 2005; Kang, 2010; Lennon & Foley, 1999; Stone, 2011), emotional and personal connection (Foley & Lennon, 1997; Mowatt & Chancellor, 2011), for unique experience (Dann, 1998), memorialization and remembrance (Pearce in Lew et al., 2014; Poria et al., 2004; Stone, 2011), education and learning (Farmaki, 2013; Keyes, 2012; Lennon & Foley, 1999; Pearce in Lew et al., 2014; Tinson et al., 2015; Yuill, 2003), to know the truth behind the stories (Dunkley et al., 2011; Lennon & Foley, 1999), interest in death and fascination with dark places (Seaton, 1996; White & Frew, 2013; Yankholmes & Mckercher, 2015), horror attraction (Ashworth & Hartmann, 2005), morbid (Wilson, 2004), confirmation of one's identity (Yuill, 2003), knowing about the country (Wight & Lennon, 2007), thrill and risk-seeking (Dann, 1998; Dunkley et al., 2011; Raine, 2013), for the sense of mortality (Dann, 1998; Farmaki, 2013; Dalton, 2015), inner purification (Blom, 2000), collecting images of historic event (Yuill, 2003), entertainment (Keyes, 2012; Lennon & Foley, 1999; Wilson, 2004), and to closely examine the past. It has observed that such researchers on tourist motivation have mostly considered popular and 'darkest' places such as holocaust sites, concentration camps, battlefield sites, etc. Lesser dark sites are not taken by academicians for empirical studies. While such places are also important as they depict historical event and a property of nation importance.

It is analyzed from past studies that not much indicates that interest in death and morbid curiosity is the main purpose for travel to dark places. Majorly visit initiated to the places of death and suffering is not due to interest in death or suffering (Seaton & Lennon, 2004; Sharpley & Stone, 2012), while the commonly reported motives are curiosity, learning, and understanding past events.

Some study concluded that visiting to dark places would result to negative emotions and destress individual mood (Austin, 2002; Miles, 2002). Many studies investigated that dark places have positively associated with tourist's emotions and attitudes (While Biran et al., 2014; Chang, 2017; Kang, 2010, Thurnell-Read, 2009; Nawjin & Fricke, 2015; Best, 2007; Lisle, 2004). Such impact of unusual dark places on tourists experience and emotions depends on various factors such as original place of the event (Miles, 2002), degree of involvement of tourists (Kang, 2010), previous knowledge about the place or tragedy (Kang, 2010; Chang, 2017), past experience of dark tourism (Chang, 2017), attraction of the place (Bittner, 2011), friendliness of tourists with dark tourism practice (Kang, 2010), and information collected about the place and its history prior to visit (Bittner, 2011). During visit to dark attraction place, tourists experience the emotions of insecurity, superiority, gratitude, and humility (Tarlow, 2005).

Understanding the tourists' motivation and behavior will help the tour operators of dark tourism sites to better understand visitors' want and provide more facilities to offer better experiences. For example, dark tourism site operators may provide more information of the event or can show firsthand information like interviews of tourists, videos of the survivors, etc. This will generate 'curiosity' about the place and the history, and can influence them to visit, as curiosity is a significant motivation to drive visitors to visit dark tourism sites.

Dark tourism possesses so many benefits in respect of revenue generation, heritage conservation, social integration, social education, job creations, indigenous community-youngsters and women empowerment, and betterment of nation, if developed effectively. Dark tourism is a best instrument to make balance between social education and leisure functions. Dark tourism holds valuable lessons for everyone that not only give us true history and initiate sensitive discussion on death but also help to examine the culture.

References

9/11 Memorial Museum welcomes more than 10 million visitors|National September 11 Memorial & Museum. (n.d.). https://www.911memorial.org/connect/blog/911-memorial-museum-welcomes-more-10-million-visitors. Accessed on December 2, 2021.

Ashworth, G. (2004). Tourism and the heritage of atrocity: Managing the heritage of South African apartheid for entertainment. In *New horizons in tourism: Strange experiences and stranger practices* (95–108). Cabi Publishing. https://doi.org/doi:10.1079/9780851998633.0095

Ashworth, G., & Hartmann, R. (2005). *Horror and human tragedy revisited: The management of sites of atrocities for tourism*. Cognizant Communication Corporation.

Austin, N. K. (2002). Managing heritage attractions: Marketing challenges at sensitive historical sites. *International Journal of Tourism Research*, *4*(6), 447–457.

Best, M. (2007). Norfolk islands: Thanatourism, history and visitor emotions. *Shima: The International Journal of Research into Island Cultures*, *1*(2), 30–48.

Biran, A., Lui, W., Li, G., & Eichhorn, V. (2014). Consuming post-disaster destinations: The case of Sichuan, China. *Annals of Tourism Research*, *47*, 1–17.

Bittner, M. (2011). Dark tourism – Evaluation of tourists experience after visiting thanatological tourist attractions. *Turizam*, *15*(4), 148–158.

Blom, T. (2000). Morbid tourism – A postmodern market niche with an example from Althorp. *Norsk Geografisk Tidsskrift*, *54*(1), 29–36.

Bowman, M., & Pezzullo, P. (2009). What's so 'dark' about 'dark tourism'?: Death, tours, and performance. *Tourist Studies*, *9*. https://doi.org/10.1177/1468797610382699

Chang, L. H. (2017). Tourists' perception of dark tourism and its impact on their emotional experience and geopolitical knowledge: A comparative study of local and non-local tourist. *Journal of Tourism Research & Hospitality*, *6*, 3.

Dalton, D. (2015). *Dark tourism and crime*. Routledge.

Dann, G. (1998). *The dark side of tourism*. International Center for Research and Studies in Tourism.

Dunkley, R. A., Morgan, N., & Westwood, S. (2007). A shot in the dark? Developing a new conceptual framework for thanatourism. *Asian Journal of Tourism and Hospitality*, *1*(1), 54–63.

Dunkley, R. A., Morgan, N., & Westwood, S. (2011). Visiting the trenches: Exploring meanings and motivations in battlefield tourism. *Tourism Management*, *32*(4), 860–868.

Farmaki, A. (2013). Dark tourism revisited: A supply/demand conceptualization. *International Journal of Culture, Tourism and Hospitality Research*, *7*(3), 281–292.

Foley, M., & Lennon, J. J. (1997). Dark tourism: An ethical dilemma. *Hospitality, tourism and leisure management: Issues in strategy and culture*. Cassell.

Handayani, B., Ivanov, S., & Korstanje, M. (2017). Smart tourism for dark sites: The sacred site of the dead, Trunyan cemetery. In *Gazing at death: Dark tourism as an emergent horizon of research*. Nova Science Publishers.

Issac, R. K., & Cakmak, E. (2014). Understanding visitor's motivation at sites of death and disaster: The case of former transit camp Westerbork, the Netherlands. *Current Issues in Tourism*, *17*(2), 164–179.

Izard, C. E., & Malatesta, C. (1987). Perspectives on emotional development I: Differential emotions theory of early emotional development. In J. D. Osofsky (Ed.), *Handbook of infant development* (2nd ed.). Wiley.

Kang, E. J. (2010). *Experience and benefits derived from a dark tourism site visit: The effect of demographics and enduring involvement*. Ph.D. Thesis, School of Tourism. The University of Queensland.

Keyes, M. (2012). War tourism: Shaping memory and perception in post-war Vietnam. In *Summer research*. University of Puget Sound.

Kusumawardhana, I., Tarigan, W. P., Komsary, K. C., & Maulana, Y. (2020). Dark tourism's products development: The cases of Fatahillah dungeon and Taman Prasasti Museum. *IOP Conference Series: Earth and Environmental Science*, *452*, 012126.

Lennon, J., & Foley, M. (1996). JFK and dark tourism: A fascination with assassination. *International Journal of Heritage Studies*, *2*(4), 198–211.

Lennon, J. J., & Foley, M. (1999). Interpretation of the unimaginable: The U.S. Holocaust Memorial Museum, Washington, D.C., and "dark tourism". *Journal of Travel Research*, *38*(1), 46–50.

Lennon, J. J., & Foley, M. (2000). *Dark tourism: The attraction of death and disaster*. Continuum.

Light, D. (2017). Progress in dark tourism and thanatourism research: An uneasy relationship with heritage tourism. *Tourism Management*, *61*, 275–301.

Lisle, D. (2004). Gazing at ground zero: Tourism, voyeurism and spectacle. *Journal for Cultural Research*, *8*(1), 3–21.

Lubbe, B. (2003). Tourism management in Southern Africa. Pearson Education South Africa.

McEvoy, E. (2016). *Gothic tourism*. Palgrave Macmillan.

Miles, W. F. S. (2002). Auschwitz Museum interpretation and darker tourism. *Annals of Tourism Research*, *29*(4), 1175–1178.

Mowatt, R. A., & Chancellor, C. H. (2011). Visiting death and life: Dark tourism and slave castles. *Annals of Tourism Research*, *38*(4), 1410–1434.

Museum Reports / Museum / Auschwitz-Birkenau. (n.d.). http://auschwitz.org/en/museum/museum-reports/. Accessed on December 1, 2021.

Nawjin, J., & Fricke, M. C. (2015). Visitor emotions and behavioral intentions: The case of concentration camp Memorial Neuengamme. *International Journal of Tourism Research*, *17*(3), 221–228.

Pearce in Lew, A. A., Hall, M. C., & Williams, A. M. (2014). *Tourism motivations and decision making in a Wiley Blackwell companion to tourism*. Wiley & Sons, Ltd.

Poria, Y., Butler, R., & Airey, D. (2004). Links between tourists, heritage, and reasons for visiting heritage sites. *Journal of Travel Research*, *43*(1), 19–28.

Poria, Y., Reichel, A., & Biran, A. (2006). Heritage site management: Motivations and expectations. *Annals of Tourism Research*, *33*(1), 162–178.

Raine, R. (2013). A dark tourist spectrum. *International Journal of Culture, Tourism and Hospitality Research*, *7*(3), 242–256.

Robbie. (2008). Touring Katrina: Authentic identities and disaster tourism in New Orleans. *Journal of Heritage Tourism*, *3*(4), 257–266.

Robinson, P., Heitmann, S., & Dieke, P. (2011). *Research themes for tourism*. CAB International.

Rojek, C. (1993). *Ways of seeing modern transformations in leisure and travel*. Macmillan.

Schneider, A., Anghel, S., Negrut, L., Gheorghe, G., & Para, I. (2021). Tourists' motivations for visiting dark tourism sites. Case of Romania. *The Annals of the University of Oradea. Economic Sciences*, *30*, 141–150. https://doi.org/10.47535/1991AUOES30(1)014

Seaton, A. V. (1996). Guided by the dark: From thanatopsis to thanatourism. *International Journal of Heritage Studies*, *2*(4), 234–244.

Seaton, A. V., & Lennon, J. (2004). Moral panics, ulterior motives and alterior desires: Thanatourism in the early 21st century. In T. V. Singh (ed.). *New horizons in tourism: Strange experiences and stranger practices* (pp. 63–82). CABI.

Sharpley, R. (2006). *Travel and tourism*. SAGE Publications Ltd.

Sharpley, R. A., & Stone, P.R. (2012). Contemporary tourist experience: Concepts and consequences. Routledge.

Stone, P. (2006). A dark tourism spectrum: Towards a typology of death and macabre related tourist sites, attractions and exhibitions. *Tourism*, *52*, 145–160.

Stone, P. R. (2011). Dark tourism and the cadaveric carnival: Mediating life and death narratives at Gunther von Hagens' Body Worlds. *Current Issues in Tourism*, *14*(7), 685–701.

Stone, P. R., & Sharpley, R. (2008). Consuming dark tourism: A thanatological perspective. *Annals of Tourism Research*, *35*(2), 574–595.

Strange, & Kempa, M. (2003). Shades of dark tourism: Alcatraz and Robben Island. *Annals of Tourism Research*, *30*(2), 386–405.

Tarlow, E. (2005). Dark tourism: The appealing 'dark' side of tourism and more. In M. Novelli, (Ed.), *Niche Tourism – Contemporary issues, trends, and cases*. Elsevier.

The main building of Hiroshima Peace Memorial Museum reopens with new exhibits that vividly convey the day of A-bombing. (n.d.). https://www.hiroshimapeacemedia.jp/blog/?p=90765. Accessed on December 1, 2021.

Thurnell-Read, T. P. (2009). Engaging Auschwitz: An analysis of young travellers' experiences of holocaust tourism. *Journal of Tourism Consumption and Practice*, *1*(1), 3.

Tinson, J. S., Saren, M. A. J., & Roth, B. E. (2015). Exploring the role of dark tourism in the creation of the national identity of young Americans. *Journal of Marketing Management*, *31*, 856–880.

Venbrux, E. (2010, April). Cemetery tourism: Coming to terms with death? *JSTOR*, *9*. https://www.jstor.org/stable/41548466

White, L., & Frew, E. (2013). *Dark tourism and place identity, managing and interpreting dark places*. Routledge.

Wight, C., & Lennon, J. (2007). Selective interpretation and eclectic human heritage in Lithuania. *Tourism Management*, *28*, 519–529.

Wilson, J. Z. (2004). Dark tourism and the celebrity prisoner: Front and back regions in representations of an Australian historical prison. *Journal of Australian Studies*, *82*, 1–13.

Yan, B. J., Zhang, J., Zhang, H.-L., Lu, S. J., & Guo, Y. R. (2016). Investigating the motivation- experience relationship in a dark tourism space: A case study of the Beichuan earthquake relics, China. *Tourism Management*, *53*, 108–121.

Yankholmes, G., & Mckercher, B. (2015). *Understanding visitors to slavery heritage sites in Ghana*. Elsevier.

Yuill, S. M. (2003). *Dark tourism: Understanding tourists' motivation at sites of death and disaster*. Texas A & M University.

Chapter 6

Battlefield Guidance From a New Angle

Seda Sökmen, Aslı Bendenay Çapa and Semra Günay

Anadolu University, Turkey

Abstract

In dark tourism, professional tourist guides are the primary intermediaries interacting with travelers. Guides provide them with an immersive and educational experience by drawing on many different fields such as history, geography, and literature in their narratives. They use a variety of sources in these fields to enrich their narratives, and literary works are one such source that can be particularly valuable. By drawing on the insights and inspiration from literary works, tourist guides can create engaging and memorable narratives that deepen visitors' understanding of the local culture and heritage. The battlefield is one area where such enriched narratives are essential. Wars that have occurred in locations with both cultural and historical significance are well documented in national and international tourism literature. The way these battlefields are shown is novel. From this perspective, this study aims to: firstly, investigate battlefields that have not received sufficient attention, utilizing novels as underexplored literary sources; secondly, to analyze these sources through geo-literary reading, a relatively novel approach. The development of tour routes in five provinces in Turkey (Ankara, Eskişehir, Bilecik, Kütahya, and Afyon), where significant battles occurred during the national struggle over a period of four years, aims to provide travel companies with innovative tour programs for the global market and to facilitate the planning of specialized battlefield training for guides in these regions.

Keywords: Dark tourism; battlefield guidance; enriched narrative; geo-literary reading; Turkey

Dark Tourism, 75–88

Published under exclusive licence by Emerald Publishing Limited
doi:10.1108/978-1-83797-336-120241006

Dark Tourism and Battlefields

In the 1990s, the relationship between tourism and mortality emerged as a major tourism topic, despite being an atypical field of study initially explored by a limited number of scholars (Light, 2017, p. 276). Traveling for the purpose of full or partial, real or symbolic interaction with death is referred to as sadness tourism in the literature (Seaton, 1996, p. 240). It is also termed variously as black spot tourism (Rojek, 1993), atrocity heritage tourism (Tunbridge & Ashworth, 1996), dark tourism (Foley & Lennon, 1996), Thana tourism (Seaton, 1996), morbid tourism (Blom, 2000), and fright tourism (Bristow & Newman, 2004). With origins dating back to the Middle Ages, sadness tourism serves as a way for tourists to discover their own meaning of life and death, or a mediator that enables association with the dead (Stone, 2012; Walter, 2009). In other words, grief tourism is a type of tourism that facilitates encounters with the idea of death and memories of mortality (Seaton, 2018). The main attraction of this type of tourism is a story that evokes death and the places, objects, activities, and people that can best represent this story (Zerva, 2021, p. 3). Dark tourism is also intertwined with cultural tourism and heritage tourism, as it occurs in places of both cultural and historical significance (Sharpley & Stone, 2009) and can occur in both real and fictionalized places all over the world. However, the places that actually witness death are considered more authentic and educational (Robb, 2009; Stone, 2006), and one of them is battlefield tourism, which is the main topic of this chapter.

The literature characterizes this kind of tourism – also known as "war tourism" – as a subset of tourism, but authors like Cohen (2011), Kidron (2013), Mowatt and Chancellor (2011), and Smith (1996) argue that because of its focus on deeper meanings, battlefield tourism can be distinguished from other forms of tourism. Battlefield tourism is important and worthy of research as it reflects the process of people recovering from the effects of war and having the courage to face history at the same time (Lee, 2016). Academic studies that provide insights into the meaning-laden nature of battlefield tourism (Lloyd, 1998; Ryan, 2007; Seaton, 1999, 2000; Smith, 1996, 1998; Tunbridge & Ashworth, 1996) highlight that battlefield tours offer opportunities to construct individual or collective memory or to validate historical events. It is even noted that they enable those who view visiting these sites as a pilgrimage to become pilgrims (Dunkley et al., 2011; Walter, 1993). In this sense, the sites that witnessed the First and Second World Wars have been frequently studied in the literature (Baldwin & Sharpley, 2009; Iles, 2001, 2008; Walter, 1993). In addition, Vietnam battlefield tours have also been the subject of many studies (Alneng, 2002; Henderson, 2000; Schwenkel, 2006, 2009).

Battlefield tourism is not a new phenomenon; what is new is the presentation of battlefields. The first battlefield to become a tourist attraction was the site of the Battle of Waterloo (1815). Visits to this site in Belgium first started in small groups and later evolved into tours involving large groups of tourists under the leadership of travel agencies (Holguin, 2005). The main motivation for visitors on these tours is to commemorate the relatives and friends they lost in the war. However, the primary goal for travel agencies is economic contribution (Holguin, 2005). Many studies have emphasized the economic benefits of tours to battlefields (Amujo & Otubanjo, 2012; Gotham, 2007; Medway & Warnaby, 2008). Battlefield tourism also contributes to building or strengthening national identities through the performance and discussion of national narratives (Buda et al., 2014; Holguin, 2005; Panakera, 2007). Cultural tools such as commemorations, monuments, and museums in battlefields are also important elements in this context (Misztal, 2003). MacCarthy and Willson's (2015) study, which analyses the museums, martyrs' cemeteries, etc., in the area where the Normandy Landings (1944) took place, confirms the importance of these cultural tools. Cooper's (2006) research, which found that younger generations of Japanese and war veterans are engaging actively with the past by visiting Pacific battlefields is important both in terms of highlighting younger generations and the impact of battlefield tours. These findings are also supported by the study claiming that collective commemoration ceremonies organized within the scope of battlefield tourism preserve collective memory and transmit it to younger generations (Dunkley et al., 2011). At this point, emotions come to the fore and are directly related to both the motivations and experiences of tourists in battlefield tourism (Braithwaite & Lee, 2006).

The emotions that battlefields evoke in visitors have been addressed in a wide range of studies (Anderson & Smith, 2001; Davidson, 2003; Knudsen & Waade, 2010; Krakover, 2005; Miles, 2002; Nawijn et al., 2016; White, 2005). Some of these emotions include disgust (Podoshen et al., 2015), fear (Buda, 2015), hope (Koleth, 2014), and pride (Cheal & Griffin, 2013). For example, in research examining the reactions of children who visited the Guba Genocide Memorial Complex in Azerbaijan, it was noted that while the majority of children felt sadness, others found the experience engaging (Israfilova & Khoo-Lattimore, 2018, p. 8). From a broader perspective, visitors to battlefield tourism can simultaneously experience positive and negative emotions, such as a transformation in which they can find new meaning for their own lives (Causevic & Lynch, 2011) or the latent pleasure of witnessing the misfortunes of others (Buda & McIntosh, 2013; Seaton & Lennon, 2004). However, the expression of such emotions is not easy due to their rich and meaningful nature (Laurier & Philo, 2006; Yankovska & Hannam, 2014). Given this, it is believed that narratives on the battlefield that aid in revealing the participant's emotions are necessary. To expand and enhance current narratives, newly formed sources are now required in addition to historical and geographical ones. Literary works can provide a distinctive and beneficial viewpoint to the investigation of battlefield narratives.

Creating and Enriching Narratives in Tour Guidance

A tourist guide is a professional who leads tourists in the language of their choice and explains the region's cultural and natural heritage. They typically hold a qualification specific to the area, which is often granted and/or acknowledged by the relevant authorities (WFTGA, 2023). In Turkey, tourist guiding was recognized as a profession with the "Tourist Guiding Profession Law" numbered 6326 published in the Official Gazette numbered 28331 on June 22, 2012. Later, on

December 26, 2014, the "Regulation on Tourist Guiding Profession" was published in the Official Gazette No. 29217 and entered into force. With the Regulation, the framework of regulations regarding the qualifications required by the profession of guidance, the conditions for fulfilling the profession, guidance certificate programs, and the execution of the profession has been established (Official Gazette, 2014). The Association of Tourist Guides (TUREB), with all its units established in accordance with the "Tourist Guiding Profession Law" numbered 6326, works to ensure that guides play a more effective and competent role in the country's tourism, to promote the country in line with cultural and tourism policies, and to prevent illegal guiding activities (TUREB, 2023).

Tour guides must receive specialized training and expertise to ensure the satisfaction of tourist groups and meet their expectations in particular subjects. TUREB is responsible for these trainings and has the authority to promote and regulate studies related to specialization training under the supervision of the Republic of Turkey Ministry of Culture and Tourism. Specialization trainings are communicated to TUREB with an official letter in line with the program prepared by the guide chambers and are carried out upon approval. For example, the Çanakkale Regional Chamber of Tourist Guides (ÇARO) provides Çanakkale Wars and Gallipoli Historic Site Specialization Training. At the end of the training, which covers both theoretical and field research, successful participants are given a Certificate of Specialization in Çanakkale Wars and Gallipoli Historical Site by TUREB and recorded in their registry (ÇARO, 2023).

Specialized tourist guides explain their area of expertise in an engaging and inspiring way, using the language of the visitor's choice (Zerva & Nijkamp, 2016). The most important element in the success of battlefield tours is the narrative of the tourist guide (Iles, 2006). These narratives should not be myths and legends, but should be founded on reliable narratives based on historical and literary sources. The main reason for this is that tour participants reject a shallow experience and want to receive reliable information from experts (Iles, 2008). However, it should not be forgotten that it is not enough for battlefield narratives to be reliable on their own. They should also appeal to the emotions of visitors, as mentioned earlier. At this point, there is a need for resources to help create and enrich the narrative. In the case of battlefields, the most frequently used sources belong to the science of history (this can also be seen in the theoretical content of battlefield specialization training). However, sources in the field of geography also offer remarkable information about battlefields. The innovative direction of this chapter is the field of historical geography, which encompasses both, and the literary works that feed this field are invaluable sources for narrative enrichment. An integrated approach to narrative is demonstrated by the incorporation of various sources, such as literary, historical, and geographical ones. These components can be combined by tour guides to create a thorough and immersive experience that goes beyond simple facts and figures, enhancing the story and drawing in more viewers. It can bring a more human dimension to the narrative and engage the emotions of the visitors.

Analyzing Literary Works and Interpretation in the Context of Battlefield Guidance

The adoption of this historical geography approach to battlefield tourism necessitates a number of methodological innovations. One such is geo-literary reading in the field of literature used for the deductive analysis of novels as secondary sources. Geo-literary reading is one of several types of reading, including biographical, political, psychoanalytic, sociological, and cinematographic. The places where events take place in literature (house, room, neighborhood, village, town, city, etc.) are important for the reader to understand the text. Analyzing the text in terms of space and the human relations analyzed, together with its spatial features, provides the reader with new perspectives. In this type of reading, the place can be determined as the place where the author was born and grew up or lived (lived geography) or a region chosen for personal, cultural, political, etc., reasons (chosen geography). The reading is also done around these two bases. Therefore, the researcher's starting points in geo-literary reading in literature can be summarized as follows (Kefeli, 2009):

Lived Geography: Real or fictional places in the texture of the text are important in terms of showing the author's observations about his or her environment and his or her understanding of art. Whether these places are real or fictional is important for the researcher. However, in both cases, the author's reason for choosing this place should be analyzed. The characteristics of these descriptions can be handled in terms of narrative techniques as well as geographical elements (vegetation, climate, mythological elements, health problems specific to the region, ethnic structure, etc.). The geography in the work(s) can be historical, economic, political, etc., analyzed with its features.

Chosen Geography: The geography of the author's life is, in a sense, a geo-literary biography. The origins of the author's family, the geography where the author was born and grew up, and the geography where the author chose for different reasons at different stages of his life are analyzed. This examination is important in terms of illuminating some motifs and themes in the author's works. A thorough analysis of the events may be given thanks to the author's depiction of the geography. It points out the explanatory aspect of the geography in which the author lives or chooses to live.

In this analysis, although reading has been done around two basic themes, it is not possible to consider it a comprehensive thematic analysis. A thematic analysis is basically carried out as a flexible and widely used research method that involves identifying, analyzing, organizing, describing, and reporting themes within a dataset. It enables in-depth analysis, provides diverse interpretations, and is useful for summarizing key features in large datasets. The method emphasizes a structured approach, contributing to clear and organized final reports. Citations are provided to credit the sources of information (Braun & Clarke, 2006). As mentioned before, geo-literary reading for literature is one of the reading types that helps to evaluate the interaction of humans and geography in the text. However, although it is stated that the themes and sub-themes of geography are used in this approach, it is obvious that these do not carry the quality of a complete thematic analysis. Whereas in this study, the sub-disciplines of geography by deductive reasoning formed the main themes, and the sub-themes under these main themes consisted of the study topics in the related disciplines.

The themes in the analysis of the artifacts considered in this case consist of the sub-disciplines of geography: physical, human, and perceptual geography. The categories are made up of the subjects that fall under these sub-disciplines, and the novel narratives serve the categories. The sub-themes of population geography, cultural geography, economic geography, military geography, and political geography are within the scope of human geography; biogeography, hydrogeography, geomorphology, and climatology are within the scope of physical geography. Perceptual geography appears as the effect of the land in which individuals live on their emotional states. Especially the effects of the war phenomenon on the emotions (grief, fear, anger, disgust, happiness, and surprise from Paul Ekman) of individuals behind the front lines are the main elements of this theme. The inclusion of the place and time of the narratives in the analysis provides a holistic knowledge of what, where, when, and how the narratives take place in the novel. As mentioned before, geo-literary reading in literature is one of the reading types that helps to evaluate the interaction of humans and geography in the text. However, although it is stated that the themes and sub-themes of geography are used in this approach, it is obvious that these do not carry the quality of a complete thematic analysis.

The Case of National Struggle From Inonu to Sakarya

In historical sources, the starting point of the National Struggle was recognized as the Armistice of Mondros signed between the Entente States and the Ottoman Empire on October 30, 1918. However, the actual start was the departure of Mustafa Kemal Atatürk, the founder of the Republic of Turkey, to Samsun on May 19, 1919. Mustafa Kemal Atatürk initiated the liberation movement in Samsun province in Turkey, and then convened congresses throughout the country. The main purpose of these congresses was to device a strategy to protect the existence of the nation. This struggle continued until the liberation of Izmir from enemy occupation on September 9, 1922. The Grand National Assembly of Turkey was inaugurated on April 23, 1920, establishing the groundwork for a new state while the battle was still ongoing (İnalcık, 2023). During the National Struggle, which extended for about four years, many battles were fought. These were the Battles of İnönü, First and Second (January 6–11, 1921/March 23–April 1, 1921) as the first victory achieved by the newly established regular army of the Republic of Turkey; the Kütahya-Eskişehir Battles (July 13–21, 1921); The Battle of Sakarya (August 23, 1921–September 13, 1921) also recorded as the Great Flood of Blood, which witnessed significant losses; The Great Offensive and the Battle of the Commander-in-Chief (August 26–30, 1922). Although the battles that took place during this period are not limited to these (Erdoğan, 2023), the case study focuses on these battles in terms of the geography it covers.

The authors of the literary works dealing with the period of the War of Independence, who witnessed the period, provide a social perspective on military and political events. Notable authors such as Halide Edip Adıvar, Yakup Kadri

Karaosmanoğlu, Ali Fuat Cebesoy, Falih Rıfkı Atay, Kazım Karabekir, Tağrık Buğra, Kemal Tahir, Reşat Nuri Güntekin, and Ahmet Hamdi Tanpınar illuminate the literature of the National Struggle period in Turkey. The most well-known authors' works were selected for examination due to their impactful subjects and experiences during the National Struggle. These authors and their works are Yakup Kadri Karaosmanoğlu's (1) Ankara (1934) and (2) Ergenekon: Milli Savaş Hikayeleri (1947); Halide Edip Adıvar's (3) Vurun Kahpeye (1970) and (4) Ateşten Gömlek (1923).

The three routes (the Battles of İnönü, the Battles of Kütahya-Eskişehir, and the Battle of Sakarya) created using Geographic Information Systems (GIS) and the five provinces in Turkey (Ankara, Afyonkarahisar, Bilecik, Eskişehir, and Kütahya) where routes are located are shown in Fig. 6.1.

The case study's main purpose is to enhance the tour guides' understanding, who are subject matter experts on the war accounts. A few examples of quotes from Karaosmanoğlu and Adıvar can be given to the narratives in the analyzed works. Yakup Kadri Karaosmanoğlu's Ergenekon: Milli Savaş Hikayeleri is one of the best examples of cultural geography in the context of human geography.

> I'm not as strong as I ever was; our home was destroyed, the village burned down, and the animals we raised one by one all died. Do you recall the grey donkey that gave birth the previous year? I'm coming back from Gecik right now; I sold it for five liras! Son, what's the point of five liras. Karaosmanoğlu (1947, p. 153)

This event took place in the village of Koçaş in Eskişehir and is inferred to have taken place in 1922, when the village was occupied. It should be noted here that some locations can be identified with certainty and some can be inferred.

Although the war at the front was fought between military forces, the people also suffered great persecution. There are very few examples of these in historical sources. However, it is possible to access the details and the vivid pleas of the people of that period through artifacts. In the same work by Yakup Kadri Karaosmanoğlu, the experiences of Kuşbaz Hüseyin, whose birds were slaughtered by Greek soldiers, are as follows:

> . . .he took each of them once by the wings and head and held them in the palm of his hand, stroking the feathers of some of them for a long time, as if with an almost transcendent emotion. Ispiro, who was talking to the officers, turned his face towards the old man. And he shouted from afar with that sly smile of his: Send them inside and let them fry them, he said. Mr. Kuşbaz Hüseyin didn't move, didn't hear, and continued to caress the blood-soaked dead birds, rubbing them on his face and eyes. Karaosmanoğlu (1947, p. 73)

In this excerpt, the exact date is 1920, but the place is inferred to be any occupied Anatolian town. This quotation is one of the findings related to the sense of surprise within perceptual geography.

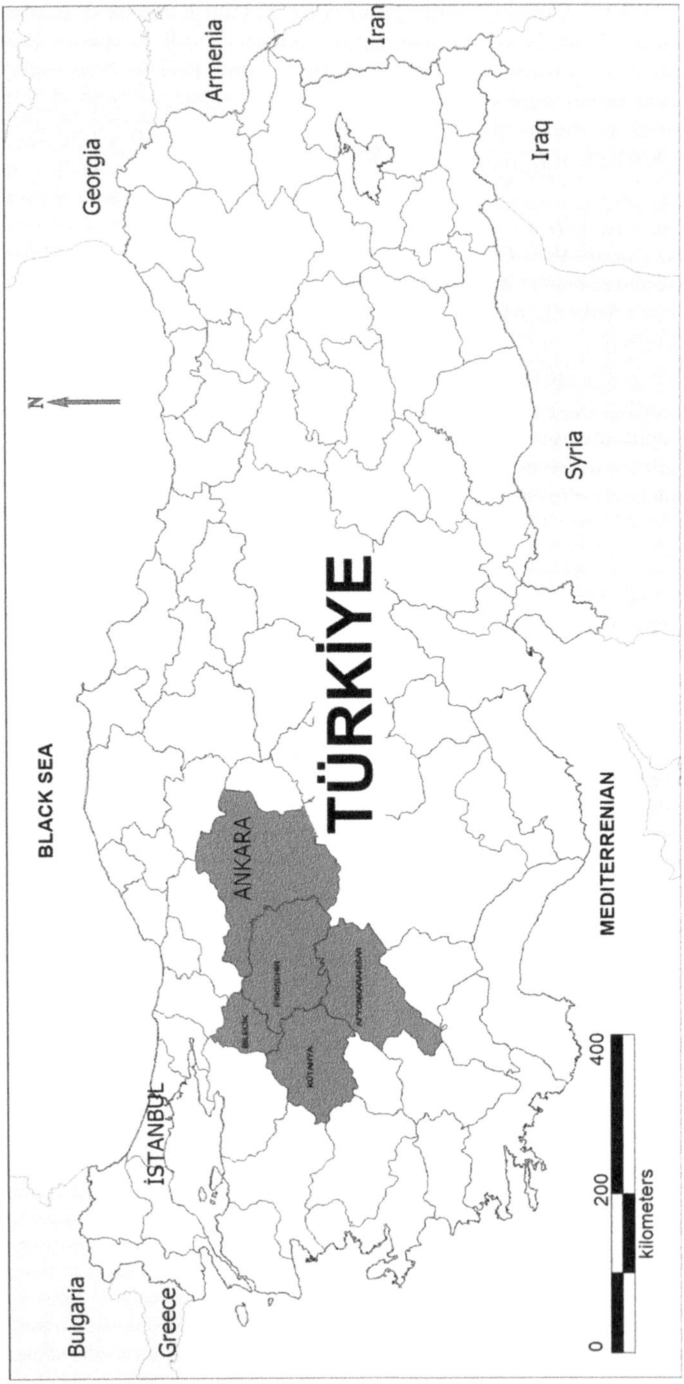

Fig. 6.1. Location Map of Routes in Five Cities. *Source:* It was created by the researchers with the help of Geographic Information Systems program.

Halide Edip Adıvar's excerpts from Vurun Kahpeye, in which she describes the situation of children in the village during the years of the War of Independence, are also noteworthy. Some of the characteristics mentioned by the author give clues about the bad conditions these children lived in the relevant period. Children were sick and had no access to medical care. The fact that all of them had tobacco in their pockets and even smoked cigarettes with their teachers emphasizes how ordinary this situation was for them.

> (. . .) The two privileged groups of children, although few in number, were the children of the gentry and the children of civil servants. The children of the gentry were usually scrawny, neglected, snotty, and dirty, but they were also bullying and immoral. All of them had a pack of tobacco in their pockets. Almost all of them would light their cigarettes with Mrs. Hatice; they would squat in the courtyard, both blowing their cigarette smoke and talking about what was going on in their homes. Miss Aliye (the teacher character in the novel) did not like these children with swollen bellies, bad breath, and being half-crippled and infirm. (p. 9)

The event in question takes place in a village near Afyon, and no exact date can be found. The fact that it takes place during the years of the national struggle is sufficient data for the analysis. This quote is an example of the economic situation within human geography.

Conclusion and Future Directions

In this case study on battlefield tourism, a type of sadness tourism, it is planned to design a tour enriched with narratives based on historical and literary texts to develop this type of tourism at the national level. In the implementation part of the designed tour, its effects on the participants' knowledge levels and emotional states regarding the national struggle are to be assessed. Thus, it is aimed to contribute to the gap in the literature that has not been filled by a few studies focusing on the outcomes of the experience, despite the large number of studies on the reasons for visitors to participate in battlefield tours (Çapa, 2022; Dunkley et al., 2011; Harmancı, 2022). The development of battlefields as a tourism product (day trips or multi-day tours) is very important to draw attention to the country's war tourism potential. Parallel to the situation in the world (Foard & Partida, 2018; Raynor, 2004), although there are many battlefields in Turkey (Doğaner, 2006; Kılıç & Akyurt, 2011; Okuyucu & Erol, 2018), very few of them are designed as a tourism product, and most of them do not offer an engaging narrative that can attract visitors.

The research process of this study can be summarized as follows: (1) designing the tour route and creating the narrative content; (2) fieldwork; and (3) monitoring the effects of the tour. The main source of data in the study, the first phase of which has now been completed, is historical and literary documents. The routes of the tour designed based on document analysis were modeled with Geographic

Information Systems (GIS). In the second phase, data will be collected from high school and university students in Eskisehir who are selected through progressive cluster sampling. The achievement test to be developed as a data collection tool will be used for the experimental design. The knowledge levels of the sample group will be determined with this test to be applied before the tour. In the last stage, this test will be applied again, and the effect of the enriched narrative on the knowledge levels of the participants about the periods of national struggle will be revealed. Furthermore, semi-structured interviews will be used to evaluate and deepen the change in their emotional states regarding the national struggle. In addition, the expectations of the participants for the development of battlefield tourism will also be examined. To summarize, this case study will theoretically add a new dimension to qualitative analysis by associating literary works with places, while in practice it will provide new material for tourism in the context of historical geography. Beyond historical information, this material will also help to communicate to the visitors to the battlefield the emotional states of the witnesses to the war. Under the leadership of this case, a good example of tour planning will be created for many battlefields around the world.

Acknowledgments

This study is funded within the scope of TÜBİTAK Innovative Solutions in Social Sciences and Humanities Research Projects Support Program, Project Number: 122G167.

References

Adıvar, H. (1923). *Ateşten Gömlek*. Teşebbüs Matbaası.

Adıvar, H. E. (1970). *Vurun Kahpeye*. Remzi Kitapevi.

Alneng, V. (2002). What the fuck is a Vietnam? Touristic phantasms and the pop-colonization of (the) Vietnam (war). *Critique of Anthropology*, *22*(4), 461–489. https://doi.org/10.1177/0308275X020220040501

Amujo, O. C., & Otubanjo, O. (2012). Leveraging rebranding of 'unattractive' nation brands to stimulate post-disaster tourism. *Tourist Studies*, *12*(1), 87–105. https://doi.org/10.1177/1468797612444196

Anderson, K., & Smith, S. J. (2001). Editorial: emotional geographies. *Transactions of the Institute of British Geographers*, *26*, 7–10. https://www.jstor.org/stable/623141

Baldwin, F., & Sharpley, R. (2009). Battlefield tourism: Bringing organised violence back to life. In R. Sharpley & P. Stone (Eds.), *The darker side of travel: The theory & practice of dark tourism* (pp. 186–206). Channel View Publications.

Blom, T. (2000). Morbid tourism a postmodern market niche with an example from Althorp. *Norsk Geografisk Tidsskrift*, *54*(1), 29–36. https://doi.org/10.1080/002919500423564

Braithwaite, D., & Lee, Y. L. (2006). *Dark tourism, hate and reconciliation: The Sandakan experience*. International Institute for Peace Through Tourism (IIPT).

Braun, V., & Clarke, V. (2006). Using thematic analysis in psychology. *Qualitative Research in Psychology*, *3*, 77–101. https://doi.org/10.1191/1478088706qp063oa

Bristow, R., & Newman, M. (2004). Myth vs. fact: An exploration of fright tourism. In *Northeastern Recreation Research Symposium* (pp. 215–221). Bolton. United States Department of Agriculture.

Buda, D. M. (2015). The death drive in tourism studies. Annals of Tourism Research, *50*, 39–51.

Buda, D. M., d'Hauteserre, A., & Johnston, L. (2014). Feeling and tourism studies. *Annals of Tourism Research*, *46*, 102–114. https://doi.org/10.1016/j.annals.2014.03.005

Buda, D. M., & McIntosh, A. J. (2013). Dark tourism and voyeurism: Tourist arrested for 'spying' in Iran. *International Journal of Culture, Tourism and Hospitality Research*, *7*(3), 214–226. https://doi.org/10.1108/IJCTHR-07-2012-0059

Çapa, A. B. (2022). *Turist rehberlerinin savaş alanları turizmi anlatılarını zenginleştirmede coğrafya merkezli okuma: Türk'ün Ateşle İmtihanı Örneği* [*Geography centered reading to improve the battlefield tourism narrations of tour guides: A sample of novel 'Türk'ün Ateşle İmtihanı'*]. Master Thesis. Anadolu University.

ÇARO (Çanakkale Regional Chamber of Tourist Guides). (2023). About TUREB. https://caro.org.tr/haberler/canakkale-savaslari-ve-gelibolu-tarihi-alani-uzmanlik-egitimi-duzenlenecekt

Causevic, S., & Lynch, P. (2011). Phoenix tourism: Post-conflict tourism role. *Annals of Tourism Research*, *38*(3), 780–800. https://doi.org/10.1016/j.annals.2010.12.004

Cheal, F., & Griffin, T. (2013). Pilgrims and patriots: Australian tourist experiences at Gallipoli. *International Journal of Culture, Tourism and Hospitality Research*, *7*(3), 227–241. https://doi.org/10.1108/IJCTHR-05-2012-0040

Cohen, E. H. (2011). Educational dark tourism at an in populo site: The Holocaust Museum in Jerusalem. *Annals of Tourism Research*, *38*(1), 193–209. https://doi.org/10.1016/j.annals.2010.08.003

Cooper, M. (2006). The Pacific War battlefields: Tourist attractions or war memorials? *International Journal of Tourism Research*, *8*, 213–222. https://doi.org/10.1002/jtr.566

Davidson, J. (2003). *Phobic geographies: The phenomenology and spatiality of identity*. Ashgate.

Doğaner, S. (2006). Savaş ve turizm: Troya ve Gelibolu savaş alanları [War and tourism: Battlefields of Trojan and Gallipoli]. *Türk Coğrafya Dergisi*, *46*, 1–21.

Dunkley, R., Morgan, N., & Westwood, S. (2011). Visiting the trenches: Exploring meanings and motivations in battlefield tourism. *Tourism Management*, *32*(4), 860–868. https://doi.org/10.1016/j.tourman.2010.07.011

Erdoğan, S. (2023). *Dakikalar içinde milli mücadele*. Kronik.

Foard, G., & Partida, T. (2018). The archaeology of medieval and early modern battlefields in Flanders. *Journal of Conflict Archaeology*, *13*(1), 12–36. https://doi.org/10.1080/15740773.2018.1551914

Foley, M., & Lennon, J. J. (1996). Special issue: Dark tourism. *International Journal of Heritage Studies*, *2*(4), 194–244.

Gotham, K. F. (2007). (Re)branding the big easy: Tourism rebuilding in post-Katrina New Orleans. *Urban Affairs Review*, *42*, 823–850. https://doi.org/10.1177/1078087407300222

Harmancı, H. (2022). *Turist rehberlerinin millî mücadele dönemi anlatılarını zenginleştirmede coğrafya merkezli okuma: Yaban Romanı Örneği* [*Geography centered *reading to improve the national struggle narrations of tour guides: A sample of novel 'Yaban'*]. Master Thesis. Anadolu University.

Henderson, J. C. (2000). War as a tourist attraction: The case of Vietnam. *International Journal of Tourism Research*, *2*, 269–280. https://doi.org/10.1002/1522-1970(200007/08)2:4<269::AID-JTR219>3.0.CO;2-A

Holguin, S. (2005). National Spain invites you: Battlefield tourism during the Spanish Civil War. *The American Historical Review*, *110*(5), 1399–1426. https://doi.org/10.1086/ahr.110.5.1399

Iles, J. (2001). *Death, leisure and landscape: British tourism to the western front deterritorialisations revisioning landscape and politics*. Black Dog.

Iles, J. (2006). Recalling the ghosts of war: Performing tourism on the battlefields of the Western Front. *Text and Performance Quarterly*, *26*(2), 162–180.

Iles, J. (2008). In remembrance: The Flanders poppy. *Mortality*, *13*(3), 201–221. https://doi.org/10.1080/13576270802181640

İnalcık, H. (2023). *Milli mücadele tarihi*. Kronik.

Israfilova, F., & Khoo-Lattimore, C. (2018). Sad and violent but I enjoy it: Children's engagement with dark tourism as an educational tool. *Tourism and Hospitality Research*, *19*(4), 478–487. https://doi.org/10.1177/1467358418782736

Karaosmanoğlu, Y. (1934). *Ankara*. Hakimiyeti Milliye Matbaası.

Karaosmanoğlu, Y. (1947). *Ergenekon: Milli savaş hikayeleri*. Varlık.

Kefeli, E. (2009). Coğrafya merkezli okuma. *Literature and History of Turkish or Turkic*, *4*(1), 423–433.

Kidron, C. A. (2013). Being there together: Dark family tourism and the emotive experience of co-presence in the holocaust past. *Annals of Tourism Research*, *41*, 175–194. https://doi.org/10.1016/j.annals.2012.12.009

Kılıç, B., & Akyurt, H. (2011). Destinasyon imajı oluşturmada hüzün turizmi: Afyonkarahisar ve Başkomutan Tarihi Milli Parkı. *Gaziantep Üniversitesi Sosyal Bilimler Dergisi*, *10*(1), 209–232.

Knudsen, B. T., & Waade, A. M. (2010). Introduction. In B. T. Knudsen & A. M. Waade (Eds.), *Re-investing authenticity: Tourism, place and emotions*. Channel View Publications.

Koleth, M. (2014). Hope in the dark: Geographies of volunteer and dark tourism in Cambodia. *Cultural Geographies*, *21*(4), 681–694. https://doi.org/10.1177/1474474013519577

Krakover, S. (2005). Attitudes of Israeli visitors towards the Holocaust remembrance site of Yad Vashem. In G. Ashworth & R. Hartmann (Eds.), *Horror and human tragedy revisited: The management of sites of atrocities for tourism* (pp. 108–117). Cognizant Communication Corporation.

Laurier, E., & Philo, C. (2006). Possible geographies: A passing encounter in a cafe. *Area*, *38*, 353–363. https://doi.org/10.1111/j.1475-4762.2006.00712.x

Lee, Y. J. (2016). The relationships amongst emotional experience, cognition, and behavioural intention in battlefield tourism. *Asia Pacific Journal of Tourism Research*, *21*(6), 697–715. https://doi.org/10.1080/10941665.2015.1068195

Light, D. (2017). Progress in dark tourism and thanatourism research: An uneasy relationship with heritage tourism. *Tourism Management*, *61*, 275–301. https://doi.org/10.1016/j.tourman.2017.01.011

Lloyd, D. W. (1998). *Battlefield tourism: Pilgrimage and the commemoration of the Great War in Britain, Australia and Canada*. Berg.

MacCarthy, M., & Willson, G. (2015). The business of D-day: An exploratory study of consumer behaviour. *International Journal of Heritage Studies*, *21*(7), 698–715. https://doi.org/10.1080/13527258.2014.1001423

Medway, D., & Warnaby, G. (2008). Alternative perspectives on marketing and the place brand. *European Journal of Marketing*, *42*(5), 641–653. https://doi.org/10.1108/03090560810862552

Miles, W. F. S. (2002). Auschwitz: Museum interpretation and darker tourism. *Annals of Tourism Research*, *29*(4), 1175–1178. https://doi.org/10.1016/S0160-7383(02)00054-3

Misztal, B. (2003). *Theories of social remembering*. McGraw-Hill.

Mowatt, R. A., & Chancellor, C. K. (2011). Visiting death and life: Dark tourism and slave castles. *Annals of Tourism Research*, *38*(4), 1410–1434. https://doi.org/10.1016/j.annals.2011.03.012

Nawijn, J., Isaac, R. K., van Liempt, A., & Gridnevskiy, K. (2016). Emotion clusters for concentration camp memorials. *Annals of Tourism Research*, *61*, 244–247. https://doi.org/10.1016/j.annals.2016.09.005

Official Gazette. (2014). Tourist guiding profession regulation. http://www.resmigazete.gov.tr/eskiler/2013/02/20130223-10.htm. Accessed on November 5, 2023.

Okuyucu, A., & Erol, F. (2018). Savaş alanları turizmine yönelik motivasyon, duygu ve deneyimler: Metristepe ve İnönü Şehitlikleri - Bozüyük örneği [Motivation, feelings and experiences towards battle fields tourism: Metristepe and İnönü war grave- The case of Bozüyük]. *Coğrafi Bilimler Dergisi*, *16*(1), 135–151. https://doi.org/10.1501/Cogbil_0000000195

Panakera, C. (2007). World War II and tourism development in the Solomon Islands. In C. Ryan (Ed.), *Battlefield tourism: History, place and interpretation* (pp. 125–141). Elsevier.

Podoshen, J., Venkatesh, V., Wallin, J., Andrzejewski, S. A., & Jin, Z. (2015). Dystopian dark tourism: An exploratory examination. *Tourism Management*, *51*, 316–328. https://doi.org/10.1016/j.tourman.2015.05.002

Raynor, M. (2004). *English battlefields*. Tempus.

Robb, E. M. (2009). Violence and recreation: Vacationing in the realm of dark tourism. *Anthropology and Humanism*, *34*(1), 51–60. https://doi.org/10.1111/j.1548-1409.2009.01023.x

Rojek, C. (1993). *Ways of escape: Modern transformations in leisure and travel*. Springer.

Ryan, C. (2007). *Battlefield tourism: History, place and interpretation*. Elsevier.

Schwenkel, C. (2006). Recombinant history: Transnational practices of memory and knowledge productionin contemporary Vietnam. *Cultural Anthropology*, *21*(1), 3–30. https://doi.org/10.1525/can.2006.21.1.3

Schwenkel, C. (2009). *The American war in contemporary Vietnam: Transnational remembrance and representation*. Indiana University Press.

Seaton, A. V. (1996). Guided by the dark: From thanatopsis to thanatourism. *International Journal of Heritage Studies*, *2*(4), 234–244. https://doi.org/10.1080/13527259608722178

Seaton, A. V. (1999). War and thanatourism, Waterloo 1815–1914. *Annals of Tourism Research*, *26*(1), 130–158. https://doi.org/10.1016/S0160-7383(98)00057-7

Seaton, A. V. (2000). Another weekend away looking for dead bodies. *Tourism Recreation Research*, *25*(3), 63–77. https://doi.org/10.1080/02508281.2000.11014926

Seaton, T. (2018). Dark tourism history. In P. Stone, R. Hartmann, A. V. Seaton, R. Sharpley, & L. White (Eds.), *The Palgrave handbook of dark tourism studies* (pp. 1–2). Palgrave McMillan.

Seaton, A., & Lennon, J. (2004). Moral panics, ulterior motives and alterior desires: Thanatourism in the early 21st century. In T. V. Singh (Ed.), *New horizons in tourism: Strange experiences and stranger practices* (pp. 63–82). CABI.

Sharpley, R., & Stone, P. (2009). *The darker side of travel: The theory and practice of dark tourism*. Channel View Publications.

Smith, V. (1996). War and its tourist attractions. In A. Pizam & Y. Mansfeld (Eds.), *Tourism, crime and international security issues* (pp. 247–264). Wiley.

Smith, V. L. (1998). War and tourism: An American ethnography. *Annals of Tourism Research*, *25*(1), 202–227. https://doi.org/10.1016/S0160-7383(97)00086-8

Stone, P. R. (2006). A dark tourism spectrum: Towards a typology of death and macabre related tourist sites, attractions and exhibitions. *Tourism: An Interdisciplinary International Journal*, *54*(2), 145–160. https://hrcak.srce.hr/161464

Stone, P. (2012). Dark tourism and significant other death: Towards a model of mortality mediation. *Annals of Tourism Research*, *39*(3), 1565–1587. https://doi.org/10.1016/j.annals.2012.04.007

Tunbridge, J. E., & Ashworth, G. J. (1996). *Dissonant heritage: The management of the past as a resource in conflict*. Wiley.

TUREB (Association of Tourist Guides). (2023). About TUREB. https://www.tureb.org.tr/.

Walter, T. (1993). War grave pilgrimage. In I. Reader & T. Walter (Eds.), *Pilgrimage in popular culture* (pp. 63–91). Macmillan.

Walter, T. (2009). Dark tourism: Mediating between the dead and the living. In R. Sharpley & P. Stone (Eds.), *The darker side of travel: The theory & practice of dark tourism* (pp. 39–55). Channel View Publications.

WFTGA (World Federation of Tourist Guide Associations). (2023). About WFTGA. https://wftga.org/about/. Accessed on November 11, 2023.

White, C. J. (2005). Culture, emotions and behavioural intentions: Implications for tourism research and practice. *Current Issues in Tourism*, *8*(6), 510–531. https://doi.org/10.1080/13683500508668234

Yankovska, G., & Hannam, K. (2014). Dark and toxic tourism in the Chernobyl exclusion zone. *Current Issues in Tourism*, *17*(10), 929–939. https://doi.org/10.1080/13683500.2013.820260

Zerva, K. (2021). Dark tourism on Netflix: From place to person-dependent. *Tourism Management Perspectives*, *38*, 100823. https://doi.org/10.1016/j.tmp.2021.100823

Zerva, K., & Nijkamp, P. (2016). Tour guides as information filters in urban heterotopias: Evidence from the Amsterdam Red Light District. *Tourism Management Perspectives*, *18*, 42–50. https://doi.org/10.1016/j.tmp.2015.12.020

Chapter 7

Exploring Marketing Strategies and Media Influences in Dark Tourism

Jeetesh Kumar, Janitha Kularajasingam and Zhang Jieyao

Taylor's University, Malaysia

Abstract

Generally, dark tourism is considered an area of segment currently being developed within the tourism industry due to its sudden popularity. Dark tourism is related to death, disaster, tragedy, and suffering. Nevertheless, the association has yet to be thoroughly researched. Thus, this study investigates the relationship between the marketing strategies employed in the dark tourism industry. This study explores the cinema or short web series/TV serials on dark tourism. In addition, the current development in entertainment media and the sudden rise of streaming platforms, social media, iCloud, content marketing, virtual influences, and bloggers in dark tourism play a pivotal role in this era. Thus, this study will investigate the relationship between these roles in dark tourism. The noticeable findings of this research are also to understand further roles within the marketing strategies that further affect dark tourism. Secondary data were explored to achieve the purpose of the study. This study has contributed to the body of knowledge regarding dark tourism and marketing strategies by providing in-depth information that will benefit scholars, tourism planners, and policymakers.

Keywords: Marketing; strategies; dark tourism; social media; content marketing

Introduction

The term "dark" describes distinctly disturbing practices and a sense of morbid products and experiences within the tourism sector. Dark tourism is a personal experience and a complex sociocultural phenomenon (Biran & Hyde, 2013). This dark tourism experience creates a difference in value perceived by tourists visiting

Dark Tourism, 89–98

Published under exclusive licence by Emerald Publishing Limited
doi:10.1108/978-1-83797-336-120241007

dark destinations (Le & Pearce, 2011). Stone (2006) described dark tourism as traveling to locations associated with death, pain, and the seemingly horrific. Dark tourism takes many forms, including prison tourism, genocide tourism, grief tourism, favela tourism, disaster tourism, and fright tourism (Shondell Miller & Gonzalez, 2013). Dark tourism has evolved into a niche tourism industry with significant growth potential in the future (Proos & Hattingh, 2022). Once dark tourism was recognized as a phenomenon, some countries attempted to incorporate dark tourism as a product into their tourism sector (Kunwar et al., 2019). Many countries worldwide have implemented structures to support this new service, and many attractions and exhibitions have emerged worldwide catering to the "dark side of travel" (Sharpley, 2022). Therefore, this study explores dark tourism's marketing strategies and media influence through reviewing and analyzing previous literature. This enables researchers to reveal dark tourism's development status and future trends and gain new insights into the application of new media in dark tourism. Additionally, some recommendations are provided for dark tourism stakeholders.

Evolution of Dark Tourism Marketing: A Historical Perspective

The phrase "dark tourism" was initially used in the middle of the 1990s by Foley and Lennon (1996) to denote death-related tourism activities. Dark tourism is also known as "morbid tourism" (Blom, 2000), "disaster tourism," "thanatourism," and "black spot tourism" (Rojek, 1993). Visit areas linked with death, pain, and tragedy. According to Niemelä (2010), dark tourism is a multilayered blend of tourism, tragedy, history, and heritage. However, it is not new (Lennon & Foley, 2000). Since the 11th century, travelers have visited places connected to death (Dale & Robinson, 2011). Dark tourism has its roots in the Middle Ages (Seaton, 1996), but it gained appeal in the late 18th and early 19th centuries thanks to trips to places like the Waterloo battlefield (Seaton, 1999). Seaton (2009) states that Christianity initiated dark tourism in the 14th and 16th centuries and was then influenced by secular sacred ideologies of national heritage, concluding that, as depicted in Romantic art and literature, by acquired the symbolic form that prevails in contemporary perceptions of death. The first death-related forms of tourism have been recognized as medieval public executions, pilgrimages to the Holy Land to visit the location of Christ's crucifixion, and Roman gladiatorial combats (Stone & Sharpley, 2008). Indeed, humans have long been captivated by places connected to conflict, tragedy, death, and atrocities (Cohen, 2011). Smith (1998) believes war-related locations may constitute the world's most significant category of tourist attractions.

Profiling Dark Tourism Audiences

Dark tourism products are presented based on educational, commemorative, or conservation perspectives (Niemelä, 2010). Historical gladiator matches in the Roman Coliseum, crowds gathering to watch public executions, visits to famous battlefields, and other activities associated with death and suffering are examples of dark tourism (Seaton & Lennon, 2004). These travel experiences offer opportunities to study the self on an intrapersonal and potentially interpersonal level (Hede & Thyne, 2010). Stone (2006) believes that the analysis of dark tourism supply is incomplete without considering tourist demand. Dark tourism is increasingly becoming an essential trend in tourism. In 2007, over 1.2 million visitors visited the Auschwitz-Birkenau massacre site, and 3.5 million people visited the Ground Zero site in New York, making it one of the most popular tourist attractions in the city (Kang et al., 2012). Extensive research on dark tourism motivations has revealed some common motivations, including curiosity, amusement, empathic affiliation, sympathy, nationalistic motives, pilgrimage, event validation, identity quest, education, and a sense of social obligation, among the several researches on dark tourism motivations (Dann, 1998; Iliev, 2021; Lewis et al., 2021; Podoshen, 2013; Rojek, 2002; Seaton & Lennon, 2004; Yan et al., 2016). The diversity of inner feelings and needs produces different forms of dark tourism and types of dark tourists. Schadenfreude (i.e., taking pleasure in other people's suffering) and thanatopsis (i.e., thinking about death) were identified by Seaton and Lennon (2004) as the two critical motivations for dark tourism. However, not all visitors who visit dark places have a dark experience (Stone & Sharpley, 2008), and not all visitors are tourists (Biran et al., 2011). In fact, rather than actively seeking to comprehend death as a significant motivation, tourists may infer the meaning of mortality from their visit (Stone, 2012). Poria et al. (2004) concur that leisure or an educational experience may induce visiting and eventually shape the experience because some tourists may not be aware of the site's dark characteristics.

The Role of Media in Shaping Dark Tourism Narratives

The media's role in shaping dark tourism is pivotal in this era. The media has an acute impact on the role of media that are being perceived, visited, and preserved. In addition to that, the media context here comes in many forms, such as cinema, short web series/TV serials on dark tourism. In addition, the current development in entertainment media and the sudden rise of streaming platforms, social media, iCloud, content marketing, virtual influences, and bloggers in dark tourism play a crucial role in this era. According to Walter (2009), there is a close link between the media and dark tourism. He stated there is a choice of visiting the places in person or watching a documentary on television.

Further, he stated an example regarding the funeral of Princess Diana, whereby several thousand people decided to watch her funeral on a giant television screen in Hyde Park. On the other hand, millions worldwide watched it at home on television, as Walter (2009) mentioned. It is evident that with increasing growth in dark tourism, especially in terms of their demand (Seaton & Lennon, 2004), death and suffering are transformed rapidly due to the current role played by media, primarily in terms of cinema, television, music, and literature as stated by Rojek (1993) and Walter (2009).

Marketing Strategies in Dark Tourism

Marketing strategies in dark tourism focus on a combination of balance of education, respect, and also toward a storytelling process. This can be achieved through capitalizing on the digital platform and providing experiences such as one of the famous destinations in terms of darker tourism, Auschwitz Memorial's virtual tour, attracts many visitors each year (Kaelber, 2007). This place allows visitors to explore the site further remotely. In addition to that, New Orleans' collaboration with Voodoo practitioners is mainly to offer educational and respectful voodoo-themed tours that eventually led to respecting religion's sacred traditions. This was mainly a strategic partnership with the local communities to take dark tourism toward a different level (Maranise, 2012).

Furthermore, Inglis and Holmes (2003) highlighted more dark tourism with a link to broader destination marketing. For example, Scotland was considered a prime tourist location, where its landscapes, cityscapes, and habitations have been recreated over the last two centuries based on the history, which has a lot of linkage with the Highland and Celtic past. In other words, this links dark history and marketing regarding destinations. Besides that, these strategies aim to satisfy the curiosity of travelers and provide an educational experience within dark tourism.

The Impact of Social Media on Dark Tourism Trends

Social media plays a crucial impact toward dark tourism trends, which can be observed via user-generated online content, as stated by Jeffrey et al. (2022). This is where it influences the consumers' attitudes and values. In addition, a few studies have looked at the role and nature of social media, whereby it gives a better understanding of the experiences and behaviors of social media users in their web exchanges, as Wight (2020) asserted. Other than that, different segments and postpurchase behaviors influence social media's impact on dark tourism trends by being able to communicate and share experiences with others, as concluded by Bizirgianni and Dionysopoulou (2013). Furthermore, sharing experiences on social media and perceiving quality in dark tourism destinations are considered a fundamental identity, as Gallarza et al. (2019) stated. This is where such experiences will drive toward dark tourism more by sharing their experiences via social networks. Additionally, moral obligations are also considered as the impact of social media on dark tourism trends, especially in obtaining comments from travelers' posts on social media, which eventually leads to visiting the dark tourism site, as identified by Wight (2020). This contributes toward personal narratives, reviews, and recommendations that can be done via social media perspective on what to expect from specific dark tourism sites.

Ethical Considerations in Marketing Dark Tourism Attractions

Dark tourism is a sociological phenomenon in which morality has been extensively questioned, particularly by the media (Stone, 2009). Dark tourism creates new venues for immorality to be depicted for modern consumption and morality to be transmitted, reconstructed, and revitalized (Stone, 2009). Dark tourism exploits taboo topics and horror for commercial purposes, stimulating moral conflicts on a collective and individual level while providing new spaces for moral exchange, reconfiguration, and revitalization (Stone, 2009). Drury (2019) noted that the proliferation of photos on social media of tourists posing on the railroad tracks leading to the Auschwitz Memorial's central entrance forced museum authorities to issue a statement on Twitter asking visitors to be more respectful when taking photos. However, it succeeded in drawing attention to the ethical problems with shooting "inappropriate" images at sensitive dark tourism destinations and the topic of transgressive tourist behavior. The moral approach of tourists to death and its various manifestations has been neglected. The method through which people decide what is morally right is becoming increasingly complex (Treviño et al., 2014). People are now more dependent on themselves for moral teaching due to the individualization process, in which old meaning systems and values lose significance in favor of individual considerations and judgments about values, norms, and behaviors (Stone & Sharpley, 2013). Dark tourism is somewhat deviant in its production and consumption of taboo topics such as death and displays of the dead. Explaining the ethics of the seemingly fleeting relationships that arise during photographic encounters between tourists and local communities (Scarles, 2013) notes that photography becomes a delicate balance – a set of circumstances that often relies on intuitive moral judgments, reasoning, and reflective argumentation.

Collaborations and Partnerships: Industry Perspectives on Marketing Dark Tourism

Dark tourism primarily involves the display and consumption (by tourists) of natural and commodified sites of death and disaster. Horodnikova and Derco (2015) pointed out that a prerequisite for the further development of dark tourism is the cooperation between the War Historical Club and all parties involved in tourism marketing campaigns supported by public finances. Dark attractions are promoted mainly by supply stakeholders such as national tourist organizations and travel operators (Farmaki, 2013). Stone (2005) indicated that the availability of dark attractions is influenced by things like attempts to gain commercial advantage or to utilize a dark historical place for political goals. The product supply-side stakeholders offer often determines what visitors consume and whether the attraction is necessary. Therefore, the development and promotion of black tourist destinations on the supply side is significant, as it can drive demand to a large extent. For the consumer side, different locations are associated with death but are not viewed as dark attractions by tourists; instead, they are viewed as heritage or religious attractions with specific cultural values. Ashworth (1991) indicated that a destination's projected image influences tourists' perceptions and actions. Tourism stakeholders need to review and enhance the image of dark tourism. A dark image is produced by projected images produced by tourism authorities, tour operators, and the media and is an important "pull" factor in attracting tourists. The projected images of dark tourism are also affected by communication channels such as the media, including the source of dark field information (i.e., the internet, travel agencies) and the passage of time since the dark event occurred. Therefore, supply and demand side understandings of projected dark images must be considered when delivering dark tourism products, as their interaction plays a role in how dark tourism is perceived and consumed.

Emerging Trends and Innovations in Dark Tourism Marketing

Tourism demand intensifies as human society transforms from a service economy to an experience economy. Nowadays, tourists focus more on their feelings, impressions, and experiences during travel. Han (2014) indicated that the general trend of dark tourism in the future should be to strengthen the dark tourism product portfolio and create a more immersive and satisfying experience for tourists. If tourists can experience different emotions through scene simulation, such as fear and sadness, immersive scenes will gradually meet tourists' growing quality needs, thus increasing the appeal of dark tourism. Dark product development requires various media, such as modern digital media technology, 3D film and television technology, and stereoscopic technology, to reproduce the event process. Wearable devices are increasingly used in the tourism industry, and virtual reality (VR) is one of the most promising tools for the tourism industry, making it possible to visit a place through VR goggles (Guttentag, 2010). Wu and Lee's (2017) research inspired a new promotion concept to introduce Taiwan's dark tourism through an augmented VR experience. Scan the QR code with a smartphone and wear the VR eyewear for visitors. They can become fully immersed in the virtual tour, which features 360-degree panorama videos, interactive audio guiding, and a self-paced journey between various locations.

Dark tourism is one of the critical trends in tourism development, and some countries have positioned themselves as national dark tourism destinations. However, there is still room for development of this type of tourism. Modern tourists have different motivations when planning their itineraries, which requires countries to provide more distinctive black tourism products. Horodnikova and Derco (2015) discovered the interconnectedness of the tourism potential of regions affected by the First and Second World Wars and, based on the connections, created itineraries with dark tourism themes, thereby gaining the international importance of the tourist destination. Such interconnections can assist destination countries in necessary cross-border cooperation and dark tourism innovation processes.

Future Directions: Navigating Challenges and Opportunities in Dark Tourism Marketing

Lennon (2017) indicated that dark tourism is a hint of postmodernity. First, the ability of global communication technology to instantaneously broadcast them and then repeat them endlessly is primarily responsible for the interest in and perception of events related to death. Second, most dark tourism destinations contest the intrinsic order, reason, and advancement of modernity (much as the idea of postmodernity does). Third, the lines between the message (educational, political) and their commercialization as tourist products have been progressively blurred at most locations.

Whether demand or supply drives dark tourism is a crucial topic of discussion. Smith (2010) suggests that from a supply perspective, the focus is on organizations that offer or deliver services or experiences. From a demand perspective, it focuses on consumers seeking travel experiences or participating in travel activities. However, Stone and Sharpley (2008) believe that whether it is supply-driven or demand-driven is still being determined based on the nature of dark tourism. Although the motivations for dark tourism seem endless, ethical considerations still need to be included (Seraphin, 2017). Stone (2007) describes how visiting places where individuals are known to have suffered or died might raise concerns about being used for commercial, educational, or recreational purposes. However, dark tourist attractions are considered to provide experiences and education in a more sophisticated way (Stokes, 2013). Garcia (2012) describes how the sensitivities of dark tourism attractions can present challenges for practitioners. Tourism markets need to be segmented because not all visitors to gloomy sites are drawn there by a fascination with death, and not all visitors have the same experience (Farmaki, 2013).

The marketing implications of dark tourism development require further research, as the marketing of dark tourist attractions and the image of dark tourist attractions remains an under-researched area. The realization of Uzzell (1998) that modern personal mobility and the internet have revolutionized the tourism business has consequences for dark tourism marketing since images of dark attractions are difficult to control. Although significant efforts have been made to maintain and exploit battle history, site management and visitor interpretation for the demands of international tourists need to be addressed (Horodnikova & Derco, 2015). Some voices suggest that dark tourism is a recovery mechanism that can help communities recover after disasters. Another sign is that dark tourism's pedagogical function is fertile ground for cultivating empathy for the suffering of others (Seraphin & Korstanje, 2021). Korstanje and Clayton (2012) concluded that there is a link between terrorism and forms of death tourism, often referred to as "dark tourism." Still, they did not address the limitations of dark tourism as a commercial activity. Because both require modern technology to function and both rely on the media and require manipulating perceptions and attitudes. Getz (2008) argued that terrorism and other catastrophic occurrences threaten tourism.

References

Ashworth, G. J. (1991). Products, places and promotion: Destination images in the tourism industry analysis. In *Products, places and promotion: Destination images in the analysis of the tourism industry* (pp. 121–142). https://www.cabidigitallibrary.org/doi/full/10.5555/19921892510

Biran, A., & Hyde, K. F. (2013). New perspectives on dark tourism. *International Journal of Culture, Tourism and Hospitality Research*, *7*(3), 191–198.

Biran, A., Poria, Y., & Oren, G. (2011). Sought experiences at (dark) heritage sites. *Annals of Tourism Research*, *38*(3), 820–841.

Bizirgianni, I., & Dionysopoulou, P. (2013). The influence of tourist trends of youth tourism through social media (SM) & information and communication technologies (ICTs). *Procedia-Social and Behavioral Sciences*, *73*, 652–660.

Blom, T. (2000). Morbid tourism-a postmodern market niche with an example from Althorp. *Norsk Geografisk Tidsskrift*, *54*(1), 29–36.

Cohen, E. H. (2011). Educational dark tourism at an in populo site: The Holocaust Museum in Jerusalem. *Annals of Tourism Research*, *38*(1), 193–209.

Dale, C., & Robinson, N. (2011). Dark tourism. In *Research themes for tourism* (pp. 205–217). CABI.

Dann, G. M. (1998). *The dark side of tourism*. CIRET-International Center for Research and Studies in Tourism.

Drury, C. (2019). Auschwitz visitors told to stop posing for disrespectful photos at Nazi death camp. *The Independent*. https://www.independent.co.uk/news/world/europe/auschwitz-selfies-visitors-posing-railway-poland-a8833746.html

Farmaki, A. (2013). Dark tourism revisited: A supply/demand conceptualisation. *International Journal of Culture, Tourism and Hospitality Research*, *7*(3), 281–292.

Foley, M., & Lennon, J. J. (1996). JFK and dark tourism: A fascination with assassination. *International Journal of Heritage Studies*, *2*(4), 198–211.

Gallarza, M. G., Arteaga, F., & Gil-Saura, I. (2019). Customer value in tourism and hospitality: Broadening dimensions and stretching the value-satisfaction-loyalty chain. *Tourism Management Perspectives*, *31*, 254–268.

Garcia, B. R. (2012). JUPTRR. *Journal of Unconventional Parks*, *4*(1), 14.

Getz, D. (2008). Event tourism: Definition, evolution, and research. *Tourism Management*, *29*(3), 403–428.

Guttentag, D. A. (2010). Virtual reality: Applications and implications for tourism. *Tourism Management*, *31*(5), 637–651.

Han, K. (2014). Research on innovation path of dark tourism based on micro-niche market. *Academic Journal of Humanities & Social Sciences*, *6*(18), 1–4.

Hede, A. M., & Thyne, M. (2010). A journey to the authentic: Museum visitors and their negotiation of the inauthentic. *Journal of Marketing Management*, *26*(7–8), 686–705.

Horodnikova, J., & Derco, J. (2015). Dark tourism, thematic routes and possibilities for innovation in the Slovak Republic. *Tourism: An International Interdisciplinary Journal*, *63*(2), 241–246.

Iliev, D. (2021). Consumption, motivation and experience in dark tourism: A conceptual and critical analysis. *Tourism Geographies*, *23*(5–6), 963–984.

Inglis, D., & Holmes, M. (2003). Highland and other haunts: Ghosts in Scottish tourism. *Annals of Tourism Research*, *30*(1), 50–63.

Jeffrey, H. L., Ashraf, H., & Paris, C. M. (2022). Hanging out on Snapchat: Disrupting passive covert netnography in tourism research. In *Qualitative methodologies in tourism studies* (pp. 148–166). Routledge.

Kaelber, L. (2007). A memorial as virtual traumascape: Darkest tourism in 3D and cyber-space to the gas chambers of Auschwitz. *Ertr, e Review of Tourism Research*, *5*(2), 24–33.

Kang, E. J., Scott, N., Lee, T. J., & Ballantyne, R. (2012). Benefits of visiting a 'dark tourism'site: The case of the Jeju April 3rd Peace Park, Korea. *Tourism Management, 33*(2), 257–265.

Korstanje, M. E., & Clayton, A. (2012). Tourism and terrorism: Conflicts and commonalities. *Worldwide Hospitality and Tourism Themes, 4*(1), 8–25.

Kunwar, R. R., Aryal, D. R., & Karki, N. (2019). Dark tourism: A preliminary study of Barpak and Langtang as seismic memorial sites of Nepal. *Journal of Tourism and Hospitality Education, 9*, 88–136.

Le, D. T. T., & Pearce, D. G. (2011). Segmenting visitors to battlefield sites: International visitors to the former demilitarised zone in Vietnam. *Journal of Travel & Tourism Marketing, 28*(4), 451–463.

Lennon, J. (2017). Dark tourism. *Oxford Research Encyclopedia of Criminology and Criminal Justice*. https://doi.org/10.1093/acrefore/9780190264079.013.212

Lennon, J., & Foley, M. (2000). *Dark tourism: In the footsteps of death and disaster*. Cassell.

Lewis, H., Schrier, T., & Xu, S. (2021). Dark tourism: Motivations and visit intentions of tourists. *International Hospitality Review, 36*(1), 107–123.

Maranise, A. M. (2012). Investigating the syncretism of Catholicism and Voodoo in New Orleans. *The Journal of Religion & Society, 14*.

Niemelä, T. (2010). *Motivation factors in dark tourism: Case: House of terror*. Bachelor's Thesis.

Podoshen, J. S. (2013). Dark tourism motivations: Simulation, emotional contagion and topographic comparison. *Tourism Management, 35*, 263–271.

Poria, Y., Butler, R., & Airey, D. (2004). The meanings of heritage sites for tourists: The case of Massada. *Tourism Analysis, 9*(1–2), 15–22.

Proos, E., & Hattingh, J. (2022). Dark tourism: Growth potential of niche tourism in the Free State Province, South Africa. *Development Southern Africa, 39*(3), 303–320.

Rojek, C. (1993). *Ways of escape: Modern transformations in leisure and travel*. Springer.

Rojek, C. (2002). Indexing, dragging and the social construction of tourist sights. In *Touring cultures* (pp. 62–84). Routledge.

Scarles, C. (2013). The ethics of tourist photography: Tourists' experiences of photographing locals in Peru. *Environment and Planning D: Society and Space, 31*(5), 897–917.

Seaton, A. V. (1996). Guided by the dark: From thanatopsis to thanatourism. *International Journal of Heritage Studies, 2*(4), 234–244.

Seaton, A. V. (1999). War and thanatourism: Waterloo 1815–1914. *Annals of Tourism Research, 26*(1), 130–158.

Seaton, T. (2009). Thanatourism and its discontents: An appraisal of a decade's work with some future issues and directions. In *The Sage handbook of tourism studies* (pp. 521–542). SAGE Publications Ltd.

Seaton, A. V., & Lennon, J. J. (2004). Thanatourism in the early 21st century: Moral panics, ulterior motives and alterior desires. In *New horizons in tourism: Strange experiences and stranger practices* (pp. 63–82). CABI Publishing.

Seraphin, H. (2017). Terrorism and tourism in France: The limitations of dark tourism. *Worldwide Hospitality and Tourism Themes, 9*(2), 187–195.

Seraphin, H., & Korstanje, M. E. (2021). Dark tourism tribes: Social capital as a variable. In *Consumer tribes in tourism: Contemporary perspectives on special-interest tourism* (pp. 83–99). Springer.

Sharpley, R. (2022). Borders as dark tourism spaces. In *Routledge handbook of borders and tourism*. Routledge.

Shondell Miller, D., & Gonzalez, C. (2013). When death is the destination: The business of death tourism–despite legal and social implications. *International Journal of Culture, Tourism and Hospitality Research*, *7*(3), 293–306.

Smith, V. L. (1998). War and tourism: An American ethnography. *Annals of Tourism Research*, *25*(1), 202–227.

Smith, S. J. (2010). *Practical tourism research*, CAB International.

Stokes, R. (2013). Can dark tourism ever be a good thing? *New Internationalist Magazine*, *21*.

Stone, P. (2005). Dark tourism-cashing in on tragedy? https://www.semanticscholar.org/paper/Review%3A-Dark-Tourism-Cashing-in-on-Tragedy-Stone/44dd0d8f3f6252239fefb79bb18f664efaaa1bc4

Stone, P. R. (2006). A dark tourism spectrum: Towards a typology of death and macabre related tourist sites, attractions and exhibitions. *Tourism: An International Interdisciplinary Journal*, *54*(2), 145–160.

Stone, P. (2007). *Dark tourism: The ethics of exploiting tragedy*. Travel Weekly.

Stone, P. (2009). Dark tourism: Morality and new moral spaces. In *The darker side of travel: The theory and practice of dark tourism*. Channel View Publications.

Stone, P. R. (2012). Dark tourism and significant other death: Towards a model of mortality mediation. *Annals of Tourism Research*, *39*(3), 1565–1587.

Stone, P., & Sharpley, R. (2008). Consuming dark tourism: A thanatological perspective. *Annals of Tourism Research*, *35*(2), 574–595.

Stone, P. R., & Sharpley, R. (2013). Deviance, dark tourism and 'dark leisure'. In *Contemporary perspectives in leisure: Meanings, motives and lifelong learning* (pp. 54–64). Routledge.

Treviño, L. K., Den Nieuwenboer, N. A., & Kish-Gephart, J. J. (2014). (Un) ethical behavior in organisations. *Annual Review of Psychology*, *65*, 635–660.

Uzzell, D. (1998). Interpreting our heritage: A theoretical interpretation. In *Contemporary issues in heritage and environmental interpretation: Problems and prospects* (pp. 11–25). Stationery Office.

Walter, T. (2009). Dark tourism: Mediating between the dead and the living. In *The darker side of travel: The theory and practice of dark tourism* (pp. 39–55). Channel View Publications.

Wight, A. C. (2020). Visitor perceptions of European Holocaust heritage: A social media analysis. *Tourism Management*, *81*, 104142.

Wu, S. T., & Lee, B. W. (2017, November). An innovative way of guided tour: A virtual experience of dark tourism. In *2017 International Conference on Information, Communication and Engineering (ICICE)* (pp. 208–210). IEEE.

Yan, B. J., Zhang, J., Zhang, H. L., Lu, S. J., & Guo, Y. R. (2016). Investigating the motivation–experience relationship in a dark tourism space: A case study of the Beichuan earthquake relics, China. *Tourism Management*, *53*, 108–121.

Chapter 8

The Impact of Dark Tourism on Local Communities

Ankit Dhiraj[a], *Sanjeev Kumar*[a] *and Divya Rani*[b]

[a]Lovely Professional University Punjab, India
[b]UUMV Bhairwa, Bisfi, Madhubani, Bihar, India

Abstract

Dark tourism is a growing phenomenon that involves visiting places associated with death, tragedy, or suffering. Dark tourism places like war memorials, concentration camps, or sites of natural disasters. While dark tourism can have several positive benefits for local communities, such as economic boost and social awareness, it also poses some challenges, such as insensitivity to the feelings of the local community and exploitation. It is essential for local communities to carefully consider the potential benefits and challenges of dark tourism before developing dark tourism attractions. This chapter explores the positive and negative effects that dark tourism can have on these communities.

Keywords: Dark tourism; local communities; impact; motivation; ethical considerations

Introduction

Dark tourism is described as travel to destinations that have historically been linked to sorrow and death. Two professors from Glasgow Caledonian University, J. John Lennon and Malcolm Foley, first used the phrase "dark tourism" in 1996 (Lennon, 2017). The phrase is now frequently used to refer to journeys to locations having a long history of tragedy and death. Dark tourism, thanatourism, morbid tourism, and mourning tourism have also been used to describe it (Seraphin & Korstanje, 2021). There are more and more possibilities for travelers to immerse themselves in horrible past events in addition to the well-known memorial and human sorrow sites.

Dark Tourism, 99–106

Published under exclusive licence by Emerald Publishing Limited
doi:10.1108/978-1-83797-336-120241008

It is now widely used to describe sights and places associated with genocide, the Holocaust, assassinations, criminal activity, or jails that have drawn tourists (Hooper & Lennon, 2016). There are following types of dark tourism:

- *Crime Scenes:* These are the locations of infamous crimes, such as the Lizzie Borden Museum in Massachusetts and the Jack the Ripper Museum in London.
- *Disaster Sites:* Tourism related to natural disasters is considered dark tourism (Rucińska, 2016) and has characteristics of soft, specialized, and distinctively different travel (Rucińska & Lechowicz, 2014). The following are some of its advantages: the mechanisms of phenomenon, risk, and presence among local victims' community is traumatized (Cioccio & Michael, 2007). Travelers are drawn to the natural "dark tourism," sometimes known as disaster tourism, when explorers, tiny groups of tourists, or drifters. These are locations where natural or artificial disasters have happened, such as the Ground Zero Memorial and Museum in the United States (Stone, 2014) and the Chornobyl Nuclear Power Plant in Ukraine (Stone, 2013).
- *Genocide Sites:* Genocide tourism (visiting places connected to or affected by genocide) is a type of dark tourism that is becoming more prevalent but whose consumption is frequently misunderstood and frequently thought to indicate voyeuristic tendencies on the part of travelers. These are locations where mass massacres have occurred. Examples are the Tuol Sleng Genocide Museum in Cambodia (Brown & Millington, 2015) and the Auschwitz-Birkenau Memorial and Museum in Poland (Griffiths, 2019).
- *Cemeteries and Graveyards:* Tourists who partake in graveyard or cemetery tours intend to learn more about the rich artistic, architectural, historical, and scenic legacy that frequently abounds in cemeteries. These are where the deceased is interred, and tourists interested in history, ancestry, or the macabre may find them attractive tourist destinations (Young & Light, 2016). Highgate Cemetery in London, UK (Ellis, 1993), Old Jewish Cemetery in Prague, Czech Republic (Greenblatt, 2002), Hollywood Forever Cemetery in Los Angeles, USA (Levitt, 2012), La Recoleta Cemetery in Buenos Aires, Argentina (Slessor, 1999), Merry Cemetery in Sapanța, Romania (Mionel, 2020) are the example of cemeteries.
- *Battlefield Locations:* Battlefield tourism is a growingly common travel trend in which tourists visit places that were the scene of famous battles to connect with history and cultural heritage (Ryan, 2007). Battle tourism can be found in many locations, particularly those with a history of ancient conflicts (Miles, 2012). Significant conflicts have occurred in these locations, such as the Battlefields of the Somme in France (Seaton, 2000) and the Gettysburg National Military Park in the United States (Cameron, 2004).

Motivation to Tourist for Dark Tourism

Travelers opt to engage in dark tourism for a variety of reasons. Some people are burning to learn more about particular historical catastrophes or events. Some people are drawn to the macabre or supernatural. Others want to know what it is like to go to a location where something dreadful has occurred (Podoshen, 2013). There are the following common motivation factors for dark tourism:

- *Education:* Dark tourism sites are frequently visited by persons who want to learn more about particular historical calamities or occurrences (Tanaya & Garry, 2006). To study more about the Holocaust, a student would visit Auschwitz-Birkenau (Niesler et al., 2010), whereas a visitor might go to Ground Zero to learn about the September 11th attacks.
- *Reflection:* Dark tourism can also catalyze individuals to consider their mortality and the impermanence of life (Stone, 2012). It may be an incredibly touching and powerful experience to visit a place where death and suffering have occurred.
- *Curiosity:* Some folks merely want to experience what it is like to go to a location where something dreadful has occurred (Lewis et al., 2021). They can be drawn to the macabre or the occult, or they might want to soak in the vibe of a creepy tourist destination (Garlick, 2022).
- *Community:* Some people use dark tourism websites to connect with a group of people who have similar interests to themselves. For instance, a group of true crime aficionados may visit a notorious crime scene, or a group of military history enthusiasts might visit a battlefield.
- *Healing:* Dark tourism destinations may help some people recover from trauma or grief (Prayag, 2016). A natural catastrophe survivor, for instance, would go to the disaster's location to process what happened.

Impacts of Dark Tourism on Local Communities

Dark tourism has become an essential branch of the tourism industry. Tourists are progressively being attracted to dark tourism sites. The areas where dark tourism attractions are located may also benefit from dark tourism. It can bring in money, produce employment, and promote knowledge of significant historical occurrences. However, it is crucial to ensure that dark tourism is done with courtesy and morality (Kang et al., 2012). There are some of the following positive impacts of dark tourism:

- *Economic Benefits:* Tourist expenditures on lodging, dining, travel, and souvenirs can bring in money for local communities due to dark tourism (Jamin et al., 2020). Dark tourism contributes to job growth and a more robust local economy.
- *Raising Awareness:* Dark tourism can aid in increasing public knowledge of significant historical occurrences and their effects on nearby communities and encourage communication and peace-making.

- *Education:* Visitors can learn about history, culture, and the human condition through dark tourism, which can be an educational experience (Jang et al., 2021). Dark tourism promotes compassion and understanding (Cohen, 2011).

Residents frequently play the role of hosts and may directly influence guests' perceptions of the place, affecting the viability of local dark tourism development. Clarifying locals' opinions on dark tourism becomes a crucial agenda item (Stone, 2006). It is significant to highlight that depending on the particular setting, the effect of dark tourism on local communities can change.

- *Exploitation:* Dark tourism may abuse nearby populations, mainly if it is not handled responsibly and with respect. Dark tourism sites, for instance, can be created without the community's approval, or they might promote a skewed or sensationalized version of history.
- *Trauma:* Local populations may experience trauma from dark tourism (Braithwaite & Lee, 2006), particularly if it is connected to recent events or catastrophes. For instance, natural disaster victims could feel uneasy about tourists visiting the disaster site.
- *Environmental Impact:* If dark tourism is managed sustainably, it may benefit the ecosystem. Dark tourist destinations, for instance, could produce much waste or harm the environment.

Here are some ways to mitigate the negative impacts of dark tourism on local communities:

- *Community Involvement:* Local groups must be involved in developing and administrating dark tourism destinations. It makes it easier to guarantee that the sites respect the locals and that the advantages of dark tourism are distributed fairly.
- *Education:* People visiting dark tourism destinations should be informed about the significance of the events or tragedies commemorated there and the site's history. Tourist fosters mutual respect and understanding.
- *Sustainability:* Dark tourism destinations should be handled sustainably to reduce adverse environmental effects (Jureniene & Radzevicius, 2022).

Dark Tourism in the 21st Century

An increasing trend in the 21st century is dark tourism. Dark tourism is brought on by a variety of factors, such as:

- *Growing Interest in History:* There is a growing interest in history in general and with regard to particular historical events and tragedies. It is brought on by several things, such as the growth of social media, the acceptance of historical documentaries, and the expansion of online historical knowledge.

- *Growing Interest in the Macabre:* There is also an increasing interest. The success of horror films and TV series and the popularity of actual crime podcasts and documentaries are evidence of this.
- *Greater Accessibility:* Sites for dark tourism are now easier to find than before due to the rise of budget travel and the increasing availability of online information about dark tourism attractions (Heidelberg, 2015).

Dark Tourism and Future

Dark tourism is likely to continue to grow in popularity in the future. This is due to several factors, including the increasing interest in history, the growing curiosity about the macabre, and the increased accessibility of dark tourism sites (Rajasekaram et al., 2022). However, it is essential to ensure that dark tourism is conducted responsibly and ethically. Dark tourism operators and communities must work together to ensure that dark tourism attractions are respectful to the victims and survivors of the events or tragedies they commemorate, accurate, educational, and sustainable. Here are some specific ways in which dark tourism can be made more responsible and ethical:

- More research is needed into the impact of dark tourism on visitors and communities. This research can be used to develop best practices for dark tourism operators.
- Dark tourism operators should work with local communities to ensure that their attractions are respectful and beneficial and involve sharing revenue with the community, hiring workers, and supporting local businesses.
- Dark tourism operators should provide visitors with information about the events or tragedies that their attractions commemorate, as well as the impact of those events on the victims and survivors. This information can help visitors understand the site's significance and reflect on their visit.
- Visitors to dark tourism sites should respect the victims and survivors (Prayag et al., 2021) of the events or tragedies the sites commemorate. Visitors avoid taking inappropriate photos or videos and be mindful of the noise level and other behaviors that may be disruptive.

Ethical Considerations

Ethical considerations must be considered when developing and managing dark tourism attractions. It is essential to ensure that these attractions are respectful to the victims and survivors of the events or tragedies that they commemorate (Fisher & Bolter, 2018). It is also essential to ensure that they are not exploitative or sensationalistic. It is vital to ensure that dark tourism attractions respect the victims and survivors of the events or tragedies they commemorate because of sensitivity to visitors' feelings and avoiding sensationalizing or trivializing the events. Dark tourism attractions should be historically accurate (Stone, 2006), providing visitors with factual information and avoiding myths and stereotypes.

Dark tourism attractions should be educational (Oren & Shani, 2012). They should help visitors learn about the events or tragedies they commemorate and reflect on their significance. Communities should have a say in how their dark tourism sites are managed and promoted.

Conclusion

Dark tourism is a complicated, multifaceted phenomenon with potential benefits and drawbacks. It may be valuable to interact with history, maintain memory, and foster empathy and tolerance with delicacy, respect, and a sincere desire to learn and understand. It also highlights ethical issues and calls for responsible engagement to prevent the exploitation of pain. Visitors should be careful about the places they visit and the reasons behind such decisions. They should also be aware of dark tourism's potential advantages and disadvantages for individuals and communities. Residents typically have a deep emotional tie to the events and aftermath of catastrophes since they own a dark tourism destination, and many are victims or family members of victims. They thus tend to be sensitive and resist the growth of dark tourism in their community. Residents nonetheless frequently serve as hosts and may directly impact visitors' perceptions of the area, affecting the viability of local dark tourism development. Locals in such circumstances often experience the dark past in different ways, leading to varying degrees of emotional attachment to the location and different perceptions of and attitudes toward the development of dark tourism.

References

Braithwaite, D., & Lee, Y. L. (2006). *Dark tourism, hate and reconciliation: The Sandakan experience*. International Institute for peace through tourism occasional paper, *8*.

Brown, C., & Millington, C. (2015). The memory of the Cambodian genocide: The Tuol Sleng Genocide Museum. *History Compass*, *13*(2), 31–39. https://doi.org/10.1111/hic3.12214

Cameron, C. M. (2004). Gettysburg National, Military Park. *Ethnology*, *43*(3), 193–216.

Cioccio, L., & Michael, E. J. (2007). Hazard or disaster: Tourism management for the inevitable in Northeast Victoria. *Tourism Management*, *28*(1), 1–11.

Cohen, E. H. (2011). Educational dark tourism at an in populo site: The Holocaust Museum in Jerusalem. *Annals of Tourism Research*, *38*(1), 193–209.

Ellis, B. (1993). The Highgate Cemetery vampire hunt: The Anglo-American connection in satanic cult lore. *Folklore*, *104*(1–2), 13–39. https://doi.org/10.1080/0015587X.1993.9715852

Fisher, J. A., & Bolter, J. D. (2018). Ethical considerations for AR experiences at dark tourism sites. In *2018 IEEE International Symposium on Mixed and Augmented Reality Adjunct (ISMAR-Adjunct)* (pp. 365–369).

Garlick, K. (2022). *Morbid curiosity: Exploring the ethics of dark tourism*. Doctoral Dissertation. West Chester University.

Greenblatt, R. L. (2002). The shapes of memory: Evidence in stone from the Old Jewish Cemetery in Prague. *The Leo Baeck Institute Yearbook*, *47*(1), 43–67.

Griffiths, C. (2019). Encountering Auschwitz: Touring the Auschwitz-Birkenau State Museum. *Holocaust Studies*, *25*(1–2), 182–200. https://doi.org/10.1080/17504902.2018.1472881

Heidelberg, B. A. W. (2015). Managing ghosts: Exploring local government involvement in dark tourism. *Journal of Heritage Tourism*, *10*(1), 74–90. https://doi.org/10.1080/1743873X.2014.953538

Hooper, G., & Lennon, J. J. (2016). *Dark tourism: Practice and interpretation*. Routledge.

Jamin, A., Zain, Z. M., Sakarji, S. R., Ahmad, N., & Beta, R. M. D. M. (2020). The benefits of dark tourism experience among visitors in Malaysia. *KnE Social Sciences*, 219–228.

Jang, K., Sakamoto, K., & Funck, C. (2021). Dark tourism as educational tourism: The case of 'hope tourism' in Fukushima, Japan. *Journal of Heritage Tourism*, *16*(4), 481–492. https://doi.org/10.1080/1743873X.2020.1858088

Jureniene, V., & Radzevicius, M. (2022). Peculiarities of sustainable cultural development: A case of dark tourism in Lithuania. *Journal of Risk and Financial Management*, *15*(6), 264.

Kang, E.-J., Scott, N., Lee, T. J., & Ballantyne, R. (2012). Benefits of visiting a 'dark tourism' site: The case of the Jeju April 3rd Peace Park, Korea. *Tourism Management*, *33*(2), 257–265.

Lennon, J. (2017). Dark tourism. https://researchonline.gcu.ac.uk/en/publications/dark-tourism

Levitt, L. (2012). Solemnity and celebration: Dark tourism experiences at Hollywood Forever Cemetery. *Journal of Unconventional Parks, Tourism & Recreation Research*, *4*(1).

Lewis, H., Schrier, T., & Xu, S. (2021). Dark tourism: Motivations and visit intentions of tourists. *International Hospitality Review*, *36*(1), 107–123.

Miles, S. T. (2012). *Battlefield tourism: Meanings and interpretations*. PhD Thesis. University of Glasgow.

Mionel, V. (2020). (Not so) dark tourism: The Merry Cemetery in Săpânţa (Romania)–an expression of folk culture. *Tourism Management Perspectives*, *34*, 100656.

Niesler, A., Górny, R. L., Wlazło, A., Łudzeń-Izbińska, B., Ławniczek-Wałczyk, A., Gołofit-Szymczak, M., Meres, Z., Kasznia-Kocot, J., Harkawy, A., & Lis, D. O. (2010). Microbial contamination of storerooms at the Auschwitz-Birkenau Museum. *Aerobiologia*, *26*, 125–133.

Oren, G., & Shani, A. (2012). The Yad Vashem Holocaust Museum: Educational dark tourism in a futuristic form. *Journal of Heritage Tourism*, *7*(3), 255–270. https://doi.org/10.1080/1743873X.2012.701630

Podoshen, J. S. (2013). Dark tourism motivations: Simulation, emotional contagion and topographic comparison. *Tourism Management*, *35*, 263–271.

Prayag, G. (2016). It's not all dark!: Christchurch residents' emotions and coping strategies with dark tourism sites. In *Business and post-disaster management* (pp. 155–166). Routledge.

Prayag, G., Buda, D.-M., & Jordan, E. J. (2021). Mortality salience and meaning in life for residents visiting dark tourism sites. *Journal of Sustainable Tourism*, *29*(9), 1508–1528. https://doi.org/10.1080/09669582.2020.1823398

Rajasekaram, K., Hewege, C. R., & Perera, C. R. (2022). "Tourists' experience" in dark tourism: A systematic literature review and future research directions. *Asia Pacific Journal of Tourism Research*, *27*(2), 206–224. https://doi.org/10.1080/10941665.2022.2046118

Rucińska, D. (2016). Natural disaster tourism as a type of dark tourism. *International Journal of Humanities and Social Sciences*, *10*(5), 1458–1462.

Rucińska, D., & Lechowicz, M. (2014). Natural hazard and disaster tourism. *Miscellanea Geographica. Regional Studies on Development*, *18*(1), 17–25.

Ryan, C. (2007). *Battlefield tourism*. Routledge.

Seaton, A. V. (2000). Another weekend away looking for dead bodies…": Battlefield tourism on the Somme and in Flanders. *Tourism Recreation Research*, *25*(3), 63–77. https://doi.org/10.1080/02508281.2000.11014926

Seraphin, H., & Korstanje, M. E. (2021). Dark tourism tribes: Social capital as a variable. In C. Pforr, R. Dowling, & M. Volgger (Eds.), *Consumer tribes in tourism* (pp. 83–99). Springer Singapore. https://doi.org/10.1007/978-981-15-7150-3_7

Slessor, C. (1999). La Recoleta Cemetery, Buenos Aires, Argentina. In *Architectural review* (Vol. 205, No. 1224, p. 90). EMAP Business Publishing Ltd.

Stone, P. R. (2006). A dark tourism spectrum: Towards a typology of death and macabre related tourist sites, attractions and exhibitions. *Tourism: An International Interdisciplinary Journal*, *54*(2), 145–160.

Stone, P. R. (2012). Dark tourism and significant other death: Towards a model of mortality mediation. *Annals of Tourism Research*, *39*(3), 1565–1587.

Stone, P. R (2013). 6 Dark tourism, heterotopias and post-apocalyptic places: The case of Chernobyl. In *Dark tourism and place identity* (pp. 79–93). Routledge.

Stone, P. R. (2014). Dark tourism as 'mortality capital': The case of Ground Zero and the significant other dead. In *Contemporary tourist experience* (pp. 71–94). Routledge.

Tanaya, P., & Garry, P. (2006). Motivations of participants in dark tourism: A case study of Port Arthur, Tasmania, Australia. In *Taking tourism to the limits* (pp. 191–198). Routledge.

Young, C., & Light, D. (2016). Interrogating spaces of and for the dead as 'alternative space': Cemeteries, corpses and sites of dark tourism. *International Review of Social Research*, *6*(2), 61–72.

Chapter 9

Dark Tourism, an Absolute Transition, a Great Reset to Propel Reconciliation: The Case of Jaffna War Zones

Suduweli Kondage Lathisha Jayangi Ramanayaka

Sri Lanka Institute of Tourism & Hotel Management, Sri Lanka

Abstract

People's fascination with historical sites led to the development of the dark tourism phenomenon. Today's tourism idea has been affected by the obsession with sorrow and death, and many tourists from across the world visit regions connected to Sri Lanka. A notable case study for the growth of dark tourism in Asia is Sri Lanka, a South Asian nation that has been involved in a 26-year civil war with the Tamil Tigers and has left behind an ingenious historical relic system. There tend to be many competing viewpoints when it comes to historical battle sites, but having a particular cause to go there gives individuals a variety of feelings and experiences. This research paper reviews academic research into dark tourism over the 1996–2016 period. The way that tourists view war memorials depends on how they understand "their reality" as it is presented by the location and its inhabitants. The structure and numerous facets of the dark areas in South Asian culture, which are distinguished by a certain sort of enlightened pilgrims and ancestral ceremonial practice, are identified using a qualitative research technique. This study aimed to address those research aims by conducting semi-structured interviews with members of the general population. The researcher examines the information obtained through the interpretation and tales of the tour leader. The study advances our knowledge of dark tourism generally and in the context of Sri Lankan war zones, where the role of dark tourism in Sri Lanka's history of war has been underappreciated.

Dark Tourism, 107–121

Published under exclusive licence by Emerald Publishing Limited
doi:10.1108/978-1-83797-336-120241009

Keywords: Ancestral ceremonial practice; dark tourism; pilgrimage; Sri Lanka; tourist experience

Introduction

The realm of dark tourism encompasses journeys to locales that have confronted or bear connections with instances of death, calamity, moral degradation, adversity, and the eerie (Lennon & Foley, 1999). This notion of dark tourism isn't a recent conception (Collins-Kreiner, 2016; Miller & Gonzalez, 2013). The presence of "darkness" in this category of tourism largely emanates from historical occurrences like wars. Other instances encompass catastrophes, purported supernatural occurrences, unconventional rituals, and assorted tragedies.

In contemporary times, dark tourism has experienced a surge in prominence, notably among travelers from Western origins. Consequently, it is commonplace to encounter multiple dark tourism sites across Europe and North America. These sites serve as repositories of diverse historical episodes, attracting visitors who are captivated by themes of mortality, misfortune, and the macabre. This attraction has propelled dark tourism into a burgeoning sphere of interest for visitors (Miller & Gonzalez, 2013; Podoshen et al., 2015).

Dark tourism stands as a venerable concept, intricately interwoven with historical epochs marked by battles, sacrificial rites, public executions, and mortality (Johanson, 2012; Podoshen et al., 2015). Its genesis can be traced to the latter part of the 1990s, notably characterized by Foley and Lennon's (1996) seminal articulation that defines dark tourism as the exhibition and consumption of authentic and commodified sites of demise and catastrophe. Visitors, seeking an unfiltered encounter with these realities and their ties to acts of horror, partake in this experiential pilgrimage (Yoshida et al., 2016).

These locales encompass places of celebrity or mass demise, denoted by Rojek (1991) as "black spots." Simultaneously, Seaton (1996) delved into the exploration of "thanatopsis" or the contemplation of mortality, elucidating its motivational role in dark tourism. Categorically, dark tourism destinations align with expansive thematic classifications, encompassing warfare, historical epochs, arcane customs, calamities, natural cataclysms, the supernatural, destitution, and other calamitous occurrences with macabre undertones (Tan & Lim, 2018).

In succinct terms, these places are inherently affiliated with the themes of demise and adversity (Reid, 2016). These nuanced perspectives and definitional frameworks constitute the bedrock of the contemporary comprehension of dark tourism, thereby facilitating the facile identification of potential dark tourism locales. In the context of this study, dark tourism is operationalized according to antecedent scholarly works as "the act of journeying to sites entwined with the underlying motifs of death, misfortune, and any other manifestations of a macabre nature."

The evolutionary trajectory of the dark tourism phenomenon aligns intricately with the proclivity of individuals toward attractions steeped in historical narratives. In the contemporary landscape, discernibly, tourists manifest a predilection for specialized tourism markets that cater to distinctive interests and bespoke needs.

The paradigm of modern-day tourism, by extension, is arguably underpinned by the ascendancy of such niche frameworks, epitomized by the emergent concept termed "dark tourism." The allure of mortality and calamity, as asserted by Reynolds (2016), has indelibly shaped the contemporary ethos of tourism. Global sojourners converge upon locales within Sri Lanka that constitute a nascent recognized niche market in the realm of dark tourism.

Curiously, the academic expanse remains conspicuously void of in-depth scrutiny and scholarly reports germane to this evolving paradigm. This research, therefore, undertakes the ambitious venture of scrutinizing the motivational underpinnings compelling individuals to visit somber destinations within Jaffna, subsequently elucidating the consequential impact on the broader tapestry of the country's reconciliation endeavors.

Within the Sri Lankan context, dark tourism materializes as an embryonic and underexplored niche market, as expounded by Fernando et al. (2016). The prospect of actualizing and refining the concept within the confines of extant geographic locales, notably in the northern precincts, specifically Jaffna, where vestiges of the war linger palpably, comes to the fore. The ensuing exploration scrutinizes the diverse attitudes exhibited by domestic tourists, delving into the motivating forces propelling them to engage with sites of darkness. The subsequent analysis extends to evaluating the repercussions on the larger ambit of dark tourism, ultimately culminating in the formulation of alternative strategies designed to intertwine tourism with the imperatives of reconciliation and sustainable tourism practices, as posited by Fernando and Jayawardena (2013).

The significance of this study is twofold, demarcating theoretical and practical dimensions. Theoretically, it seeks to bridge a discernible void in extant research, thereby positioning itself as a pivotal intervention within the scholarly discourse. Substantively, this endeavor holds promise in augmenting the collective reservoir of knowledge pertinent to the intricacies of dark tourism. Jaffna, post the tumultuous period of conflict, emerges as an enticing destination, holding latent potential for exploration by the inquisitive traveler, poised to glean not only aesthetic gratification but also educational edification.

The meticulous delineation of tourist motivations and behaviors is poised to furnish invaluable insights for industry stakeholders. In the multifaceted tapestry of the tourism sector, replete with diverse service providers and stakeholders, opportunities abound for strategic engagement with this burgeoning niche market. Specifically, the Jaffna region stands as a repository of tourism destinations that can be strategically curated and promoted through targeted market segmentation initiatives.

Concomitantly, the outcomes derived from the research findings proffer a potential catalyst for governmental intervention. New regulatory frameworks and policies, informed by empirical evidence, may be instituted to govern tourist activities, ushering in a nuanced era of governance aligned with the exigencies of this nascent tourism paradigm. It is incumbent upon the government to leverage these insights for experimental endeavors and implementations.

However, the discerning researcher must grapple with certain limitations inherent in this scholarly pursuit. Notably, the dearth of antecedent studies within the Sri Lankan context poses a constrictive parameter, constraining the depth of contextual understanding. Simultaneously, the intricacies of data collection are exacerbated by the formidable language barrier, impeding direct communication with participants and precipitating challenges in extracting nuanced responses. The direct impact on participants' sentiments and cognitions, pivotal facets for comprehensive comprehension, is subtly obfuscated by this linguistic constraint.

In summation, this research endeavors to unravel the latent potentials and challenges intrinsic to the dark tourism landscape within the Jaffna region, casting a discerning gaze upon the intersectionality of historical narratives, tourist motivations, and the imperatives of reconciliation and sustainable tourism practices.

Review of Literature

The advent of fatalities and calamities within the realm of tourism has emerged as an increasingly pervasive facet in the contemporary tourist landscape, thereby ushering in the phenomenon of dark tourism (Singh, 2014). Despite the escalating scrutiny bestowed upon dark tourism, a global cohort of individuals persists in journeying to locales marked by mortality, a trend that has not gone unnoticed (Chang, 2014). This proclivity is notably conspicuous in Sri Lanka, a nation etched with the indelible imprints of a 20-year civil conflict that concluded in 2009 and the abrupt onslaught of a devastating tsunami in 2004 (Euromonitor International, 2018).

In the aftermath of these harrowing episodes, tourists have inundated the country, converging upon sites emblematic of the postwar and posttsunami epochs, thereby paying homage to the vicissitudes of the affected regions (Fernando & Jayawardena, 2013). The government, cognizant of the potential sociopolitical and economic ramifications, strategically leverages this trend to bolster stability, embarking upon multifaceted initiatives ranging from the construction of opulent resorts to the encouragement of tourist influx from the southern regions to the northern and eastern fringes of the nation (Burns & Novelli, 2007). Simultaneously, infrastructural enhancements, exemplified by the development of transport corridors leading to these sites, constitute instrumental endeavors aimed at rendering the overall tourist experience more amenable and comfortable (Burns, 2004).

Upon scrutinizing the remnants of the war, a vivid tapestry of scars emerges, bearing witness to the tumultuous 20-year phase of the civil conflict – a landscape replete with ruins, the remnants of tanks and ships, and desolate towns haunted by the spectral echoes of the past (Miles, 2013). War memorials, erected in solemn tribute to the valiant soldiers who sacrificed their lives, further punctuate this historical tableau (Ratnayake & Hapugoda, 2016).

In a rather disquieting juxtaposition, individuals evince a palpable interest in visiting these evocative sites. This inclination stems, on one hand, from an appetite for the thrill intrinsic to such ventures and, on the other, a profound appreciation for the historical narratives embedded within these events.

A comprehensive overhaul of the tourism industry is imperative for Sri Lanka to establish competitiveness within the global travel market, taking all factors into consideration (Kamble & Bouchon, 2014). In the Sri Lankan context, tourism holds the rank of the third-largest contributor to export earnings, with the past seven years witnessing unprecedented growth in visitor numbers (Birna et al., 2013). The escalating influx of visitors, particularly toward "dark" sites in the northern hinterlands and select regions within the eastern, southern, and western provinces, intensifies the imperative to effectively manage environmental ramifications and sustain the quality standards of tourist services and products (Samaranayake et al., 2013).

Globally, tourism emerges as a swiftly burgeoning industry, underscored by its overarching importance and economic significance. Within the Sri Lankan milieu, the emphasis on tourism resides in its potential contribution to future developmental objectives, serving as a pivotal mechanism for income generation, foreign exchange accrual, and the creation of employment opportunities within a holistic framework (Jayawardena, 2013).

The nuanced tapestry of Sri Lanka's tourism narrative is marked by diverse influences, one of which is the ambiance engendered by the Liberation Tigers of Tamil Eelam (LTTE) a decade ago, exerting a formative impact on the tourism sector and positioning Sri Lanka as a significant player in the domain of dark tourism (Senaratne, 2010).

Notably, the Sri Lankan tourism industry finds itself intricately entwined with travel agents and tour operators, particularly concerning inbound tourism. In light of this, expanding the value proposition and formulating strategic decisions becomes imperative to fortify and nurture the essence of destinations, thereby adding economic value to the invaluable assets surrounding these destinations marked by historical remnants (Sri Lanka Tourism, 2015).

Amid the transformational themes, enhancing governance regulations emerges as a salient facet, particularly in association with postwar affected locales, cities, and towns that bore the profound impacts of the 2004 tsunami (SLTDA, 2016).

Research Methodology

The present investigation employed the qualitative research methodology, grounded in an inductive approach. As elucidated by Selvanathan (2007), qualitative research is predicated upon the observation and interpretation of individuals' perceptions of diverse events, capturing a snapshot of their perceptual nuances within a natural setting. This method, rooted in the exploration of people's perceptions, proves instrumental in discerning respondents' perspectives regarding the phenomena encapsulated by dark tourism. The research philosophy underpinning this inquiry adopts a "constructivist" stance, entailing the acceptance of human behaviors and mindsets. This philosophical orientation closely aligns with a pragmatism derived from human experiences, rendering the qualitative method the most apt for the study's thematic terrain.

The study endeavored to amass multidimensional data from both domestic Sinhala people and Tamil residents in Jaffna. The methodological approach comprised semi-structured interviews, encompassing a diverse sample of Tamil residents and domestic tourists as respondents.

Primary Objective

The primary objective was to unravel the motivations propelling domestic travelers to engage with dark sites in Jaffna and assess the ramifications of these motives on the reconciliation dynamics between Sinhala and Tamil residents in the region.

Sampling

The author's purposive sampling strategy comprises domestic travelers who visited postwar Jaffna and Tamil residents in the locale.

Primary Data Collection Methods

Employing semi-structured interviews, seven conducted with Tamil residents in Jaffna and five with domestic travelers, the inquiry incorporated pretest surveys among domestic tourists by Jaffna researchers before their visits. The interview questions meticulously probed into the locations visited in Jaffna, the rationale behind selecting this locale, and the emotional responses during the visit, including considerations of revisiting memorials. For Tamil residents, inquiries delved into their interactions with Sinhala people, soliciting their perspectives on reconciliation and attitudes toward establishing mutual connections with the Sinhala populace.

The authors complement these interviews with firsthand observations during their visits to Jaffna. The observed locales encompassed a spectrum of significant sites, including the Old Kachcheri building, Nallur Kovil, Jaffna Museum, Library, Dambakola Patuna, Jaffna University, Dencil Kobbekaduwa Memorial, Thilipan Memorial, Naga Vihara, Kadurugoda Bhikshu Burials, and the Kirimale Premise. This selection is informed by previous studies and the evidentiary tapestry of postwar memories.

Data Analysis Method

The method employed for data analysis herein is thematic analysis, as delineated by Braun & Clarke, 2006. Following the transcription of interviews, a meticulous process of familiarization with the data ensued, wherein coding was systematically executed to distil and synthesize emergent themes reflective of the respondents' opinions and attitudes. The denouement of this analytical endeavor culminated in the identification of four overarching themes, each further delineated by 11 subcodes, meticulously derived from the nuanced reservoir of research findings.

The Cynicism of Dark Tourism: Dark Tourism and Mythology Belief Systems, Values, and Traditions of Asian Countries

In the intricate ballet between morose fascination and cultural profundity, the exploration of dark tourism within the milieu of Asian countries assumes a poignant resonance. This essay embarks on a cerebral odyssey, navigating the labyrinthine interplay between the cynicism inherent in dark tourism and the profound mythological belief systems, cultural values, and time-honored traditions woven into the very fabric of diverse Asian societies. Dark tourism, with its predilection for sites steeped in the narratives of mortality and historical lament, becomes an anthropological canvas upon which the complexities of human motivations and cultural entanglements are painted. The symbiotic dance between dark tourism and the mythopoeic constructs unique to the Asian continent beckons discerning scrutiny, unraveling layers of emotional intricacy and philosophical resonance. As we probe the depths of this dialectic, we seek to illuminate not only the shadows cast by somber locales but also the nuanced hues of cultural identity, inviting contemplation on the delicate balance between reverence and commodification in the tapestry of dark tourism across Asia.

Sri Lanka has emerged as an increasingly favored destination among both local and international travelers, gaining significant recognition from various acclaimed sources. Notably, *Lonely Planet* consecutively listed Sri Lanka as the premier destination to visit in both 2013 and 2019. *Forbes* magazine also bestowed upon the island the honor of being among the top 10 coolest countries to explore in 2015. Additionally, prominent publications like *The Guardian* and *The New York Times* identified Sri Lanka as a premier travel location in 2016 (SLTDA, 2016).

The island's allure stems from its natural splendor, strategically nestled in the heart of the Indian Ocean and renowned for its rich cultural heritage (Embuldeniya & Rajapaksha, 2013). Despite these accolades, several aspects of Sri Lanka's tourism sector remain untapped, holding immense potential for further development (Seoighe, 2017).

Dark tourism, notably significant for economic growth, gained prominence, particularly following the decade-long civil strife (Nisthar et al., 2017).

Initially perceived as an economical travel destination due to its remarkable climate and affordable travel rates, Sri Lanka's reinvestment potential was limited. However, post the cessation of the civil war in 2009, the country's economy, particularly within the tourism sector, underwent rapid expansion (Nisthar et al., 2017). This transformation presents an opportunity for continued economic growth. Dark tourism, an emerging facet within the broader tourism industry, stands as an avenue through which Sri Lanka can make substantial contributions and further its economic prowess.

The enduring strife between the Sri Lankan government and the LTTE stands like a behemoth, casting its shadow over nearly three decades, earning the title of the most protracted civil war in Asia (Jayashree, 2009). Known as the Tamil Tigers, the LTTE yearned to etch a separate destiny for the island's Tamil minority. After an unyielding triad of years from 2006 to 2009, the Sri Lankan military triumphantly dismantled the insurgent LTTE, culminating in the final chapter of Eelam War IV, an odyssey that commenced in 1983 and etched itself as Asia's longest internal strife (Hashim, 2020).

Within this tapestry of ethnic-separatist dissent, the LTTE wielded a formidable expertise across a myriad of adversarial tactics aimed at the Sri Lankan government. Across the initial trio of conflicts, christened as Eelam Wars I to III from 1983 to 2002, eventual discussions brokered a comprehensive peace pact. In these tumultuous chapters, the LTTE deployed an intricate strategic ballet, intertwining diverse methodologies encompassing a symphony of technologies, tactics, and artistic maneuvering, all poised to outwit or stymie the Sri Lankan military (Satgunarajah, 2016). The enduring saga of this conflict was fueled by deep-rooted disparities in the divergent creeds of the opposing camps. While the LTTE fervently advocated for the liberation of the northern Tamil minority, aspiring to carve out dominion over a third of the island and the lion's share of its coastline, the government pursued a path of self-rule. These foundational contrasts reignited armed hostilities, even after attempts at reconciliation (Biswas, 2006). This saga, marked by its intricate threads, strategic complexities, and protracted chronicle, stands as a testament to the Herculean challenges ingrained within the tapestry of ethnic and territorial discord. The remnants of the bulldozer stand as a poignant exhibit at Elephant Pass, accompanied by an imposing plaque and billboard dedicated to honoring Corporal Gamini Kularatne, the valiant soldier who foiled the audacious endeavor. A sign inscribed in Sinhala conveys, "For the first time in the world, the Vibhusanaya award [a prestigious bravery medal] was conferred upon Corporal Gamini Kularatne for jeopardizing his life at this camp by leaping onto the Tiger bulldozer on July 1, 1991, at Elephant Pass." Kularatne sacrificed his life in thwarting the assault, preventing the bulldozer from reaching its intended target. His legacy reverberates throughout Sri Lanka, and his modest village near Mahiyangane has attained prominence owing to his extraordinary deeds (Steward, 2014).

The inaugural heritage site of significance resides in the remnants of the railway system that once traversed the Yaldevi route leading to Jaffna in the northern expanse. A tragic episode unfolded in 1985 when the Tigers orchestrated a bomb detonation on the train, resulting in the loss of 11 lives. After this catastrophic event, the railway line ceased its operations. Astonishingly, the wreckage of the train has now been repurposed as a site accessible to visitors. A sign inscribed in Sinhala delineates, "This is the area where the terrorists destroyed the Yaldevi railway" (Steward, 2014).

Moving further away from the ruins of Yaldevi lies Kilinochchi, the erstwhile bastion and de facto capital of the Tamil Tigers. A commemorative memorial, meticulously preserved by the Sri Lankan government, stands as a testament to the military triumph at Kilinochchi. This memorial encompasses a colossal water tower adorned with the scars of bullet craters. Notably, the environs surrounding the tank, in stark contrast to the dilapidated and war-ravaged structures nearby, exude a semblance of meticulous upkeep and cleanliness (Ramakrishnan, 2016).

Venturing northwards from Kilinochchi, one encounters Elephant Pass, situated in proximity to the Jaffna peninsula. Elephant Pass has borne witness to several pivotal conflicts, including an audacious yet failed suicide strike targeting a Sri Lankan army camp. In this striking episode, the Tamil Tigers ingeniously crafted a counterfeit tank from a commandeered bulldozer, fortified with armored infantry plates. Operated by dedicated Cadres, this makeshift tank was designed to breach the camp's defenses before detonating within the premises, causing extensive damage.

Since the cessation of the war, Sri Lanka has experienced a transformation in its tourism landscape, witnessing the emergence of a new facet of internal tourism (SLTDA, 2016). The government, military, and potential stakeholders are strategically aiming to capitalize on this trend by establishing luxury resorts around both dark and nondark sites (Light, 2017). Numerous memorials have been erected to attract travelers to various cities. The government has been actively involved in developing existing infrastructure, including roads, and has undertaken initiatives to renovate buildings (Fallon & Robinson, 2017).

In terms of war sites, there still exist evident scars, including ruins, tank wrecks, and ship remains, contributing to the characterization of some areas as "ghost towns." War memorials commemorate Sri Lanka's victory and honor those who lost their lives during the conflict. Additionally, these sites preserve relics associated with the Tamil LTTE (Shanmugaratnam, 2005).

The concept of dark tourism has swiftly gained traction in the country, with groups of people visiting these sites to pay homage to the fallen and displaying a sudden interest in the cultural and historical significance of these relics. From an economic perspective, this trend proves beneficial. Further investments in developing comprehensive policies and strategies could enhance Sri Lanka's attractiveness to tourists worldwide (Mullins, 2016).

Regarding dark sites related to natural disasters, they are considered supplementary attractions for tourists. An example is Yala on the southern coast, where the beach suffered considerable damage after the 2004 tsunami. Sri Lanka's east and south coasts bore the brunt of natural disasters, with some memorials erected but later removed. Approximately, 95% of the country's destruction and damage resulted from floods and tsunamis, while the rest was attributed to landslides and animal attacks. With well-crafted marketing strategies, Sri Lanka has the potential to lead the tourism industry, provided a meticulous approach is undertaken (Stone, 2006). Various policies could be implemented to promote and influence dark tourism across different sectors for the country's benefit.

Findings and Discussion

Utilizing respondent data and guided by an intricate master coding hierarchy table, the author employed qualitative techniques to extract pertinent themes. The primary focus revealed four overarching themes: Dark Tourism Destinations, Tourist Gaze on Post-War Sites, Reconciliation between Sinhala and Tamil People, and Sustainable Tourism in Jaffna. The elucidation of the Dark Tourism Destinations theme involves three subcodes: war memorials, war attractions, and war sites. Within the second theme, tourist curiosity is dissected, revealing an inclination to explore war sites fueled by inquisitiveness and religious interests.

The Reconciliation theme unfolds with two subcodes: interactions among civil groups and the attitudes of Tamil residents toward tourist behaviors. The final theme, Sustainable Tourism in Jaffna, is parsed into three subcodes: the removal of war attractions, the preservation of cultural attractions, and satisfaction with government initiatives.

The primary objective of this study is to discern the varied motivations underpinning visits to dark destinations. Foremost among these motivations is curiosity, a pivotal driver for attending war attractions. The study delineates the nuanced tourist gaze upon dark sites, emphasizing the prevalence of curiosity as participants share and discuss their fascination with these sites. On a global scale, millions of tourists engage in journeys to locations such as the Vietnam Cu Chi Tunnel and Auschwitz campsites in Germany. Notably, the Thilipan memorial in the northern region of Jaffna remains a prominent attraction, particularly for those still traversing the area. A key finding underscores that domestic tourists predominantly focus their gaze on religious attractions (Sharpley, 2005).

The second objective centers on identifying the motives influencing the achievement of reconciliation between Sinhala and Tamil people. Critical findings underscore divergent attitudes among both groups. Sinhala domestic travelers express a proclivity for visiting dark places to experience historical veracity, driven by distinct motives. Identified motives encompass education, history, novelty, curiosity, photography, leisure, emotions and feelings, and myths and stories.

Curiosity emerges as the predominant motivation, propelling domestic tourists to dark destinations. Despite these motives, there appears to be no discernible impact on the reconciliation between Sinhala and Tamil people. Domestic travelers exhibit a proclivity for visiting, whereas Tamil individuals, plagued by lingering sorrowful memories and emotional scars, are disinclined. Both groups possess dual attitudes toward this concept, necessitating strategic interventions to elucidate the value of the dark tourism concept within the tourism industry for both parties.

Conclusions

Dark tourism emerges as a burgeoning niche market within the expansive realm of the tourism industry, a phenomenon not novel in the global landscape with established footprints in countries such as Vietnam, Japan, Korea, and Italy. Within the Sri Lankan context, however, dark tourism remains an untapped and unexplored niche market. Yet, there exists a plausible prospect of conceptual development within the existing geographical confines of the northern region. In evaluating its impact on reconciliation, the study discerns no substantial influence on the achievement of reconciliations, prompting the formulation of strategic initiatives to fortify appropriate tourism practices in Jaffna through robust collaboration with host participants.

The crux of this study revolves around unraveling the nuanced impact of visiting dark destinations on the reconciliation dynamics between Sinhala and Tamil people in the country. Both communities exhibit dual attitudes toward this concept. However, a discernible impact on reconciliation is notably absent. Noteworthy locations and memorials, such as Tholepin's memorial, evoke profound emotional responses among the Tamil populace, with some expressing reluctance to visit these sites, given the traumatic memories of war. In contrast, Sinhala individuals, often bereft of direct wartime experiences, demonstrate a pervasive curiosity to explore these locales. A substantial contingent of Tamil respondents articulates a disinterest in visiting sites laden with dark memories, emphasizing the inconsequence of such visits to them.

The study culminates in the identification of diverse motives propelling individuals to engage with dark destinations: education, history, novelty, curiosity, photography, leisure, emotions and feelings, and myths and stories. Curiosity emerges as the predominant motivator, driving domestic tourists to explore these dark destinations.

Despite these motivations, the dark tourism concept appears to wield minimal impact on the broader reconciliation efforts between the two nations. Notwithstanding this, strategic interventions are posited as imperative to enlighten both parties about the intrinsic value of the dark tourism concept within the tourism industry. This underscores its potential utilization as an asset for the sustainable future of the younger generation by imparting knowledge about the nation's historical narrative.

The Northern Province has gained prominence in the domain of dark tourism owing to its association with the Sri Lankan War, a conflict notorious for its intense and violent clashes.

Former battlegrounds from the Sri Lankan civil war, marked by the echoes of bloody confrontations, draw the attention of both foreign and domestic travelers to Jaffna. The northern province of Sri Lanka has thus evolved into a renowned tourism destination, attracting visitors seeking commemorative experiences and engaging in pilgrimages.

This distinctive incentive for exploring historical locales manifests as a diverse spectrum of tourist experiences. The intricate connections visitors establish with these sites, often rooted in personal and familial histories, serve as illuminating lenses through which one can comprehend their cognitive processes, emotional responses, and behavioral patterns. Consequently, the act of visiting war memorials becomes a nuanced and reciprocal process, wherein visitors undergo psychological transformations as a direct consequence of their immersive encounters.

Suggestions

The ongoing investigation delves into a distinctive Sri Lankan interpretation of the enduring ramifications of the civil war, wherein on-site guides assume multifaceted roles encompassing political and educational functions. These guides serve as mediators, navigating the intricate terrain between historical events and the contemporary context, concurrently advocating for peace. The exploration of dark sites, as unveiled by the study, unfolds as a narrative more intricate than the conventional dichotomy of the dark-light spectrum, involving nuanced considerations of motivation, perception, and experiential dimensions.

The research findings not only furnish novel perspectives on the structure and characteristics of dark tourism in Asia but also illuminate the intricacies of the demand–supply imbalance. This imbalance, notably observed in Sri Lanka's Northern Province, holds significant consequences for the evolution of dark tourism products. Moreover, the study extends its contributions to the broader understanding of dark tourism, both in a general context and within the nuanced backdrop of Sri Lanka. In doing so, it rectifies an oversight in the existing tourist literature by providing a more comprehensive comprehension of the conflict from a distinctly Sri Lankan perspective.

Future Implications

The principal pragmatic implication emanating from this research study lies in its prospective utility for tourism boards and travel agencies, posited as a potential instrument for cultivating cognizance and fostering the advancement of dark tourism within specified categories contingent upon discernible motivations. As an illustrative instance, these entities may amplify their promotional initiatives directed toward dark tourism attractions characterized by historical profundity, exemplified by sites steeped in the annals of warfare in Jaffna, Sri Lanka.

Moreover, a strategic emphasis could be placed on elucidating the experiential and educational dividends inherent in these locales, thereby accentuating the multifaceted value proposition they confer upon the discerning tourist.

References

Birna, A., Hyde, K. F., & Farmaki, A. (2013). Dark tourism revisited: A supply/demand conceptualization. *International Journal of Culture, Tourism and Hospitality Research*, *7*(3), 281–292.

Biswas, B. (2006). The challenges of conflict management: A case study of Sri Lanka. https://www.researchgate.net/publication/233253068_The Challenges of_Conflict_Management_A_Case_Study_of_Sri_Lanka. Accessed on September 11, 2021.

Braun, V., & Clarke, V. (2006). Using thematic analysis in psychology. *Qualitative Research in Psychology*, *3*(2), 77–101.

Burns, P. M. (2004). Tourism planning: A third way? *Annals of Tourism Research*, *31*(1), 4–43.

Burns, P. M., & Novelli, M. (2007). *Tourism and politics*. Routledge.

Chang, T. Y. (2014). Dark tourism the effects of motivation and environmental attitudes on the benefits of experience. *Revista Internacional de Sociología*, *72*(2), 69–86.

Collins-Kreiner, N. (2016). Dark tourism as/is pilgrimage. *Current Issues in Tourism*, *19*(12), 1185–1189. https://doi.org/10.1080/13683500.2015.1078299

Embuldeniya, A., & Rajapaksha, W. (2013). The significance of tourism in Sri Lankan economy. https://www.researchgate.net/publication/326543634_The_Significance_of_Tourism_in_Sri_Lankan_Economy

Euromonitor International. (2018). Travel in Sri Lanka. https://www.euromonitor.com/travel-in-sri-lanka/report. Accessed on July 26, 2019.

Fallon, P., & Robinson, P. (2017). 'Lest we forget': A veteran and son share a 'warfare tourism' experience. *Journal of Heritage Tourism*, *12*(1), 21–35.

Fernando, S., Bandara, J. S., & Smith, C. (2016). Tourism in Sri Lanka. In C. M. Hall & S. Page (Eds.), *The Routledge handbook of tourism in Asia* (pp. 271–284). Routledge.

Fernando, P., & Jayawardena, C. (2013). Focusing on realistic tourism targets for post-war Sri Lanka. *Worldwide Hospitality and Tourism Themes*, *5*(5), 495–504.

Foley, M., & Lennon, J. J. (1996). JFK and dark tourism: A fascination with assassination. *International Journal of Heritage Studies*, *2*(4), 198–211. https://doi.org/10.1080/13527259608722175

Hashim, A. S. (2020). 10 lessons of modern war: A case study of the Sri Lankan War. In *Learning the lessons of modern war* (pp. 181–196). Stanford University Press. https://doi.org/10.1515/9781503612518-014

Jayashree. (2009). *The Sri Lankan conflict*. Council on Foreign Relations. https://www.cfr.org/backgrounder/sri-lankan-conflict. Accessed on September 11, 2021).

Jayawardena, C. (2013). Innovative solutions for future tourism development in Sri Lanka (2013-2026). *Worldwide Hospitality and Tourism Themes*, *5*(5), 512–531.

Johanson, M. (2012, May 5). Dark tourism: Understanding the attraction of death and disaster. http://www.ibtimes.com/dark-tourism-understanding-attractiondeath-and-disaster-696604

Kamble, Z., & Bouchon, F. (2014). Tourism planning and a nation's vision: A review of the tourism policy of Sri Lanka. *Procedia-Social and Behavioral Sciences*, *144*, 229–236.

Lennon, J. J., & Foley, M. (1999). Interpretation of the unimaginable: The U.S. Holocaust Memorial Museum, Washington, D.C., and "dark tourism". *Journal of Travel Research*, *38*(1), 46–50. https://doi.org/10.1177/004728759903800110

Light, D. (2017). Progress in dark tourism and thanatourism research: An uneasy relationship with heritage tourism. *Tourism Management*, *61*, 275–301.

Miles, P. (2013). Sri Lanka criticized over 'war tourism'. https://www.telegraph.co.uk/travel/news/Sri-Lanka-criticised-over-war-tourism/. Accessed on July 26, 2019.

Miller, D. S., & Gonzalez, C. (2013). When death is the destination: The business of death tourism – Despite legal and social implications. *International Journal of Culture, Tourism and Hospitality Research*, *7*(3), 293–306. https://doi.org/10.1108/IJCTHR-05-2012-0042

Mullins, D. (2016). What is dark tourism? https://culturaltourism.thegossagency.com/what-is-dark-tourism/. Accessed on July 26, 2019.

Nisthar, S., Vijayakumar, S., & Nufile, A. A. M. (2017). Dark tourism and its potential impacts on tourism industry in Sri Lanka. http://ir.lib.seu.ac.lk/handle/123456789/2191

Podoshen, J. S., Venkatesh, V., Wallin, J., Andrzejewski, S. A., & Jin, Z. (2015). Dystopian dark tourism: An exploratory examination. *Tourism Management*, *51*, 316–328. https://doi.org/10.1016/j.tourman.2015.05.002

Ramakrishnan, T. (2016). Post-civil war, Sri Lanka's tourism industry having one great run. *The Hindu*. https://www.thehindu.com/news/international/postcivil48war-sri-lankas-tourism-industry-having-one-great-run/article8541224.ece. Accessed on July 3, 2021.

Ratnayake, I., & Hapugoda, M. (2016). Tourism under military: A critique on land utilization and tourism in Postwar Sri Lanka. *Sabaragamuwa University Journal*, *15*(1), 18–35.

Reid, R. (2016). Is 'dark tourism' OK? http://www.nationalgeographic.com/travel/features/is-dark-tourism-ok-chernobyl-pripyat-disaster-sites/

Reynolds, D. (2016). Consumers or witnesses? Holocaust tourists and the problem of authenticity. *Journal of Consumer Culture*, *16*(2). https://doi.org/10.1177/1469540516635396

Rojek, C. (1991). *Ways of escape: Modern transformations of leisure and travel*. Doctoral dissertation. http://theses.gla.ac.uk/2507/1/1991rojekphd.pdf

Samaranayake, H. M. S., Lantra, N., & Jayawardena, C. (2013). Forty-six years of organized tourism in Sri Lanka (1966-2012). *Worldwide Hospitality and Tourism Themes*, *5*(5), 423–441.

Satgunarajah, M. (2016) *"Never again": The case study of Sri Lanka and the collective action theory of genocide* (p. 82). Master Thesis.

Seaton, A. V. (1996). Guided by the dark: From thanatopsis to thana tourism. *International Journal of Heritage Studies*, *2*(4), 234–244. https://doi.org/10.1080/13527259608722178

Selvanathan, S. (2007). The effect of war and other factors on Sri Lankan tourism. *Applied Economics Letters*, *14*(1), 35–38. https://doi.org/10.1080/13504850500425576

Senaratne, K. (2010). Sri Lanka tourism - Marketing and manpower. http://www.sundaytimes.lk/100926/BusinessTimes/bt30.html. Accessed on July 26, 2019.

Seoighe, R. (2017). War, denial and nation building in Sri Lanka - After the end. In R. Mac Ginty (Ed.), *Palgrave studies in compromise after conflict*. Springer Nature.

Shanmugaratnam, N. (2005). Challenges of post-disaster development of coastal areas in Sri Lanka. In *Consultative Workshop on Post-Tsunami Reconstruction Experiences of Local NGOs* (Vol. 23).

Sharpley, R. (2005). Travels to the edge of darkness: Towards a typology of dark tourism. In *Taking tourism to the limits: Issues, concepts and managerial perspectives*. Elsevier.

Singh, R. (2014). Development of dark tourism. *International Journal of Management Research and Review*, *4*(8), 830–833.

SLTDA (2016). *Sri Lanka tourism strategic plan 2017-2020*. Ministry of Tourism Development and Christian Religious Affairs.

Sri Lanka Tourism. (2015). Half year tourism earning at $1.3bn - The informal sector continues to drive tourist arrivals. https://www.srilanka.travel/travel-news&news=332. Accessed on July 26, 2019.

Steward, J. (2014). After the slaughter: War tourism in modern Sri Lanka. https://www.e-ir.info/2014/02/28/after-the-slaughter-war-tourism-in-modern-sri-lanka/

Stone, P. R. (2006). A dark tourism spectrum: Towards a typology of death and macabre related tourist sites, attractions and exhibitions. *Turizam: Međunarodni Znanstveno-Stručničasopis*, *54*(2), 145–160.

Tan, G., & Lim, S. (2018). The "Pearl of the Orient" is a dark tourism destination in Malaysia. *International Journal of Tourism Cities*, *4*(1), 68–80. https://doi.org/10.1108/IJTC-04-2017-0023

Yoshida, K., Bui, H. T., & Lee, T. J. (2016). Does tourism illuminate the darkness of Hiroshima and Nagasaki? *Journal of Destination Marketing & Management*, *5*(4), 333–340. https://doi.org/10.1016/j.jdmm.2016.06.00

Chapter 10

Issues and Challenges of Dark Tourism

Birendra Kishore Roy

Department of Learning and Development, Suryagarh palace, Rajasthan, India

Abstract

According to the definition, visiting any specific destination to explore and unfold the darkest event of human history is known as dark tourism. This journey involves much of the darker side of the travel which includes massacre, genocide, harlotry or flesh trade, ethnic cleansing, war, or disaster which is either natural and accidental or created by humans. Many international and domestic tourists travel around different states of India to explore its culture and community, foods and festivals, art, crafts and music, and the climate. The idea behind "dark tourism" is highly sensitive in the case of the Indian tourism context as Indian culture is much more possessive and protective compared to Western culture. The idea behind dark tourism and its utilization started in 1996; the phenomenon of this tourism mainly exists within a destination and is mainly controlled by societies and political regimes. This augmented exploration either encourages or discourages a tourist from exploring a destination for solo or group travel or restricts them due to psychological and social aspects. Many people think that dark tourism can increase interest in committing crimes and may increase the criminal mindset among individuals. The main objective of this research is to explore the major constraints and challenges of dark tourism.

Keywords: Dark tourism; tourist; destination; culture; human

Introduction

Traveling and staying outside of one's normal area for recreational, professional, or other objectives for a maximum of one year is referred to as tourism. One of the most significant sources of revenue is now provided by the service industry, which includes tourism. Over the past 10 years, India's tourism industry has grown

Dark Tourism, 123–133

Published under exclusive licence by Emerald Publishing Limited
doi:10.1108/978-1-83797-336-120241010

quickly (Biran et al., 2011). Over time, specialty tourism has also developed; one such niche sector is gloomy travel. Dark tourism is defined as travel to places that are typically associated with death, misery, and suffering (Dann & Seaton, 2001). Dark tourism is described as tourism that involves visiting locations traditionally connected with death and misery. Dark tourism has been more popular over the past 10 years, mostly due to people's intense interest in death. It's still a novel idea for the travel sector (Dunkley et al., 2011). It may be characterized as a synthesis of tragedy, history, heritage, and tourism. The concept of heritage tourism was essentially enlarged to include locations with a tragic or morbid past. They are areas that have seen social unrest or political unrest, as well as memorials and places having sad historical relevance (Robinson et al., 2011).

Dark tourism is a subgenre of tourism that deals with pain, violence, graveyards, worldwide tragedies, death, and dormant locations associated with history, culture, and heritage tourism. This type of leisure, which may be compared to morbid curiosity, has great growth potential, even though it is recognized that education is required to avoid the formation of incorrect impressions about these specific sites (Foley & Lennon, 1996). India is known for having some of the most well-liked destinations for tourists on the planet, including the Taj Mahal, the Qutub Minar, the Rajasthan desert, and numerous more breathtaking locations (Lennon & Foley, 2000). However, there is limited information available regarding gloomy and somber tourism destinations that people interested in this niche can visit and use to enhance their dark encounters (Kamat, 2023; Verma & Jain, 2013). The practice of dark tourism is the deed or usage of sending tourists to locations where there has been a catastrophe, death, or suffering. This sector of tourism (Jain et al., 2020) is connected to gloomy occurrences, such as suffering populations, violence, war zones, graveyards, international tragedies, death, or dormant locations connected to historical, cultural, and heritage tourism (Light, 2017). Such a form of leisure, which is similar to morbid curiosity, has great growth potential, even if it is recognized that formal learning is required to avoid the formation of incorrect impressions about these specific spots (Miller & Gonzalez, 2013). Given the enormous increase in dark tourism visitation numbers, this development potential is becoming extremely noticeable. In 2018, for example, approximately 2.1 million tourists went to the Holocaust monument, while 3.2 million people went to the 9/11 memorial at Ground Zero annually. Researchers claim that even though dark tourism is becoming more and more popular, little is known about this complex phenomenon (Wilson, 2011). Many academics have indicated that it is necessary to comprehend its empirical study from the perspective of tourists to fully comprehend their motives and experiences as its thorough and more in-depth comprehension is absent in current research (Miller & Gonzalez, 2013). Dark tourism can be seen for a variety of reasons, including political, educational, nostalgic, economic growth, or amusement. It can also be observed as a reminder of one's mortality and fragility (Rojek, 1993). One would think that by engaging in dark tourism, visitors will value life more and develop compassion for victims (Petr, 2015). Dark tourism (Sharma & Nayak, 2019) is a multifaceted phenomenon with its own merits. Owing to the emotional connection created by the encounter and the desire to understand more about the unknowable, people frequently express enthusiasm for "extinction," "death," and any "disaster" (Sarkar, 2022). Another type of illicit travel is "battlefield tourism," which entails visiting locations and cities connected to conflict areas or taking part in simulations of warfare (Preece & Price, 2005). War museums use terminology to describe previous battles and wars to visitors to elicit strong emotions in those who are unfamiliar with such conflicts and wars (Sharma, 2021). The purpose of this essay is to assess how local communities view dark tourism. This study also seeks to determine how India may gain from underground tourism for its purposes as well because it is clear that more research is needed in this area.

The Genesis of Dark Tourism and Its Numerous Manifestations

Roman gladiator matches, escorted visits to see hangings in England, and medieval executions are all examples of early-evening tourism (Ryan, 2007). Researchers found that travel has long been related to mortality and conflict. Tourists are still captivated by death and violence; consequently, they visit places like Ground Zero 9/11 (Sharma, 2018) or the John F. Kennedy assassination site (ROBB, 2009) in Dallas (Seaton & Lennon, 2004). Tourists in pursuit of unique attractions frequently visit abandoned locations such as jails and places of punishment. But it wasn't until 1996 that the phrase "dark tourism" was made known to academics, which spurred a flurry of additional research on the subject (Lennon, 2017). We may group seven different sorts of dark tourism into the prior research that is currently available, including:

- Cemetery or Graveyard tourism (Sallay et al., 2022).
- Battleground tourism (Buntman, 2008).
- Slavery tourism (Powell & Kennell, 2016).
- Monumental travel (Banaszkiewicz, 2017).
- Inmate travel (Richard, 2006).
- Tourism in undesirable locations (Seaton, 1999).

We can categorize and see that the term "dark tourism," as it is known, contains an extensive variety of different types of excursions and venues. To accurately estimate what visitors will need, it is crucial to comprehend their primary driving forces and goals. According to a survey, tourists drawn to Genocide or WWII sites are driven by their enthusiasm for learning the truth about the period and their understanding of history (Sharpley & Stone, 2009; Stone, 2012). In a study conducted for the museum of the massacre (Walby, 2011), it was discovered that visitors are eager to learn the truth about an event and, as a result, decry the catastrophe (Martini, 2020). Researchers observed in another case study about jail tourism that visitors are drawn to the institution by their curiosity, desire to learn about its history, and availability of educational programs (Soulard et al., 2023). In contrast, Wight and Lennon's study on tourism to museums in Lithuania revealed that visitors are curious to learn about the nation.

The vast majority of the illicit tourism unbiased was educational and commemorative. Death tourism (Ryan & Kohli, 2006) is the ultimate manifestation of "dark tourism," according to academic researchers (Pratt et al., 2019). This extreme kind of black tourism happens when a person goes to locations to commit suicide, frequently with medical assistance (Sharma, 2021). Though it may not be directly tied to dark tourism, given that it is a vacation activity and a product of nature, one may argue that it is. This study might conclude that illicit tourism is a multifaceted phenomenon (Lennon, 2017) based on the literature (Stone, 2012). Dark tourism was practiced by people for a variety of reasons, but mostly for educational and memorial ones. Visitors can elicit emotional reactions that accomplish their most fundamental needs by engaging in dark tourism events. Old regions, like battlegrounds, can also benefit commercially from dark tourism, making it beneficial for safeguarding them.

Dark Tourism Destinations Include India

Due to India's extensive past, dark tourism is bound to exist. Natural disasters, disputes, civilizations, poverty, cultural diversity, monarchy and dynasties, belief systems, and yearning for independence are only a few of the examples of the horrible events that India has experienced. Therefore, India has an enormous amount of potential for dark tourism. Several locations can be promoted called "dark tourist destinations," which will help strengthen the economy of the country by bringing lots of domestic and foreign tourists. Since plenty of locations throughout the nation have historically encountered a certain kind of phenomenon, the places might be advertised to tourists as being uncomfortable (Dandotiya & Aggarwal, 2022).

Some Major Dark Tourism Destinations and How They Impacting in Niche Tourism Market in India

This study rests on descriptive and exploratory analyses that drew on secondary data retrieved and collected from publicly available material online, in newspapers, in magazines, journal articles, and books and book chapters. After gathering pertinent information, conclusions were reached after interpreting the results. This study has also taken into consideration several working papers and reports produced by renowned authors and institutes on Indian tourism, as well as pieces from renowned publications and the Indian Ministry of Tourism. It might be challenging to embrace the idea of "dark tourism" in India since it seems strange to visit places where people have died or experienced the loss of someone close to them. One can hardly picture going to such a place as a tourist destination. The researchers claim that there is always a chance that a gloomy tourist attraction may shock people since it deals with the agony or death of survivors. However, it appears that dark tourism is becoming a sizable niche trend in contemporary India. Numerous kings, protracted combat, massacres, and a range of acts of terror that are frequently decried rather than glorified have all occurred throughout India's history. People who aren't fond of devouring history books can take advantage of the possibilities to head to these historical locations and get a firsthand glimpse of history in action. By doing so, they can affirm the adage that "Seeing is believing" by experiencing the suffering of a previous generation directly. There are several excellent prospective dark destination spots in India. Here are a few excellent examples. Beginning with Amritsar (Mehta et al., 2023), British soldiers with weapons slaughtered hundreds of innocent Indians in 1919 (Jamalian et al., 2020).

Another excellent example is Kuldhara in Hamlet in Rajasthan when 83 people perished in the course of a single night. This location, which has been abandoned since the beginning of the 19th century, is thought to be haunted. Local government officials have increased their efforts in recent years to market this location as a tourism destination.

Another excellent example is the Eerie Savoy Hotel in Missouri. This hotel, where guests can stay, has a lengthy history of being haunted, with several strange occurrences having been noticed and documented.

Last but not least, the Taj Mahal Palace Hotel, a landmark hotel in the city of Mumbai, gained notoriety across the world after being attacked by terrorists in 2008 (Machold, 2017).

Also noteworthy is the fact that illicit tourism is becoming a smaller and smaller specialized business. Dark tourism, according to *The Times of India* (India, 2017), has contributed to a 10% over the previous year increase in Indian tourists to unusual European sites (India, 2018). According to one of the owners of a popular newspaper in India, there is currently a growing demand for creepy characters, mysterious places, and unpleasant history (Hindu, 2017), which some people find enjoyable, while others are more interested in discovering the truth behind old stories and confirming what is true and folklore. Visitors between the ages of 35 and 50 are drawn to the trend in recent years (Lewis et al., 2022). The areas of death and devastation associated with dark tourism are perceived negatively by society. Due to its abundance of spiritual and cultural attractions, India has always highlighted the good aspects of the locations and monuments. It's tough for people to embrace this idea since the psychology underlying tourists' motivations is in doubt. Due to a lack of understanding and awareness of the concept as well as a contradiction against Indian conventional principles, values, and ethical standards, dark tourism has never been properly promoted (Dresler, 2023).

Major Challenges and Issues in Promoting the Dark Tourism Market in India

Despite the nation's extensive history and numerous cultural assets, establishing the dark tourism sector in India brings several difficulties and problems. Dark tourism, which entails visiting locations connected to sorrow, tragedy, and death, can be a fragile challenging sector to advocate for. Here are some significant problems and difficulties.

Compassion and Responsiveness

There are a lot of dark tourism destinations in India that are connected to sad occurrences like natural catastrophes, terrorist attacks, and historical wars. It is extremely difficult to strike an equilibrium between the obligation to respect those who died and their loved ones and the need to draw tourists. It is vital to make sure that tourism-related activities are carried out respectfully and sensitively (Stone, 2013).

Preservation and Conservation

Dark tourism sites may deteriorate over time, either due to the passage of time or from the increased foot traffic from tourists. It's essential to strike a balance between preserving these sites for historical and educational purposes while allowing controlled tourism (Baloch et al., 2023).

Development of Infrastructure and Accessibility

India includes numerous dark tourist attractions, but getting to them might be difficult because of their rural or underdeveloped locations. To draw visitors and give them a secure and enjoyable background, improvements to the infrastructure are required, such as changes to the roads, lodgings, and visitor facilities (Stone, 2011).

Lack of Wisdom Among the Community

India has a relatively low level of dark tourism expansion when compared to other nations. Many prospective visitors may be unaware of the physical presence or historical importance of these locations. To promote these locations, successful advertisement and awareness efforts are required.

Protective Measures

Additionally, there may still be certain black tourism destinations in India that are situated in regions in which there are possible security risks. Establishing security measures and ensuring tourist safety while wanting to draw tourists is essential.

Regulatory Obstacles

Regulations, such as getting permission and approvals, may present obstacles to the establishment of dark tourism destinations. It is crucial to streamline the licensing procedure while being sure that it serves the objectives of promoting safe travel.

Insight and Instruction

It is crucial to provide a sympathetic and correct presentation of these locations. The historical and cultural relevance of these locations should be communicated without sensationalizing or exploiting them, according to experienced guides and educational materials.

Cultural Sensitivity

India is a culturally diverse country with a complex history. Understanding the cultural sensitivities of the local population and involving them in the development and management of dark tourism sites is crucial to avoid cultural clashes (Sharma, 2020).

Comparing Economic Gains and Exploitation

It is very challenging to strike an acceptable compromise between the likelihood of exploitation and commodification and the economic advantages of dark tourism. It is crucial to ensure that local communities capitalize on tourism while limiting commercialization (Sharma, 2022).

The Environmental Consequences

In ecologically fragile locations, the inflow of tourists may hurt the environment. Plans for developing the dark tourism industry should take into account sustainable tourism practices, garbage prevention, and environmental conservation initiatives (Kamat, 2023).

Conclusion

In this chapter, the author examined Indian perceptions on this delicate subject while discussing the importance of a new tourist niche called "dark tourism." This segment of the tourist industry concentrates on previous and historical events that occurred in India, primarily during the reign of the Mughal Empire, but also afterward, when the British controlled. This judgment brought about several things, including the establishment of what are now regarded as "dark" locations since they are close to the spots where many deaths and invasions have taken place in the past. Even while they offer great financial possibilities, these locations and events may also have a highly detrimental impact on people's attitudes. In this sense, governments and the media have a key responsibility to inform the public about such locations since they should also be viewed as a source of cash that may assist in addressing the local community's socioeconomic problems. Numerous restrictions and difficulties exist for the study of dark tourism. The main impediment to this paper's completion was the dearth of thorough research on the topic. However, there is sufficient research and literature on religion, leisure, and amusement, nevertheless is a dearth of literature because the idea of dark tourism seems new. By writing this essay, the author hopes to have filled in part of this gap. The subject itself has its restrictions. One may contend that governments are not doing sufficient promotion given the key barriers. Since many local populations are unaware of dark tourism along with its commercial implications, especially in the context of India, that might be due to a concern about how the local populace will react to their feelings. Consequently, it is advised that governments aid in encouraging this subset of travel. A further barrier can be the absence of amenities, essential services, advice, conservation, unsuitable infrastructure, and upkeep of those gloomy tourism historical and abandoned locations. Due to dread, sentiments, emotions, and psychological obstacles, local populations' cooperation is at a minimum. In conclusion, India has a lot of unexplored places that have the potential to enhance the tourist industry and are of tremendous significance historically. However, an efficient marketing plan alongside increased awareness is necessary to change how people perceive these sites. If put into practice, it will benefit local communities by creating new employment possibilities, fostering the expansion of local firms already in place, and bringing in foreign exchange reserves – essential components of any economy. In addition, if these monies are appropriately governed, they may be utilized to fight poverty and aid those who have been hurt or negatively affected by traumatic occurrences. Dark tourism may also be used as an avenue to teach people about their misconceptions of such locations, to help them gain clarity in their thinking, and to assist them in intellectualizing and capitalizing on the positive aspects of unfavorable circumstances.

References

Baloch, A. G., de Almeida Leite, E. M., Ribeiro, H. N., & Alves, S. P. (2023). Dark tourism in India: Significance and challenges. In *Economic and Social Development: 98th International Scientific Conference on Economic and Social Development* (pp. 312–318). https://www.esd-conference.com/upload/book_of_proceedings/Book_of_Proceedings_esdPorto2023_Online.pdf

Banaszkiewicz, M. (2017). Dissonant heritage and dark tourism at Lenin's Mausoleum. In M. Frihammar, & H. Silverman (Eds.), *Heritage of death* (pp. 78–91). Routledge.

Biran, A., Poria, Y., & Oren, G. (2011). Sought experiences at dark heritage sites. *Annals of Tourism Research*, *38*, 820–841. https://doi.org/10.1016/j.annals.2010.12.001

Buntman, B. (2008). Tourism and tragedy: The memorial at Belzec, Poland. *International Journal of Heritage Studies*, *14*(5), 422–448. https://doi.org/10.1080/13527250802284867

Dandotiya, R., & Aggarwal, A. (2022). An examination of tourists' national identity, place attachment and loyalty at a dark tourist destination. *Kybernetes*. https://doi.org/10.1108/K-08-2021-0756

Dann, G., & Seaton, A. (2001). Slavery, contested heritage and thanatourism. *International Journal of Tourism and Hospitality Administration*, *2*, 1–29.

Dresler, E. (2023). Multiplicity of moral emotions in educational dark tourism. *Ourism Management Perspectives*, *46*, 101094. https://doi.org/10.1016/j.tmp.2023.101094

Dunkley, R., Morgan, N., & Westwood, S. (2011). Visiting the trenches: Exploring meanings and motivations in battlefield tourism. *Tourism Management*, *32*, 860–868.

Foley, M., & Lennon, J. (1996). JFK and dark tourism: A fascination with assassination. *International Journal of Heritage Studies*, *2*(4), 198–211.

Hindu, T. (2017, October 26). 'Dark' tourism: Walk through the valley of the shadow of death. https://www.thehindu.com/life-and-style/travel/walk-through-the-valley-of-the-shadow-of-death/article19918095.ece

India, T. (2017, July 17). Indians take a ticket to dark destinations. https://timesofindia.indiatimes.com/india/indians-take-a-ticket-to-dark-destinations/articleshow/59624548.cms

India, T. (2018, February). Dark tourism in India—Walking through the alleys of India's dark past. *Travel Trends India*. https://timesofindia.indiatimes.com/travel/destinations/dark-tourism-in-indiawalking-through-the-alleys-of-indias-dark-past/articleshow/66107504.cms

Jain, D., Madireddy, M., & Rao, R. N. (2020). Potential assessment of dark tourism in India. *International Journal of Management*, *11*(8), 1284–1296. https://doi.org/10.34218/IJM.11.8.2020.117

Jamalian, M., Kavaratzis, M., & Saren, M. (2020). A happy experience of a dark place: Consuming and performing the Jallianwala Bagh. *Tourism Management*, *81*, 104168. https://doi.org/10.1016/j.tourman.2020.104168

Kamat, S. B. (2023). Decoding the dark tourism Traumascape: An auto ethnographical study of the tourist destination - Goa (India). *International Journal of Hospitality & Tourism Systems*, *16*(2), 43–51.

Lennon, J. (2017). *Dark tourism*. Oxford University Press. https://researchonline.gcu.ac.uk/en/publications/dark-tourism

Lennon, J., & Foley, M. (2000). *Dark tourism: The attraction of death and disaster*. Continuum.

Lewis, H., Schrier, T., & Xu, S. (2022). Dark tourism: Motivations and visit intentions of tourists. *International Hospitality Review*, *36*(1), 107–123. https://doi.org/10.1108/IHR-01-2021-0004

Light, D. (2017). Progress in dark tourism and thanatourism research: An uneasy relationship with heritage tourism. *Tourism Management*, *61*, 275–301. https://doi.org/10.1016/j.tourman.2017.01.011

Machold, R. (2017). Militarising Mumbai? The 'politics' of response. *Contexto Internacional*, *39*, 477–498. https://doi.org/10.1590/S0102-8529.2017390300002

Martini, A. (2020). Dark tourism and affect: Framing places of death and disaster. *Current Issues in Tourism*, *23*(6), 679–692. https://doi.org/10.1080/13683500.2018.1518972

Mehta, D., Gupta, S., & Singla, H. (2023). Dark tourism intentions: Moderating effect of xenophobia. *An International Journal of Tourism and Hospitality Research*, *34*(3), 401–413. https://doi.org/10.1080/13032917.2022.2048403

Miller, D., & Gonzalez, C. (2013). When death is the destination: The business of death-tourism–despite legal and social implications. *International Journal of Culture, Tourism and Hospitality Research*, *7*(3), 293–306.

Petr, C. (2015). How heritage site tourists may become monument visitors. *Tourism Management*, *51*, 247–262.

Powell, R., & Kennell, J. (2016). Dark cities? Developing a methodology for researching dark tourism in European cities. In *Tourism and Culture in the Age of Innovation: Second International Conference IACuDiT. Athens 2015*. Springer International Publishing. https://doi.org/10.1007/978-3-319-27528-4_21

Pratt, S., Tolkach, D., & Kirillova, K. (2019). Tourism & death. *Annals of Tourism Research*, *78*, 102758. https://doi.org/10.1016/j.annals.2019.102758

Preece, T., & Price, G. (2005). Motivations of participants in dark tourism: A case study of Port Arthur, Tasmania, Australia. In C. Ryan, S. Page, & M. Aicken (Eds.), *Taking tourism to the limits: Issues, concepts and managerial perspectives* (pp. 191–198). Elsevier.

Richard, S. (2006). Travels to the edge of darkness: Towards a typology of "dark tourism". In C. Ryan, J. P. Stephen, & M. Aicken (Eds.), *Taking tourism to the limits* (pp. 215–226). Routledge.

ROBB, E. M. (2009). Violence and recreation: Vacationing in the realm of dark tourism. *Anthropology and Humanism*, *34*(1), 51–60. https://doi.org/10.1111/j.1548-1409.2009.01023.x

Robinson, P., Heitmann, S., & Dieke, P. (Eds.). (2011). *Research themes for tourism*. CABI.

Rojek, C. (1993). *Ways of seeing-modern transformations in leisure and travel*. Macmillan.

Ryan, C. (2007). *Battlefield tourism: History, place and interpretation*. Elsevier.

Ryan, C., & Kohli, R. (2006). The Buried village, New Zealand – An example of dark tourism? *Asia Pacific Journal of Tourism Research*, *11*(3), 211–226. https://doi.org/10.1080/10941660600753240

Sallay, A., Mikházi, Z., Gecséné, I. T., & Takács, K. (2022). Cemeteries as a part of green infrastructure and tourism. *Sustainability*, *14*(5), 2918. https://doi.org/10.3390/su14052918

Sarkar, A. (2022). People's perception on dark tourism: A quantitative exploration. *Current Issues in Tourism*, *25*(13), 2042–2047. https://doi.org/10.1080/13683500.2021.1889483

Seaton, A. V. (1999). War and thanatourism: Waterloo 1815-1914. *Annals of Tourism Research*, *26*(1), 30–158.

Seaton, A., & Lennon, J. J. (2004). Thanatourism in the early 21st century: Moral panics, ulterior motives and ulterior desires. In T. Singh (Ed.), *New horizons in tourism: Strange experiences and stranger practices* (pp. 63–82). CABI Publishing.

Sharma, N. (2018). Self-care for the researcher: Dark tourism in Varanasi, India. *Journal of Teaching in Travel & Tourism*, *18*(1), 41–57. https://doi.org/10.1080/15313220.2017.1403801

Sharma, N. (2020). Dark tourism and moral disengagement in liminal spaces. *Tourism Geographies*, *22*(2), 273–297. https://doi.org/10.1080/14616688.2020.1713877

Sharma, N. (2021). Dark tourism and moral disengagement in liminal spaces. In R. S. Bristow, & I. S. Jenkins (Eds.), *Liminality in tourism* (pp. 55–79). Routledge.

Sharma, N. (2022). Acknowledging the shades of grey: The past, present and future of dark tourism in India. In N. Chowdhary, P. Tiwari, & S. Billa (Eds.), *Indian tourism: Diaspora perspectives* (pp. 125–142). Emerald Publishing Limited.

Sharma, P., & Nayak, J. K. (2019). Dark tourism: Tourist value and loyalty intentions. *Tourism Review*, *74*(4), 915–929. https://doi.org/10.1108/TR-11-2018-0156

Sharpley, R., & Stone, P. (2009). *The darker side of travel: The theory and practice of dark tourism*. Channel View Publications.

Soulard, J., Stewart, W., & Samson, E. (2023). Dark tourism and social mobilization: Transforming travelers after visiting a Holocaust Museum. *Journal of Travel Research*, *62*(4), 820–840. https://doi.org/10.1177/00472875221105871

Stone, P. R. (2011). Dark tourism: Towards a new post-disciplinary research agenda. *International Journal of Tourism Anthropology*, *1*(3–4), 318–332. https://doi.org/10.1504/IJTA.2011.043713

Stone, P. R. (2012). Dark tourism and significant other death: Towards a model of mortality mediation. *Annals of Tourism Research*, *39*(3), 1565–1587. https://doi.org/10.1016/j.annals.2012.04.007

Stone, P. (2013). Dark tourism scholarship: A critical review. *International Journal of Culture, Tourism and Hospitality Research*, *7*(3), 307–318. https://doi.org/10.1108/IJCTHR-06-2013-0039

Verma, S., & Jain, R. (2013). Exploiting tragedy for tourism. *Research on Humanities and Social Sciences*, *3*(8), 9–13.

Walby, K. (2011). The polysemy of punishment memorialization: Dark tourism and Ontario's penal history museums. *Punishment & Society*, *13*(4), 451–472. https://doi.org/10.1177/1462474511141478

Wilson, J. (2011). Australian prison tourism: A question of narrative integrity. *History Compass*, *9*, 562–571.

Chapter 11

Dark Tourism in Vietnam: A Case Study of Hoa Lo Prison Relic

Hanh Thi Pham

Vietnam Women's Academy, Vietnam

Abstract

This chapter clarifies the "dark tourism" in Vietnam; a case study of Hoa Lo Prison Relic. Development orientation of "dark tourism" in Vietnam contributes to healing the wounds of war. The author mainly uses a system approach, method of analysis and synthesis, and method of comparison. "Dark tourism" in Vietnam has not really developed as a specific type of tourism like ecotourism or meetings, incentives, conferences, and exhibitions (MICE) tourism. At Hoa Lo Prison historical site, "dark tourism" has been effectively exploited: (1) in the first 6 months of 2023, Hoa Lo Prison welcomes nearly 295 thousand visitors, charging fees of 7 billion VND, the situation of "sold out" takes place regularly; (2) many unique tourism products, one of the successful and widespread tourism products must be mentioned the "night tour" experience; (3) many creative forms of sightseeing with automatic explanation system so that visitors can experience 35 historical stories in their own way; and (4) always create the best conditions for visitors' sightseeing experience. To promote dark tourism in Vietnam, it is necessary to execute some schemes: (i) enhance the sense of responsibility for peaceful tourism in the operation of monuments; (ii) stakeholders coordinate the organization of the tour program and tourist routes with appropriate contents and forms for each target group, especially foreign tourists; (iii) there is cooperation between national and international experts in organizing the "dark tourism" program aimed at peace, preventing the risk of war and conflict, and overcoming the consequences of war; and (iv) application of scientific and technological achievements in "dark tourism" activities.

Keywords: Dark tourism; Hanoi; Hoa Lo Prison Relic; Vietnam; war and conflict

Dark Tourism, 135–148

doi:10.1108/978-1-83797-336-120241011

Introduction

Although "dark tourism" in Vietnam has been implemented but has not really developed and been researched, "Dark tourism" is known by many other names such as "thana tourism," "black tourism", and hoặc "grief tourism." "Dark tourism" is a type of tourism where visitors will visit places where some dark events, often related to death and tragedy of human history, have occurred. These dark events can originate from natural causes such as earthquakes, tsunamis, or originate from humans such as genocide, assassination, or can be prisons, war scenes, terrorism, etc., with a rich history of resistance against invaders. In Vietnam, there are many places marking the painful years but expressing the indomitable spirit of resisting foreign invaders of the nation. Some typical places that can exploit the type of "dark tourism" can be mentioned as Pile dump on Bach Dang River, Go Dong Da, Hoa Lo Prison (Hanoi), War Remnants Museum (Ho Chi Minh City), or Hang Duong Cemetery, Con Dao Prison. International tourists are also particularly interested in destinations such as Cu Chi Tunnels, Dang Thuy Tram Hospital, My Lai Village, or places reminiscent of ancient battlefields such as the Quang Tri Citadel area.

"Dark tourism" is really necessary when studying in Vietnam. Tourism is no longer just about going on vacation, enjoying the beauty of nature, going to amusement parks but sometimes it is also the desire to experience the authentic past of the destination. Therefore, in recent years, memorial tourism or travel to the past is becoming a trend both internationally and domestically.

One of the typical sites is Hoa Lo Prison Historical Site. This prison was built by the French colonialists in 1896 on the land of Hoa Lo village (specializing in making earth kettles, super earth, and earthen furnaces sold throughout the period).

It used to hold the majority of political prisoners, patriots who opposed the French colonial government, and a few were ordinary and foreign prisoners. For those sentenced to a maximum of 5 years or guillotine, the French colonialists detained them in Hoa Lo Prison.

The study focuses on clarifying the concept of "dark tourism" and the factors affecting "dark tourism," and studies the case of the historical site of Hoa Lo Prison, in Hanoi, Vietnam. This study proposes a number of solutions to heal the wounds of war and aim to effectively exploit the relatively new type of "dark tourism" in Vietnam with the participation of stakeholders including the state, businesses, communities, and tourists themselves using "dark tourism" services.

"Dark Tourism"

Rationale for "Dark Tourism"

According to authors John Lennon and Malcom Foley, the Canadian Tourism Recovery Association, "dark tourism" is a form of tourism that has not been well studied, especially in North America. The authors argue that "dark tourism" does not stop at visiting the locations where crimes take place. They argue that modern politics, sociology, and technology also contributed to the creation of "dark tourism" venues.

People gradually gravitate toward dark travel, mainly because of their curiosity about death. Lennon and Foley (2000) and Yuill (2003) point out that "dark tourism" is interesting and emphasizes that when people are concerned about issues of death and disaster. When these types of tourism do not satisfy people's curiosity about the world, they expect to experience the death or disaster of dark monuments. "Dark tourism" is increasingly accepted by tourists and has become a hot topic in the tourism industry. Foley and Lennon (1995) define "dark tourism" as the development of places or landmarks associated with historic human transportation into tourist attractions.

Dann and Seaton (2001) argue that events involving death, disaster, violence, tragedy, or crime as opposed to humanity can be defined as "dark tourism." Stone (2005) points out that pain and tragedy are touring themes of "dark tourism," in which real death sites or related exhibits will be visited. However, "dark tourism" is still a new concept in tourism studies. Sharpley (2005) identified the limitations of consumer psychology and motivation in the study of "dark tourism." Lennon and Foley (2000) proposed that the study of "dark tourism" should focus more on culture, society, and psychology, and emphasized that the motivations of visitors would benefit from the management of "dark tourism" venues (Lennon & Foley, 2003). The conceptualization of "dark tourism" is important (Avital et al., 2011), and the motivation for visiting "dark tourism" sites and the benefits obtained is the additional significance of attracting tourists.

As a result, numerous studies over the past decade have looked at the motivations of "dark tourism" guests through various case studies, where concentration camps or historical battlefields were hotspots (Lennon & Foley, 1999; Miles, 2002). Research shows that tourists visit "dark tourism" locations for various reasons, such as curiosity (Biran et al., 2014; Isaac & Çakmak, 2014), eager to educate and learn about what happened at that location (Kamber et al., 2016; Yan et al., 2016), interested in history or death (Yankholmes & McKercher, 2015; Raine, 2013), and connected to one's personal or family heritage (Mowatt & Chancellor, 2011; Le & Pearce, 2011). Drawing from the literature, four common themes (i.e., dark experiences, fascinating entertainment, unique learning experiences, and common interests) emerged serving as foundational pillars for research.

The main attraction of dark tourism sites is their historical value rather than their connection to death and suffering. It contains aspects of both "dark tourism" and heritage tourism. "Dark tourism" refers to the philosophical interrogation of death. Visitors who are interested in these spaces demonstrate an intention to understand the pain of others or simply an educational goal. Dark tourists often imagine their own finiteness through the silhouettes of others. Therefore, "Dark tourism" helps improve the reception capacity of society as well as leave lessons for the next generation.

Factors Affecting "Dark Tourism"

Positive Effects

Culture is one of the important elements that constitute tourism. The formation of a trip stems from our desire to explore, learn, and experience for ourselves unique cultural features that we have never known before. Culture is the constituent element as well as attracting tourists. Thanks to the need to learn and explore people; culture brings the tourism industry a great source of revenue. So what about "dark tourism" positively impacting?

Referring to the positive impacts of "dark tourism," it is first impossible not to mention that this is an extremely effective means to recreate the image of the country and people of previous generations to tourists. In addition, dark tourism also brings new cultural "winds," not only to visit but also to learn and sympathize with what previous generations have encountered. "Dark tourism" also helps us to know and be conscious in preserving and promoting national culture. The beauties of lifestyle, customs, and valuable experiences will be known, sympathetic and conscious of preservation and development by descendants.

Negative Effects

The limitations of this form of tourism can be mentioned as: *First*, psychological problems of tourists. Not everyone likes the experience of death. *Secondly*, as the knowledge of the tour guide, these tour guides must have a profound knowledge of history and consistent perspectives along with the ability to explain to customers. In many foreign destinations such as S21 Prison or the Killing Fields, Cambodia allows visitors to use audio guides – customers wear a guide transmitter in multiple languages be different and follow the guided journey to be able to hesitate to explore and listen. It also focuses heavily on travelers who have been to or experienced loss and grief, so it's not for everyone.

To develop "dark tourism" in Vietnam, it is necessary to identify customer files and have reasonable marketing forms. The form of promotion must arouse the curiosity of customers, especially young and adventurous customers.

Methodology

Study Area

Geographical Location

Hoa Lo Prison is a famous landmark because it was a place to hold many major Vietnamese revolutionaries during the Indochina War and American pilots during the Vietnam War. The place was recognized as a historical relic by the Ministry of Culture, Sports and Tourism No. 1543-QD/VH dated June 18, 1997.

Hoa Lo Prison or Hoa Lo Pha House is a prison built by French colonialists on the old land of Hoa Lo village, currently located at No. 1 Hoa Lo Street, Tran Hung Dao Ward, Hoan Kiem District, Hanoi.

It has a total area of 12,908 m^2.

Today, Hoa Lo only has a part of more than 2,000 m^2 adjacent to Hoa Lo street preserved and embellished into Hoa Lo Prison Souvenir Area.

Built in 1896, Hoa Lo Prison is designed into four zones A, B, C, and D with different purposes: Zones A and B for inmates under investigation, misdemeanors, and violations of prison regulations. Area C is for prisoners who are Lagoon, foreigners. Area D is for felony prisoners, awaiting the death penalty (Figs. 11.1 and 11.2).

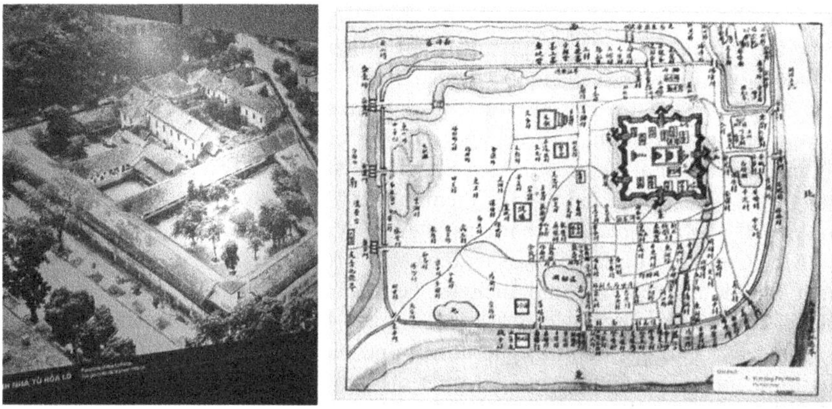

Fig. 11.1. Diagram of Hoa Lo Prison. *Source:* Hoa Lo Prison Relic (2015a).

Fig. 11.2. Main Gate of Hoa Lo Prison. *Source:* Hoa Lo Prison Relic (2016).

Hoa Lo Prison Wards

Hoa Lo is one of Vietnam's most notorious prisons. This prison was used by the French authorities and later the Vietnamese authorities during the war. Hoa Lo Prison is divided into four main wards, A, B, C and D, each with different characteristics and forms of torture.

Area A of Hoa Lo Prison was built in the early 20th century and was used by French authorities to hold political prisoners. Area A covers an area of about 450

m^2 including cells, trial waiting rooms, interrogation rooms, private cells, and hospital rooms. In Area A, prisoners are often tortured by beatings, hanging upside down, or placed in small wooden crates.

Area B of Hoa Lo Prison is used to hold female political prisoners. Zone B covers an area of about 150 m^2 and includes cells, interrogation rooms, courtrooms, and hospital rooms. In Area B, prisoners endure torture such as beatings, hanging upside down, and placing them in small wooden crates; however, torture in Area B is often not as barbaric as in Area A.

Area C of Hoa Lo Prison was used to hold prisoners of war. Zone C covers an area of about 2,000 m^2 and includes cells, courtrooms, interrogation rooms, media rooms, classrooms, and hospital rooms. In Area C, prisoners were often tortured by beatings, hung upside down, and placed in small wooden crates. In addition, prisoners endure other forms of torture such as being tied to poles, placed in water tanks, and closed for long periods of time.

Area D of Hoa Lo Prison is used to hold criminal prisoners and emotional prisoners. Zone D covers an area of about 2,500 m^2 and includes cells, courtrooms, interrogation rooms, restrooms, and patient rooms. In Area D, prisoners are also subjected to torture such as beatings, hanging upside down, and placing in small wooden crates. In addition, inmates in Area D work in labor groups and often suffer from hunger, fatigue, and debilitating health.

All Hoa Lo Prison complexes have some common features in the way prisoners are held and tortured. It is one of the harshest prisons in Vietnam's history. Here, prisoners endured brutal forms of torture and lived in harsh living conditions. However, these actions could not defeat the spirit of those who were once detained here and they became a symbol of the perseverance and sacrifice of the Vietnamese people in the fight for independence and freedom.

History

French Colonial Period. This prison was built by the French colonialists in 1896 on the land of Hoa Lo village (specializing in making earth kettles, super earth, and earthen furnaces sold throughout the period).

In 1899, Hoa Lo Prison was not completed, but had to be put into use immediately because of the French colonialists arresting a series of patriots to fight against them. In the following years, the prison was regularly renovated, repaired, and increased the area of cells.

It used to hold mostly political prisoners, patriots who opposed the French colonial government, and a few were ordinary and foreign prisoners. For those sentenced to a maximum of 5 years or guillotine, the French colonialists detained them in Hoa Lo Prison.

During the Vietnam War. It was a place where American pilots were held alive until after the 1973 Paris Agreement. American prisoner of war (POWs) and pilots are called Hoa Lo Prison "Hanoi Hilton." Of the American POWs captured, the most famous of the latter was US Senator John McCain.

In addition, Hoa Lo Prison was also a detention ground for alleged spies sent by the Republic of Vietnam to undermine the North during the division of the country and also a detention ground for those allegedly involved in post-1954 anti-Party Revisionist cases in the North.

Post-War. In 1993, in order to meet the development needs of the Capital, the State of Vietnam decided: part of Hoa Lo Prison to be used to build a commercial building with the name Hanoi Tower – "Hanoi Tower."

Currently, Hoa Lo Prison ruins have become "Red Addresses," a place to educate patriotic and revolutionary traditions to all classes of people, especially the young generation of the Capital and the whole country; a place that attracts a large number of domestic and international visitors to visit, research, and study.

Hoa Lo Prison architecture is part of Vietnam's cultural heritage. It was built in 1894 by the French, and was used to hold important prisoners during this period. Hoa Lo Prison is built in classical French architecture, with some distinctive features. It was designed by a French designer, and built of stone and porcelain. Hoa Lo Prison is in rectangular shape, with several windows throughout the inner wall. The exterior wall has some large windows, and there are also some smaller windows on the wall. These windows are designed to allow light into the prison.

Hoa Lo Prison Historical Site also has a number of different characteristics. It has a large entrance door and a smaller entrance door at the rear. The prison also has some windows throughout the interior wall, and some smaller windows are on the walls. The prison also has some large windows at the front and some smaller windows at the back.

At the time of completion, Hoa Lo Prison has the largest and most solid scale in Indochina with a total area of 12,908 m^2, including the following construction items: a guard house; two infirmaries; an alms-giving hospital; two defendant jails; a workshop house; and five prison houses.

Besides, the exterior architecture has a very seamless and sturdy design to prevent defections. On the four corners of the prison are four watchtowers for observation inside and outside the prison, surrounded by a solid stone wall with bottle fragments and high-voltage wires. At the base of the wall is the patrol line. Especially, construction materials were carefully selected by the French colonial government and transported from France to meet the highest quality requirements.

In Hoa Lo Prison, there are some special rooms designed in the classic French architectural style. These rooms are decorated with unique and distinctive shapes. In addition, Hoa Lo Prison also has several windows on all four sides, designed in the classic French architectural style.

Hoa Lo Prison was built with a modern information system. The system includes security cameras, door control systems, and light control systems. The system is designed to help thwart any attack from outside. Built with a state-of-the-art escape system, this system consists of hidden emergency exits, door control systems, and light control systems. The system is designed to help prevent any attack from within. And most especially, Hoa Lo Prison is built on a large area with an area of more than 20 hectares. The area is divided into three different sections including prison cells, practice rooms, and public areas. All prison cells are designed in a circular shape, with each room having a central window. These windows are designed to allow natural light into the prison cell.

In 1954, after the liberation of North Vietnam, the government administered Hoa Lo Prison, which served as a detention ground for lawbreakers and detained American pilots shot down or American soldiers captured while bombing North Vietnam.

Today, the area retained for conservation and tourist sightseeing is only about 2,400 m^2.

Results and Discussion

Exploiting "Dark Tourism" at Hoa Lo Prison Historical Site, Hanoi, Vietnam

Forms of Brutal Torture at Hoa Lo Prison

Hoa Lo is nicknamed "*hell on earth*" in the heart of Hanoi and prisoners. They endured fiery slaps as they stepped through heavy ironwood gates, then were shackled, brutally beaten in solitary confinement cells or guillotine cells.

Cachot is used to hold prisoners who are punished for violating prison rules or committing acts of resistance (organizing struggles, organizing prison escapes, revolutionary propaganda). Cachot in Hoa Lo is "hell of hell;" the cell is cramped, dark, and the floor is sloping upside down so that prisoners cannot lie down. People imprisoned in Cachot after only a short time will suffer from edema, blurred vision, and scabies full of people due to lack of hygiene, light, and oxygen.

The colonial prison had its own extremely vicious torture methods for female prisoners in the form of electric shock with electric cameras at the Hanoi Secret Intelligence Department or rudimentary batoong to use corporal punishment on the weak spots of female political prisoners.

Along with the most barbaric weapon of the French colonialists is the giant guillotine, also known as the medieval guillotine, Hoa Lo Prison is known as one of the 10 most notorious prisons in the world, topping the top five creepiest places in Southeast Asia. This guillotine is designed by two wooden pillars 4 m high with blades held high above by pegs. Below is an iron compartment to drop the death row prisoner's head, next to which is a rattan barrel that weaves the body. In January 1930, the guillotine was transported to Yen Bai to execute 13 Vietnamese Nationalist soldiers captured during the Yen Bai Uprising (headed by Nguyen Thai Hoc). This barbaric weapon in the prison is in constant operation, being used to circulate from prison to prison throughout Tonkin.

In addition to torture and brutal beatings, prison guards also forced prisoners to do hard labor: repairing houses, pounding rice, working in prison guards' living quarters, and working on battlefields.

Hoa Lo Prison is known as one of the harshest detention grounds in Vietnam's history, where many Vietnamese leaders and soldiers experienced months of hardship and suffering. However, during those years, many stories of indomitable patriotism of the incarcerated became inspirations for the next generation and became legends of noble sacrifice.

Outstanding Stories of Indomitable Patriotism Inside Hoa Lo Prison

General Do Muoi (Hoa Lo Prison Relic, 2015b). In the 1940s, General Do Muoi was arrested by the French and imprisoned at Hoa Lo Prison. During that time, he never gave up his indomitable patriotism. He even managed to contact comrades outside prison to spread messages of independence and freedom to the nation. He also helped other prisoners by providing little food and medicine, and never betrayed his patriotism.

Nguyen Thi Quang Thai – First Wife of Legendary General Vo Nguyen Giap (Hoa Lo Prison Relic, 2017). Nguyen Thi Quang Thai was the first wife of General Vo Nguyen Giap, with whom she had a daughter, Nguyen Hong Anh; In 1942, during a search, the French arrested her and comrade Nguyen Duy Trinh, sentenced her to 16 years in prison and transferred Quang Thai to Hoa Lo Prison. At the end of 1943, there was a typhoid epidemic in Hoa Lo prison, many sisters were infected. Having medical knowledge, Quang Thai wholeheartedly took care of her sick sisters, but later, due to her poor health due to the torture of French secret agents, she became infected. Knowing that she would not survive, Quang Thai sent her family members to bring her daughter Hong Anh in so that the mother and daughter could see each other. Receiving the news, her mother-in-law took Hong Anh away, but was only halfway there when the train was bombed by allied aircraft, and the two grandmothers had to return. Quang Thai died without seeing her daughter and husband for the last time.

The Legendary General Vo Nguyen Giap. In 1942, General Vo Nguyen Giap was arrested by the French and imprisoned at Hoa Lo Prison. During that time, he demonstrated his indomitability by refusing to talk to other detainees if they were not his comrades. This has left French officials worried about his perseverance and courage.

Actor Trinh Cong Son. Trinh Cong Son is a famous Vietnamese actor who was arrested and imprisoned at Hoa Lo Prison in the 1960s for his revolutionary ideology. During that time, he wrote many poems and songs to express his patriotism and desire to live in a free and just society. His works have become an inspiration for many Vietnamese in their struggle for independence and freedom.

Le Duc Tho (Hoa Lo Prison Relic, 2017). He was a revolutionary activist and Vietnam's representative in the Paris peace negotiations in the 1960s. He was briefly detained in Hoa Lo Prison, before being released as part of the Paris peace agreement.

Soldier Nguyen Van Troi. Nguyen Van Troi was a Vietnamese revolutionary soldier who was arrested at Hoa Lo Prison and then executed by firing squad in 1964. During his imprisonment, Nguyen Van Troi committed many daring actions such as whitewashing other prisoners with powdered sugar, writing stone stelae to remember the revolutionaries who had sacrificed, and writing poetry to convey the message of patriotism.

Prospects of Exploiting "Dark Tourism" at Hoa Lo Prison, Hanoi, Vietnam

For a long time, Hoa Lo Prison Ruins have been an indispensable destination for tourists when coming to Hanoi. This is a place containing historical values, high cultural and spiritual values of the Vietnamese people with many potentials for economic and tourism development.

Firstly, according to the Department of Culture, Sports and Tourism, only from the beginning of 2023, Hoa Lo Prison Ruins welcomed nearly 295 thousand visitors, collecting fees reaching VND 7 billion. A representative of Hoa Lo Prison Monuments Management Board said that in March, the monument welcomed 60,000 visitors equivalent to 2,000 visitors per day; as of the evening of April 11, the number of visitors to the monument during the 3-day public holiday was 4,300 visitors, up 53.8% over the same period last year. Young people coming to Hoa Lo Prison Ruins during the holidays increased much compared to previous years. This proves that history education with practical knowledge has become attractive to many students. Thereby, to arouse national pride and pride for them the monument is constantly innovating tours and focusing on conveying messages and historical stories in new and youthful forms, so the situation of "sold out" takes place regularly.

Secondly, Hoa Lo Prison has many tourism products; one of the successful and widespread tourism products is the "night tour" experience of Hoa Lo Prison Relics. During the 70-minute journey back in time, visitors will witness the harsh life in hell on earth that former political prisoners have endured. This place is also known as the first and only historical site to deploy this new tourism product.

Third, Hoa Lo Prison has many creative forms of sightseeing with an automatic explanation system so that visitors can experience 35 historical stories in their own way. To serve visitors who have not had the opportunity to visit this place because of geographical distance, the monument also launched an online tour through the "Online Exhibition Space." In addition, visitors can visit the monument on two platforms Spotify and Apple Podcasts. Here, three online exhibitions named "Light the Fire of Love," "Iron – Son," and "Oath of Death" are conveyed through emotional voices that make it easier for audiences to access the historical story.

Fourth, Hoa Lo Prison always creates the best conditions for visitors' sightseeing experience. When visiting directly, visitors will be supported with free or discounted tickets if they belong to special subjects such as students, the elderly, people who have contributed to the revolution, and people with disabilities. In addition, the thoughtfulness right from the simplest things such as preparing umbrellas, lockers, and weather forecasts is the reason for leaving an unforgettable impression on visitors from the first time they visit.

In the midst of a modern, bustling capital, always interwoven with ancient spaces, historical works bring many ups and downs, nostalgia, evoking strong emotions in each person. Hoa Lo Prison with many unique and diverse tourism products, creative forms of sightseeing along with the thoughtfulness in the way of service, thereby increasing the value of the destination, attracting a large number of domestic and international tourists to visit to help develop the tourism economy, contribute to creating livelihoods for local residents.

Some Suggestions for Using "Dark Tourism" to Contribute to Healing the Wounds of War

Tourism is considered an important and effective activity contributing to healing the wounds of war. Through tourism activities, it will help enhance mutual understanding and promote peace among nations and peoples.

One visitor said:

> The world has suffered many painful wars, and now humanity is still constantly fighting new wars that have not ended. You have to go to Hoa Lo to see that peace is precious to any people. That lesson needs to be passed down for generations. We have been through pain, what we need for the future is peace.

"Weekend prison visit" and "prison outing" – the funny saying of citizens of the Capital and tourists every time they come to Hoa Lo makes tourists excited. Prison visits become a good option when families want to bring children closer to the nation's history.

Foreigners who want to decipher the aspirations for freedom and national independence of Vietnamese people can also come here. The younger generation also wants to come, basically because Hoa Lo Prison Ruins are now unlike any remnants of war left in Vietnam.

The relics have been digitized in tourism conservation and exploitation, are different products that are put on a new type of tourism thinking, understand the desires of young people, and meet the needs of historical tolerance with many levels of emotions.

Using national memory for tourism is not new in nations, war and the obsession with death is something no nation wants. Faced with the historical evidence of Hoa Lo Prison in Hanoi, this painful time inspired the young generation to act for the love of the country, for a civilized and peaceful life.

Although the war has receded, international friends have been contributing to healing the wounds of the war in Vietnam, extending the bridge of peace for humanity.

On the basis of domestic and international conditions, I would like to contribute a number of recommendations to develop "dark tourism" and at the same time contribute to healing the wounds of war, specifically as follows:

First, to strengthen the sense of responsibility for peaceful tourism in the operation of monuments. Every monument, every attraction, every museum must play the role of a messenger carrying a message of peace. Hoa Lo Prison is not only a place to display artifacts and documents about the war, but also needs to become a space for cultural activities, organizing exchange activities, connecting through culture and art so that visitors can clearly feel the losses in the war and join hands to build peace. The tour guides must also have specific professional training, in addition to basic knowledge. It is necessary to exchange skills in community and international communication, appropriate and professional discourse so that visitors can clearly feel the historical and human value of the heritage.

Second, management agencies and tourism enterprises shall coordinate in organizing tour programs and tourist routes with appropriate contents and forms for each target group, especially foreign tourists. When designing and building a "dark tourism" program, it is necessary to ensure science, meet the balance between the needs of understanding historical events and enjoying landscapes and cultural experiences. The journey of visitors comes alive when they are truly "living" in historical memory, not only in a monument but also expanding the sociocultural space in that land. Avoid building formal programs, imposing sociopolitical views on tourists. The historical story is skillfully led and reproduced objectively depending on the level of awareness and learning needs of each group of visitors.

Third, there is cooperation between national and international experts in organizing the "dark tourism" program toward the goal of peace, preventing the risk of war and conflict, overcoming war consequences. The research works will shed more light on historical issues and highlight the value of relics. Consulting and applying theories, research results and international tourism trends will help us orient and develop tourism effectively, overcome war consequences, and move toward peace. The creation of a cooperative exchange environment between domestic and international experts will bring useful and practical works, contributing to improving the quality of relics and enhancing the experience for visitors. Cultural and educational exchange programs between domestic and international youth are a way to enrich cultural values and introduce the beauty of Vietnam to the world.

Fourth, the application of scientific and technological achievements in "dark tourism" activities. With the fourth industrial revolution (4.0) taking place rapidly in the world, digitalization is an inevitable trend in the operation of monuments and museums. Renovate heritage displays and explanations to attract visitors offline and online. Travelers now have the habit of researching on the internet about destinations (or travel services) before making a decision. Stories, events, or historical figures are created by the monument's media team in many forms, both topical and catching up with trends that are prevalent on social networks. Typically, it can be mentioned the integration of puzzle wit, lyrics, and illustrations with information tied to history that seems "extremely dry." To date, the homepage "Hoa Lo Prison Relic" has attracted more than 125,000 followers, receiving thousands of likes, interactions, and shares every day. In addition to Facebook, the management board of the monument also focuses on developing content on other social networking platforms such as Instagram and Spotify by "time travel." It is known that during the forced closure due to the Covid-19 pandemic, Hoa Lo monuments still constantly connect with visitors on online spaces. It can be said that rarely has the communication channel of a historical site successfully interacted and emerged into such a social network "phenomenon." Applying technology to tourism activities enhances the experience for tourists, adds value to relics, and promotes and creates advantages for Vietnam's tourism industry in regional and international markets.

Conclusion

"Dark tourism" is a type of tourism that is directly associated with heritage and war relics. The goal of "dark tourism" is to vividly recreate the resistance war against the United States (1954–1975) exactly as it took place, giving visitors an authentic experience of historical events, thereby conveying the message of peace and solidarity among peoples around the world, responsibility to preserve the country's independence for future generations. War relics in Vietnam in general and Hoa Lo Prison in particular have not been properly recognized and exploited with the role and meaning of "dark tourism." Continuing to study cases that have successfully developed "dark tourism" in the world will make an important contribution to the research and practical application in tourism activities, helping to effectively exploit rich and potential tourism resources in the country. "Dark tourism" is also an effective solution to attract international tourists to Vietnam, creating a special attraction for tourists in distant markets such as the United States, France, Europe, Australia, and Russia.

Research Limitations and Further Research Directions

This study was only qualitative research at Hoa Lo Prison Historical Site and has not organized quantitative research here; for example, using a plan behavior theory (TPB) and "dark tourism" structure to demonstrate that there is a relationship between TPB and "dark tourism" structures. At the same time, this study did not study motivation for "dark tourism" and the intensity of participation. And the ethic of "dark tourism" has not been mentioned. These are the topics the authors may conduct research on in the future.

References

Avital, B., Poria, Y., & Oren, G. (2011). Sought experiences at (dark) heritage sites. *Annals of Tourism Research*, *8*(3), 820–841.

Biran, A., Liu, W., Li, G., & Eichhorn, V. (2014). Consuming post-disaster destinations: The case of Sichuan, China. *Annals of Tourism Research*, *47*, 1–17.

Dann, G., & Seaton, A. V. (2001). Slavery, contested heritage and thanatourism. *International Journal of Hospitality and Tourism Administration*, *2*(3/4), 1–29.

Foley, M., & Lennon, J. (1995). JFK and dark tourism: A fascination with assassination. *International Journal of Heritage Studies*, *2*(4), 198–211.

Hoa Lo Prison Relic. (2015a). The process of formation and construction. https://hoalo.vn/Articles/38/48/qua-trinh-hinh-thanh-va-xay-dung.html. Accessed on November 05, 2023.

Hoa Lo Prison Relic. (2015b). Comrade Do Muoi escaped from Hoa Lo prison, March 1945. https://hoalo.vn/Articles/34/49/đong-chi-đo-muoi-vuot-nguc-hoa-lo-thang-3-1945.html. Accessed on April 05, 2024.

Hoa Lo Prison Relic. (2016). The arduous exile journey of Hoa Lo prisoner (part 1). https://hoalo.vn/Articles/15/186/Hanh-trinh-luu-day-gian-kho-cua-nguoi-tu-Hoa-Lo-phan-1.html. Accessed on November 05, 2023.

Hoa Lo Prison Relic. (2017). When the Homeland needs it, they know how to live apart. https://hoalo.vn/Articles/12/9410/khi-to-quoc-can-ho-biet-song-xa-nhau.html. Accessed on April 05, 2024.

Isaac, R. K., & Çakmak, E. (2014). Understanding visitor's motivations at sites of death and disaster: The case of former transit camp Westerbork, The Netherlands. *Current Issues in Tourism*, *17*(2), 164–179.

Kamber, M., Karafotias, T., & Tsitoura, T. (2016). Dark heritage tourism and the Sarajevo siege. *Journal of Tourism and Cultural Change*, *14*(3), 155–269.

Lennon, J. J., & Foley, M. (1999). Interpretation of the unimaginable: The US Holocaust Memorial Museum, Washington, DC, and "dark tourism". *Journal of Travel Research*, *38*(1), 46e50.

Lennon, J., & Foley, M. (2000). *Dark tourism: The attraction of death and disaster*. Continuum.

Lennon, J., & Foley, M. (2003). Dark tourism and sites of crime. In D. Botterill & T. Jones (Eds.), *Tourism and crime: Key themes* (pp. 215–228). Goodfellow Publishers.

Miles, W. (2002). Auschwitz: Museum interpretation and darker tourism. *Annals of Tourism Research*, *29*(4), 1175–1178.

Raine, R. (2013). A dark tourism spectrum. *International Journal of Culture, Tourism and Hospitality Research*, *7*(3), 242–256.

Sharpley, R. (2005). Travels to the edge of darkness: Towards a typology of dark tourism. In C. Ryan, P. Stephen, & A. Michelle (Eds.), *Taking tourism to the limits: Issues, concepts and managerial perspectives* (pp. 217–228). Elsevier.

Stone, P. (2005, January). "Dark tourism-an old concept in a new world" tourism magazine by the Tourism Society (UK). 20. http://www.works.bepress.com/philip_stone/26

Yan, B.-J., Zhang, J., Zhang, H.-L., Lu, S.-J., & Guo, Y.-R. (2016). Investigating the motivation-experience relationship in a dark tourism space: A case study of the Beichuan earthquake relics, China. *Tourism Management*, *53*, 108–121.

Yankholmes, A., & McKercher, B. (2015). Understanding visitors to slavery heritage in Ghana. *Tourism Management*, *51*(12), 22–32.

Yuill, S. M. (2003). *Dark tourism: Understanding visitor motivation at sites of death and disaster*. Doctoral dissertation. Texas A&M University.

Chapter 12

Chambal Valley: the Myth, the Curse, the Legend, the Challenges, and the Opportunities for Dark Tourism – Case Study

Shaheed Khan[a], *Swarna M. Freeda Maria*[a] *and Ajoy K. Bhattacharya*[b]

[a]Dharthi NGO, India
[b]Madhya Pradesh Ecotourism Development Board, India

Abstract

Call it the curse, the myth, the legend, the dacoits, the flora and fauna, the River, and the valley remains untouched and pristine and is home to many endemic species and is blessed with stark but beautiful landscape, the surreal world of an era gone by, the ravage of the ravines, which have been preserved, protected, and conserved, albeit for the 'ancient curses,' and yes being a home to the 'dreaded outlaws' and 'modern dacoits.' Whether it was the fear for one's life considering it was a dacoit area or also known as 'gun powder area,' or the curses that none wanted to garner to themselves, Chambal is slowly but steadily ebbing out of the dark shadows that it once was. The river itself passes through Madhya Pradesh, Rajasthan, and Uttar Pradesh, before joining the river Yamuna. Declared a Protected Area (PA) in 1978, the nomenclature of dark tourism will be ubiquitous to the Chambal River valley, considering that the area is dotted by a multitude of forts, and heritage destinations, that have frozen in the sands of time; either on account of the stories that meander, as the river does, with its deep ravines, ravines that invite the adventurer. It is the last bastion for the nation's endangered wildlife, viz., Gharial, Maggar, Turtles, Otter, and the fresh water Dolphin, which has encouraged the Government to get the local community to be part of the special purpose vehicle promoting tourism. The authors known for their contribution to community-based ecotourism (CBE) have visited the

Dark Tourism, 149–173

doi:10.1108/978-1-83797-336-120241012

Chambal area and worked with the communities and bureaucracy to ensure conclusions.

Keywords: Dark tourism; chambal; heritage tourism; ecotourism; flora; fauna

Introduction

A drive from Gwalior to the Chambal River takes about 90 minutes. Gwalior is a historical place which has today grown to be an important city in the state of Madhya Pradesh. Gwalior is a good base to start the trip toward probing Chambal River (Fig. 12.1), which was once known for armed dacoits, who were also known locally as *baghis* (rebels) and spelt terror in the region (Banerjee, 2017). As one arrives in the Chambal region, it is the undulating landscape, with mud hillocks and thorny shrubs that greet a visitor. Every Indian, who has heard about Chambal, will recall something about the dacoits who had made the Chambal area their home. *Baghis* for some, and dacoits for many, we remember the names of Maan Singh, Paan Singh Tomar, Phoolan Devi, Malkhan Singh, Jagjivan Parihar, and Nirbhay Singh Gujjar as we eulogize the era of a region (Dabas, 2016) that had its tryst with destiny, albeit as a rebellion against the societal system that existed or the *baghis* becoming 'robin hoods' to many, celebrated by others, hated by the Government machinery, the *baghis* carried a bounty, which ensured either they surrendered and live/lived a life in politics or got to the oblivion or they were killed in police action or even assassinated as in the case of Phoolan Devi in broad daylight at the age of 37. Phoolan Devi, popularly known as the Bandit Queen, had surrendered in 1983, and served prison

Fig. 12.1. Picture of the Chambal. *Source:* Parameswaran (n.d.).

time for 11 years. She even got elected as a Parliamentarian to the Lower House (*Lok Sabha*) from Mirzapur in Uttar Pradesh with more than 300,000 votes on a *Samajwadi* Party ticket. The Indian film world brought forth the life of Phoolan Devi in celluloid titled Bandit Queen (1994) (https://www.youtube.com/watch?v=iERIJL3UGvI), and received acclaim at the Cannes Film Festival and was India's official Foreign Language Film entry at the 67th Academy Awards held in 1995 (Borealis, 2021). The story of Paan Singh Tomar, killed by the police, was eulogized through a phenomenal 2012 Bollywood biopic (https://www.youtube.-com/watch?v=tly2nDuB4es) rendered by late Irfaan Khan which has been considered as one of the best films on the *baghis* (The Lallantop, 2021). Paan Singh Tomar was a solider and an athlete, who was a seven-time national steeplechase champion and even represented India at the 1958 Asian Games that were held in Tokyo (Deni, 2022).

The Chambal River and its primary tributaries of Parbati, Kali Sindh, Shipra, Banas, and Mej Rajras (n.d.) (see Figs. 12.2 and 12.2a) spread over the region of north-central of India, and the engulfs within itself, southeastern Rajasthan, southwestern Uttar Pradesh, and northern Madhya Pradesh creating a multitude of opportunities of tourism promotion in an area that had been declared as badland topography. Sutikshan (2020) speaks of how badland topography (Fig. 12.3) is characterized by an undulating floodplain, gullies, and ravines. It is a type of dry terrain where softer sedimentary rocks and clay-rich soils have been extensively eroded by wind and water.

Badlands are terrains with intense past or present erosion that create extremely high impressive erosion landforms on earth. They have poorly consolidated sediments and regolith (a layer of unconsolidated solid material covering the bedrock of an area (Rafferty, 2018), scarce vegetation, and intense drainage, all of

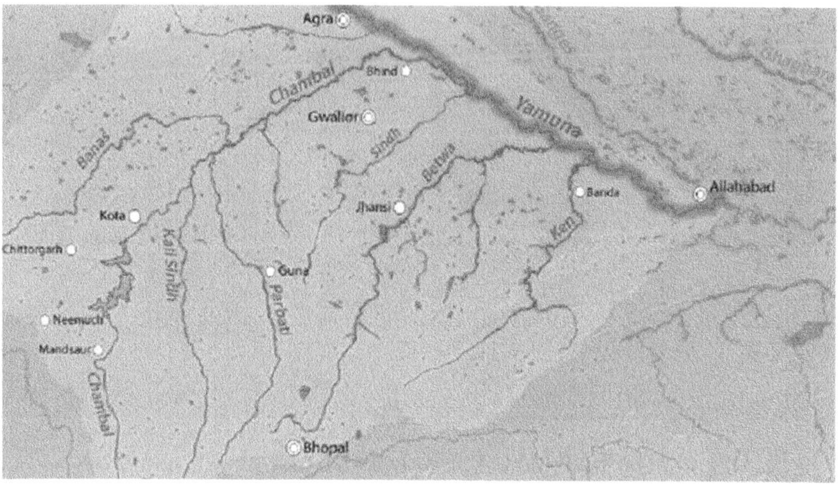

Fig. 12.2. The Chambal River Ecosystem. *Source:* Sutikshan (2020).

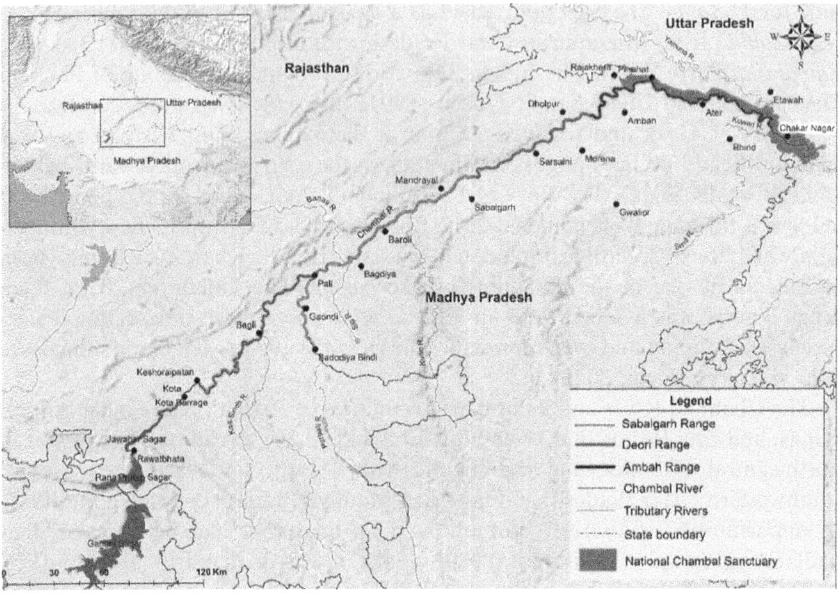

Fig. 12.2a. The Chambal River Ecosystem. *Source:* Management Plan (n.d.).

Fig. 12.3. 'The Badlands,' the Ravines of Chambal Region. *Source:* Images by the Authors.

which make agriculture and grazing impossible. The place/area therefore becomes synonymous with dacoits, who made use of the geography and created a home for themselves (Business Standard, 2019). For the local community in Chambal, the dacoit or *Baghi* culture is something of the past, the populace has moved on, no doubt. The authors got to interact with one, Raghuveer (not his full name) (Raghuveer, 2023), who enjoys taking visitors on guided trips, stated,

> . . .no doubt, our area is known for all the bad aspects, but that is our past, a past, that we cannot forget. It lives with us. Recently a visitor stayed here for about ten-days and moved around in all modes of transport and got a feel of what is the 'so called' dreaded area all about.

Abhishek Chaubey (Business Standard, 2019), who has thrived writing stories and even screenplays for films about the Badlands, mentions "the place has moved on, there are no dacoits now, but the issues of caste and gender discrimination exist. The local community members still remember the culture of the *Baghis* of the area though."

Nair (2017) speaks about the fact that the unique ecosystem of the Chambal River the region has ensured the creation of the National Chambal Sanctuary (NCS) (Fig. 12.4) (also known as the National Chambal Gharial Wildlife Sanctuary), the first and only tristate protected area (PA) in India. With a PA of NCS that moves with the river for nearly 425 km and is 2–6 km wide (Water Database, n.d.), it is perhaps the longest riverine PA in the world. Nair and Krishna (2013) mention that NCS was established on September 30, 1978, by the Government of

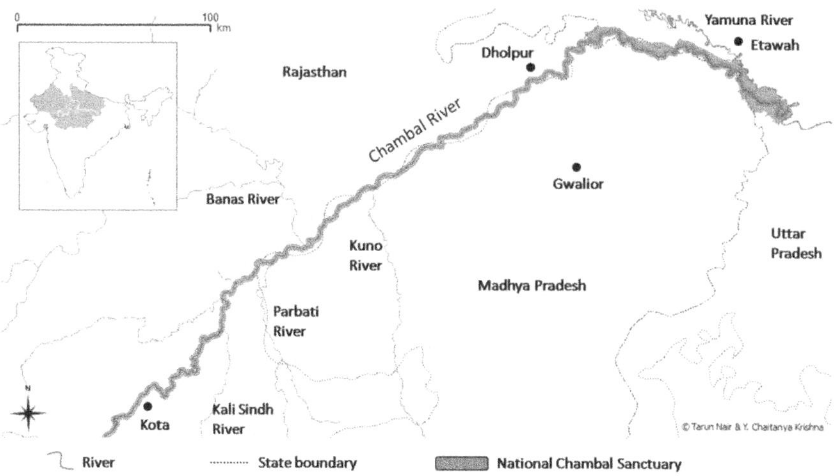

Fig. 12.4. The Area of the National Chambal Sanctuary. *Source:* Nair and Krishna (2013).

India, Notified by States: Madhya Pradesh (December 20, 1978); Uttar Pradesh (January 29, 1979); and Rajasthan (December 7, 1979), and lies between latitude and longitude 240 55′–26°50′N & 750 34′–790 18′E. NCS primarily came into existence with a charter to conserve, preserve, and protect the Gharial and the unique Chambal River ecosystem, which covers an area of 1,800 km^2 (Figs. 12.4–12.6). The Government of India, Ministry of Environment and Forests (MoEF) (MoEF, 2011) even constituted the National Tri-State-Chambal Sanctuary Management and Coordination Committee for conservation of Gharial (NTRIS-CAS MACC), which was done with an intent of having better coordination among the three states sharing the Chambal river for ensuring the protection, preservation, and conservation of the biodiversity of the area. The coordination committee was to have members from the three states' Departments of Water Resources, Irrigation and Power, besides, Wildlife Institute of India, Madras Crocodile Bank, The Gharial Conservation Alliance, Development Alternatives, Ashoka Trust for Research in Ecology and the Environment, Worldwide Fund for Nature, and the DFOs (related to NCS) of the three states.

Chambal River

The Chambal River runs for 885 km (550 miles) (Water Database, n.d.) in the states of Rajasthan, Madhya Pradesh, before joining the River Yamuna in Uttar

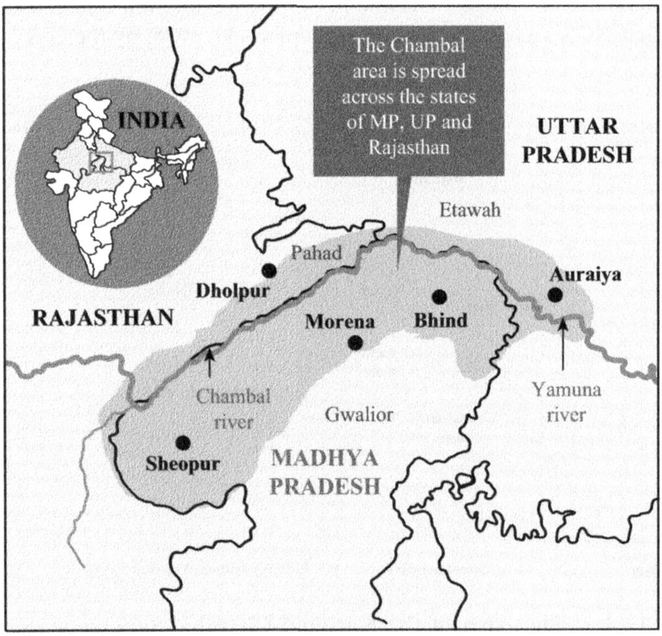

Fig. 12.5. River Chambal and Its Environs. *Source:* Adapted from Banerjee, 2017.

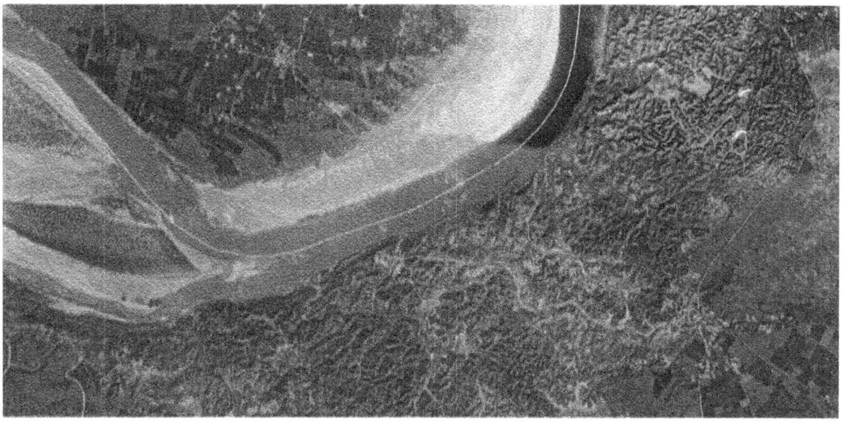

Fig. 12.6. Chambal River and the Badlands Near Morena. *Source:* Satellites Pro Morena (n.d.).

Pradesh (Travel Trends, 2021) has its origin at Janapav (Singar Chouri peak) west-south-west of Mhow, Madhya Pradesh in the Vindhya Mountain range at an elevation of 843 m and joins the River Yamuna at Pachnada near Bhareh in Uttar Pradesh, after flowing Rajasthan (Rajras, n.d.). Hussain and Badola (2001) have considered the Chambal River as one of the last remnant rivers in the greater Gangetic Drainage Basin which has retained significant conservation values.

Rivers in India no doubt have been worshiped and considered as divine; however, the Chambal River has a different story to speak about itself from historical times. Referred as Charmanvati (River of Hides: Charman-skin, hide) in the historical times (Fitzgerald, 2004), it is believed that during the times of Mahabharata, King Rantideva of Dasapura (Mandasaur in Madhya Pradesh) is said to have distributed beef along with food grains to Brahmins; and thence, the river oozed out from the heaps of the animal hides (Charmanyavati) (Rajras, n.d.) and came to be known as the River of Hides (Charmanvati) (Chambal Safari, n.d.). The irony of the river was that it was further cursed by Draupadi who was disgraced following the game of dice that was played between the Kauravas and Pandavas. She had cursed that anyone who drinks water from the river will be filled with a thirst for vengeance that would be unquenchable for the river stood testimony to the humiliation of Draupadi (Travel Trends, 2021). By reading about the curse by Draupadi, when anyone thinks of the Chambal River, it evokes a scene of the Chambal *Behaad* (ravines), which became linked to the dacoits, the *Baghis* (rebels) who were in search of revenge and justice for a long time, during the contemporary period. Writing about Chambal, Mohan (2010) mentions that, during the days of the Mahabharata, the area was part of the kingdom of Shakuni, who was with the Kauravas.

Vinod (personal communication, October 5, 2022) a tour guide who operates in the Gwalior and Khajuraho destinations mentions, "may be because of the Chambal River curse, the river was quietly flowed without much human intervention, and no industries around the river, the river and the water are clean and unpolluted." He even reiterates, "the Chambal no doubt joins the Yamuna River first and then the Yamuna mergers into the Holy Ganga River; but Chambal is unpolluted, while we can see too much pollution in the Yamuna and Ganga." Vinod laughs and mentions, "the curse of Draupadi, the curse of the river oozing out of the skin of animals, and the dacoits menace ensured the Chambal ecosystem was kept clean without any interference." Das (2016) reiterates, due to the dreaded dacoits of Chambal region, folks of Morena, Sheopur, and Bhind, along Rajasthan-Madhya Pradesh border, believe that one should not consume Chambal's water. The actual fear was that if one were to consume the water of Chambal, it would make the person aggressive, thence, the proof that the curse in a way was real. Parameswaran (n.d.) in his blog succinctly concludes, the curse of the Chambal Valley on account of two facets:

(1) the river has been a recipient of many ancient curses;
(2) the home of the dreaded modern dacoits.

Chambal and Tourism

Nandagopal (personal communication, January 4, 2023) a domestic and inbound tour operator mentions, "the name of Chambal ushers in a sense of fear in the heart and minds of people. It became famous for the natural ravines, and the badlands that were ruled by the dacoits." Nandagopal recalls his travels to Delhi from Chennai as a youngster by train and states,

> ...we would sit up and watch the Chambal ravines go by, and see the river in a few places, it would bring about an eerie sense of fear. I even got to watch a film on the most famous dacoit of Chambal, Phoolan Devi and realized why Chambal came to be known as a place of fear and dead.

Nandagopal adds, "today of course, we speak of Chambal as a tourist destination, for its natural beauty, the flora and fauna and the heritage sites the visitors get to see and understand the ecosystem." Vinod (personal communication, October 5, 2022) states,

> ...as a child I grew up the stories of fear when one spoke about Chambal River and the environs. Sometimes my mother would scare my siblings and me to sleep stating that the 'badman' (the dacoit) would come and take me away, and yes, we would fall asleep immediately. But today, when we speak of the Chambal environs, we see an opportunity for tourism promotion. I have

> been to the Bateshwar area where many temples were resurrected by the Archaeological Survey of India (ASI) (https://asi.nic.in), and believe me, the visitors would like the destinations that has never seen visitors for years.

"Besides, Culture and Heritage tourism, one can also plan for visits to places of nature, which are very pristine and the Chambal River provides a phenomenal ecosystem that can be visited," he adds.

S. P. Singh (personal communication, December 3, 2021), a Retired Deputy Director General with the Ministry of Tourism, Government of India, mentions,

> ...during my early days as an officer with the Ministry, when I was stationed in Agra, visitors would ask about the Chambal and its environs, but then in the 70's and 80's the dacoit menace ensured that there was an element of fear in the mind of people, and hence, no visitations occurred. But then once the conditions were normalized, roads, hotels and even resorts coming up has ensured, the place becomes an attraction under the garb of 'dark-tourism,' only on account of the fact of the stories of the Historical curse as we get to hear and yes the presence of the dacoits.

S. P. Singh states,

> ...as I read about the Chambal area today, I feel it will become an important destination where the Tourism, Travel and Hospitality (TTH) business segments can promote tourism and travel in a big and get visitors to see the offbeat destination, which will also help the local communities that are spread over three states of Madhya Pradesh, Rajasthan and even Uttar Pradesh.

S. Tomar (personal communication, February 9, 2024) a Naturalist on the Chambal River, near Deori in Madhya Pradesh, who engages visitors on the Chambal states,

> ...the opportunities to promote tourism is phenomenal, Chambal has everything within itself, the history, the intrigue, the darkness, the Wildlife, the flora, the geology, the climate, and the visitors. If more local community members like me participate in the ecotourism endeavors, it will be an asset to the to one an all.

He further added, "it may be tough, as people are unlettered, but it will not be an obstacle to initiating something good for ecotourism."

Historical Heritage

The Restored Bateshwar Group of Temples

Blogger Lakshmi Sharath (Sharath, 2018) who visited and got an opportunity to interview the restorer (Dr KK. Muhammed of the Archeological Survey of India (ASI) (https://asi.nic.in)) of the ninth century Bateshwar (Fig. 12.7) temple complex speaks about how the area was flattened on account of an earthquake and for centuries over 200 temples of Lord Shiva and Lord Vishnu, built by the Gurjara Pratihara Dynasty (ruled from Ujjain and later Kanauj), a greater part of Northern India from the 8th to 11th centuries (Britannica, 2009), were lying in the place with entwined roots of trees.

Located to the north of Gwalior (31 km) and 34 km east of Morena, the Bateshwar Group of Temples spread over 25 acres, are a treat to the eye. Bhot (2022) calling Bateshwar a ‘hidden gem’ writes the writ of Dacoit Nirbhay Singh Gujjar who had made the ruins of Bateshwar his home, was so phenomenal, that no one ventured to the temple complex which had temples dedicated to Shiva, Vishnu, and Shakti. Pezarkar (2018) mentions, though the Bateshwar temples were mentioned by Alexander Cunningham, and even considered as a protected site in 1924, the site was ruins and was only in 2002 that Archeologist KK. Muhammed began his tryst to bring back to life the fallen temples. True, he had to use his acumen to get permission of the dreaded Dacoit, and he did not fail in his endeavor. Once restored, Nirbhay and his gang also would visit the local Hanuman temple for prayers, and they are believed to have felt proud that their ancestors had built the temple complex that was being restored under their watchful eyes. Nirbhay Singh Gujjar was killed in an

Fig. 12.7. Glimpses of the Bateshwar Temple Complex After Restoration. *Source:* Images by Authors.

encounter in 2005, but the legacy of the temples now stands tall, and visitors not only get to see a heritage of the past, but also get to the know the stories of the way the temple complex was restored.

Ater Fort of the Bhadauria Kings

Located north-east of Gwalior (100 km), south-east of Agra (106 km) and north-west of Bhind (35 km), Ater fort (Fig. 12.8), located on the ravines of Chambal, was built by Bhadauria rulers Badan Singh, Maha Singh, and Bakhat Singh between 1664 and 1668. The area where the fort lies is called Badhwar after the Bhadauria rulers (Madhya Pradesh Tourism, n.d.). The main attractions of the fort are 'Khooni Darwaza,' 'Badan Singh Ka Mahal,' 'Hathiapor,' 'Raja Ka Bangla,' 'Rani Ka Bangla,' and 'Barah Khamba Mahal' (District Bhind, n.d.). The Chambal flows about 4-km from the Ater fort and after one cross the Badlands of Chambal and reaches the river there are opportunities for boating and even seeing the wildlife that exists in the Chambal. One needs to make an effort to enroll the community members and the concerned departments should play a role of a facilitator to ensure qualitative promotion of the destination. The location at Ater is also a gateway for the visitors from Uttar Pradesh to come and enjoy the Chambal in its pristine.

Chausath Yogini Temple, Morena

At 34 km to the east of Morena (District Morena, n.d.) and 34 km to the north of Gwalior and just 4 km from the Bateshwar temple complex is the Chausat Yogini

Fig. 12.8. Glimpses of the Ater Fort. *Source:* Pictures and satellite images Authors.

Temple also known as Ekattarso Mahadev Mandir in Mitaoli (Mitawali or Mitavali) near Padaoli. Yadav (2019) states, the Chausath Yogini temple (Fig. 12.9), which stands out as a replica of the Parliament House, is situated in the desolate Chambal Valley. It is a temple of Lord Shiva and many consider the Parliament House in New Delhi as a replica of the structure, an argument which a section of archaeologists and historians do not agree. However, KK. Muhammed, the Archeologist who resurrected the Bateshwar complex, mentions the temple was built by Kachchhapaghata King Devapala in 1323, and it was between the 5th and 13th centuries when the semi divine Goddesses were widely recognized and worshiped in the religions landscape of the time. Dr Muhammed views that the design of the Parliament House was inspired from the circular Chausath Yogini temple. He said the Vedic mantras engraved on the walls of the Parliament House is a vindication that the British architects did consider the Indian building traditions. The Chausath Yogini temple was a center for 'tantra vidya' (black magic). The circular structure has 64 temples of Lord Shiva. It has images of Lord Shiva with 64 yoginis, which in tantric tradition connote feminine spirit. The yoginis are also said to represent the 64 *kalas* or arts.

A flight of 100 steps ensures that the visitor reaches the Chausat Yogini temple, perched atop a hill. Gupta (2022) mentions that, a demon King Raktabija, whose blood if spilt could create thousands of evil offspring, lived in the area. When Mother Durga clashed with the invincible *asura* (demon), she is said to have unleashed 64 *Yoginis* – feminine beings – who drank all the blood of the *asura* even before it touched the ground. It is from this event that we have the story of the Chausath Yoginis (Chausath, meaning 64). The Chausath Yoginis of tantrism were one of the most esoteric cults of India – feared or venerated, depending on who you asked. The yoginis, they say, were a tight clan of sorceresses, and what

Fig. 12.9. Pictures of the Chausath Yogini Temple. *Source:* Images by the authors.

transpired in their haven will remain a mystery forever, save for their sculpted carvings on the temple walls. Much like the yoginis, their temples wherever found are distinct from the mainstream, in form if not in function. Rongmei (2023) confirms the existence of the other three Chausath Yogini Temples in Hirapur, Ranipur-Jharial in Odisha, and another at Khajuraho in Madhya Pradesh. As esoteric as it sounds, the temple is a must visit when one visits the Chambal area.

Kakanmath Temple

Maheshwari (2022) writes about 55 km north of Gwalior and 32 km east of Morena is Sihoniya which is famous for the Kakanmath temple (Fig. 12.10). Standage (2019), a photo journalist and blogger, states, "Kakanmath temple, is defying the laws of gravity," and mentions "after visiting the place, one wonders how is the temple still standing." The Kakanmath Temple was commissioned by the Kachchhapaghata ruler Kirttiraja sometime between 1015 and 1035 AD. An inscription found at the Sas Bahu temple in Gwalior records a Shiva temple being built at Siṁhapānīya which is known as Sihoniya today.

Temples of India (n.d.) have professed that the name of the temple derives from the *kanak* (gold) and *maṭha* (shrine). Originally, the site had a temple complex, with a central temple surrounded by four subsidiary shrines. Only the ruins of the central temple stand now; its outer walls, balconies, and a part of its spire have fallen. This damage probably happened during an earthquake. From any visitor to Chambal area, the Kakanmath temple brings an understanding of the way the ancient kingdoms looked at art and architecture.

Fig. 12.10. Pictures of the Kakanmath Temple Complex. *Source:* Images by the authors.

Gwalior as a center for the Scindia Royalty has a multitude of forts and palaces in Gwalior, which can be visited by the tourists, depending on the duration of stay in the area.

Natural Heritage

Considered to be a river free of pollution, Chambal River is home to a variety of riverine fauna, which includes two types of crocodiles, the Gharials (*Gavialis gangeticus*) and Mugger (*Crocodylus palustris*), Nilgai (*Boselaphus tragocamelus*), Jackals (*Canis aureus*), Hyaena (*Hyaena hyaena*) smooth coated otters (*Lutrogale perspicillata*), eight varieties of freshwater turtles, jungle cats, Indian Skimmer (Indian Scissors-bill) (*Rynchops*), Pallas's fish eagle, Gangetic River dolphins (*Platanista gangetica*), black-billed terns, sarus cranes, bar-headed goose, and black-necked storks (*Ephippiorhynchus asiaticus*) (Rajras, n.d.; Nair & Krishna, 2013). Approaching from a scientific perspective, Singh and Rao (2017) concur that the Chambal River in India supports a rich variety of fauna and flora including the endangered Ganges River Dolphin and Crocodile as well. If the highest density of the endangered Gharial which are endemic to the Gangetic System of which Chambal River is an important feeder (Choudhury et al., 2007) and Reeves (2009) writes about the fact that the Dolphins of the River Ganga have received national recognition and have been declared the Indian National Aquatic Animal. It has been also reported that there are seven species of freshwater turtles, and 78 species of wetland birds (Hussain, 1993). Nair and Krishna (2013) through their Scientific research, study, and reporting state that they were able to record the following faunal diversity (Fig. 12.11) in the Chambal area,

Fig. 12.11. Wildlife (Gharial, Sarus, Turtle, and Tern) From the Chambal. *Source:* Images by the authors.

which included 147 fish species comprising 32 families, 56 reptile species comprising 19 families, 308 bird species comprising 64 families, and 60 mammal species comprising 27 families. This includes six critically endangered, 12 endangered, and 18 vulnerable species, as categorized by the International Union for Conservation of Nature (IUCN) (https://www.iucn.org) Red List of Threatened Species.

The Natural heritage in the Chambal (NCS) area has become a business opportunity for the local populace, as they have resorted to indiscriminate sand mining by the sinister sand mafia, which has led to depletion of the resources for the local fauna to survive. Ghose (2018), in a researched article on the dacoits of Chambal, writes, *behaad* or the Chambal ravines are the deadly sand mountains that stretch along the river, like a gargantuan labyrinth, sitting in a terrifying treacherous silence, as if to lure human beings to death. If the *behaad* was a place of fear, today, the sand miners have ensured that the fear has disappeared. Sand mining has been banned in the NCS since 2006 (Mazoomdar, 2022), but the local contractors and the politicians see it as a business bonanza, as illegal sand mining continues. Mazoomdar (2022) mentions that, the Madhya Pradesh Government in December 2021 suggested to the Ministry of Environment, Forest, and Climate Change (MoEFCC), Government of India to open five stretches of land on the Chambal River and its tributary Parvati, which amounts to 292 hectares. Whether the state Government is making efforts to end the conflict and please the lobbies, argue and state that the local Forest Department is over-burdened in managing and maintaining the forest resources, or to gain local support and also get revenue for the state; it puts all those involved and those who read and understand Chambal and its pristine options in a quandary. The environmentalists would call this hoarse and would say, it is a dark day for the ecology, environment, the flora, fauna, and the communities. The National Green Tribunal (NGT) (NGT, 2022) too jumped into the fray, after seeing media reports of the illegal sand mining and spoke about the blanket clearances are given for mining projects in violation of Sustainable Sand Mining Guidelines, 2016 and 2020. The NGT even created a committee headed by the Director General Forests and Special Secretary, MoEFCC, involving various officers of agencies across the three states to address the issue of sand mining. Despite all the push toward commercialization of the Chambal River and the ensuing issues that are to be found, the Chambal meanders silently, with the curses of the past and those that are being accumulated for the present and the future.

Opportunities for Tourism

The visitors of today depend on the various information sources, including the Government run Tourism offices, websites, social media, and its influencers to get to information and plan their travel to destinations. The authors of the chapters too looked to various sources and found that Native Planet (Native Planet, n.d.) had put forward the following to promote the Chambal destination spread across three states, which included the following important postulates for a visitor to travel: (i) birding (bird watching), (ii) camel safari, (iii) chambal safari, (iv) jeep safari, (v) culture/eritage tourism-Bateshwar Temples, (vi) village safari, and (vii) river rafting. True to its ecosystem, a variety of tourism products can and have been created by the TTH stakeholders in the Chambal River arena, which will ensure a sustainable traction for the destination. Vinod (personal communication, October 5, 2022), the tour guide, mentions, "we need lot of infrastructure, and all the stakeholders of tourism should come and contribute, engage the local community and bring about a change in the tourism ecosystem." He further adds, "boating, rafting, interpretation center about the history of Chambal, Photography expeditions, will add a lot of value to Chambal as a destination." "Dark tourism, and its ethos will lead us to a lot of light and it will ensure that the community which has been bereft of opportunities will see the light of day through tourism and its service offerings."

Tourism of various hues, ecotourism, community-based tourism (CBT), tent tourism, and there are even prospects of promoting tourism based on the impact assessment on the river via luxury boats. It can be said that the options for Sustainable Tourism is an opportunity that the Chambal River provides considering that is pollution free, unlike the other rivers. Kumar (2023) makes a mention of how Ram Pratap Singh, through the Mela Kothi Chambal Safari Lodge (https://www.chambalsafari.com), has promoted Sustainable and Responsible Tourism, mentions, "our lodge is carbon negative, water positive by 15 million liters, zero plastic, zero chemicals. We are completely glued to sustainable and responsible tourism and have ensured that no new inventory is added to the existing structures."

It is good to note that, in 2011, the Madhya Pradesh Ecotourism Development Board (MPEDB) (https://mpforest.gov.in/ecotourism/) along with Responsible Tourism Society of India (RTSOI) (earlier known as Ecotourism Society of India) (https://rtsoi.org) made an effort to introduce nature-based tourism (NBT) as an alternative income source in the formerly bandit ridden area of the Chambal and also make efforts to declare the Chambal River Ecosystem as a World Heritage Site under the UNESCO norms (Ibex Expeditions, n.d.).

An earlier effort in 2007 was undertaken by the same team to understand the facets of conservation and biodiversity and what activities could be rolled out for the local communities. The leaders at RTSOI believed that Chambal was no doubt once the abode of the outlaws (the dacoits), who operated in an ecologically sensitive region and when it comes to transformation of the region, it is the duty of the tourism ecosystem and the stakeholders to play a dynamic role along with the Government and ensure for the communities in the area. By training the local community members as tour guides, naturalists, boatmen, camel safari managers and providing opportunities to work in the rest house and resorts would not only help the community to eke out livelihoods, but would help the tourism ecosystem to reveal the many hidden treasures of Chambal through their work (RTSOI, 2021). Working with the MPEDB, the team also suggested to have a Chambal Mega Circuit (Figs. 12.12 and 12.13) by the river, which will help visitors to understand the destination better and provide ample opportunities for the local community with livelihood options and opportunities (MPEDB, 2011).

Fig. 12.12. The MPEDB and RTSOI Expedition of 2011 in and Around the Chambal River. *Picture 1:* Rakesh Mathur (center) and Mandip Singh Soin of RTSOI; *Picture 2:* Sujit Banerjee IAS, Mandip and Dr AK Bhattacharya IFS, CEO, MPEDB; *Picture 3:* Discussion with the Local Community. *Source:* Ibex Expeditions (n.d.A).

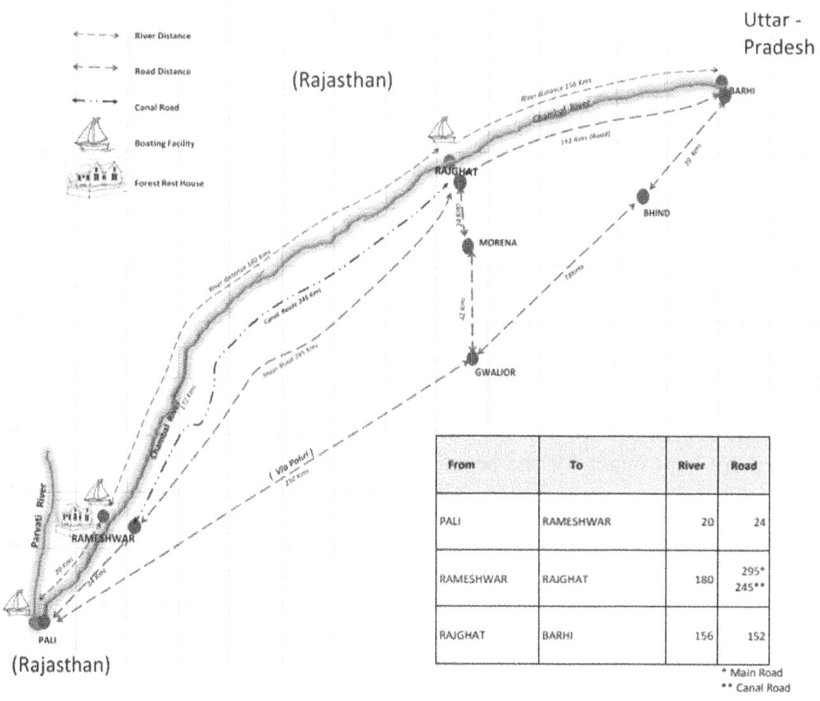

Fig. 12.13. Chambal Mega Circuit. *Source:* MPEDB (2011).

When one assesses the dynamics of NCS, being a tristate Sanctuary, it covers various districts (Fig. 12.14) of the three states and hence, the interventions, the thought processes, and the tourism products that can be evolved will surely add value to the way tourism/ecotourism can be promoted in the NCS area.

	States	Districts
1	Madhya Pradesh	Bhind, Morena, Sheopur, and Gwalior
2	Rajasthan	Bharatpur, Karauli, Sawai Madhopur, Kota, Baran, and Dholpur
3	Uttar Pradesh	Agra, Etawah, Jhansi, Firozabad, Auriya, and Jalaun

Fig. 12.14. Districts Coverage in the NCS Area of Operation.
Sources: NCS Conservation Plan (2023) and NGT (2023).

The authors made their best of efforts to ensure the district footprints where the Chambal and how they are making efforts to preservation, conservation, and protection of the sites; however, Yadav et al. (2016), while researching on the Sustainable Tourism Management (STM) impediments and opportunities in NCS, mention that though there are multiple districts that cover the Administrative mechanism of NCS across the three states, only three districts of Morena in Madhya Pradesh, Agra in Uttar Pradesh, and Sawai Madhopur in Rajasthan see tourism operations being done in a way desirable. One needs to get to the detailing of the tourism options that are available in each district of the tristates and create a plan to ensure, as has been done by the Forest Department of Madhya Pradesh, who have created the Management Plan (MP) that provides for a plan mechanism from 2023–2024 to 2027–2028 (NCS Conservation Plan, 2023). The District Forest Officer (DFO), Morena, has been designated as the Director for NCS and hence much of the orientation to plan and strategize rests with the DFO and the team.

Arun Rai (Rai, 2022), a former Manager at MPEDB, mentions "Chambal River, the Riverine, the flora and fauna that abounds and yes, an element of secrecy it has on account of it not being promoted as a tourism destination earlier, will create a multitude of options for promoting Sustainable Tourism and its development. MPEDB along with the other stakeholders, the Forest Department, the private sector should join in and play a role which will ensure the local communities participate and ensure the flora, fauna, the cultural and natural heritage are kept intact for the benefit of our generation and the future ones to cherish. Chambal in a way is a fresh wave to promote 'dark tourism,' juxtaposed with Ecotourism, which will attract multitude of visitors from near and far. By organizing events like Birding, and watching the Gharials and muggers from boats, it will be a treat to the visitors."

It is worth noting that the Bombay Natural History Society (BNHS), Mumbai (https://www.bnhs.org), has identified Chambal as one of the Important Bird Areas (IBA), which helps to identify the area of Chambal as an excellent indicator of the ecosystem on account of the good bird count (WII Envis, n.d.). This clearly means, besides the Gharial and the Mugger, it is an opportunity to promote Bird Festivals based on the seasons and ensure serious visitors come to study the Birds and document them. Arun Rai (Rai, 2022) also mentioned that, "during the migratory season in winter one can proceed to Rajghat (near Morena) and Barhi (near Bhind), where the Madhya Pradesh Forest Department (MPFD) had organized facilities for boating and nature trails."

C. S. Dubey (personal communication, March 17, 2021) former Deputy Conservator of Forest in the Government of Madhya Pradesh mentions that Chambal region is a space that can be promoted for a variety of tourism activities. We always think that Chambal River and its tributaries are no doubt a phenomenal base for ecotourism and Sustainable Tourism activities. The region also has opportunities for Cultural and Heritage tourism, which will attract the visitors from different places. It is an area loaded with legends, and stories that can keep visitors at the destinations for a long time. Considering that the region had issues of dacoits and came to be known as badlands, in an era bygone, the present and future offers opportunities that the tourism system should enable itself and ensure the local communities participate and use the opportunities for livelihood generation and become part of the tourism/ecotourism set up. A few entrepreneurs have set up resorts and accommodation facilities, but more needs to be done. The Forest Departments across the Chambal region, be it Madhya Pradesh, Rajasthan, and Uttar Pradesh, must work with the communities and create options for Eco-Development Committees (EDC) and facilitate wherever the forests are involved to promote Community-based ecotourism (CBE). This will ensure the management of the local resources, the flora, and the fauna. Further by training the local community members as Guides, and as associates in the resorts to manage the same, one can see opportunities for evolving a robust environment for conservation, preservation, and protection of the Chambal Region, which following the fall of the Dacoit (*Baghi*) era has seen a new type of mafias emerging, viz., the sand mafia, the stone mafia, and others.

R. J. Rao (personal communication, April 5, 2022), former Vice Chancellor of Barakatullah University, and a Scientist of Eminence, who is known for his work on the Ecology of aquatic animals in the NCS and who has worked with the Forest Department in a multitude of ways, mentioned that "the opportunity for Tourism, with community participation in the Chambal River Ecosystem will help not only the community, but also in preserving and conserving the fragile ecosystem. One needs to ensure systems such as the EDCs that will work along with the Gram Sabhas, which will go a long way to help augment not only quality in the tourism product, but provide for livelihood options to the locals." He further added, "the local community is aware of the presence of the Gharials, the muggers, the turtles, the skimmers, their nesting places, and the patterns they follow; hence it will be pertinent to take the support of the community and promote tourism in what otherwise was a dark area for tourism for more than fifty years."

Singh and Rao (2017), in a seminal research paper on conservation, mention success will depend on the acceptance and participation of the local population. Implementation of conservation initiatives must be achieved on a long-term basis through awareness, education, and the involvement of local people with an interest in the area. Under river basin management policy, as proposed by the Ramsar Convention, water abstraction should be such that it maintains the normal flow of the river in low flow seasons without affecting the hydrological functioning of the river system and aquatic life. Under the Indian Wildlife (Protection) Act (1972), any form of natural resource extraction from the protected areas is banned and an Environmental Impact Assessment (EIA) of all development projects close to these areas is mandatory. Strict implementation of this Act, coupled with integrated conservation planning, will save this important river system from imminent disaster. For sustainable biodiversity conservation, it is important to reduce dependence of local people on natural resources in terms of fuel wood, fodder, and reduction of the number of feral cattle in the region. This could be achieved by adopting the process of eco-development through people's participation, development of alternatives to natural resources, and an increase in the supply of biomass resources outside the PAs. In the NCS, micro plans for eco-development had been prepared but could not be implemented due to lack of coordination among stakeholders. But the environment today is such that the stakeholders due to various pressures want to create opportunities wherein the local community participates and ensures for the Chambal.

The Eco Park Centre (Fig. 12.15) at Deori (45 km from Gwalior on the highway toward Agra) is part of the NCS and a place where rejuvenation efforts for the Chambal River is happening in a scientific way. The MPFD centre has ensured that the Gharials, Crocodiles, and other species are taken care of and ensured protection. The center in fact helps to bread and rehabilitate the species and ensure protection and conservation. Speaking about the Eco Park Centre, a senior Forest Officer of Madhya Pradesh, from another division in the southern part of the state, mentioned, "I am here with my team to understand the way the Chambal fauna is being preserved and it is interesting to note that my colleagues in Deori are adding value to the finer aspects of preservation and conservation."

Fig. 12.15. Deori Eco Park Centre. *Source:* Images by the Authors.

A Scientist who has worked for three decades with MPFD while wanting to remain anonymous states, "the Eco Park is a labor of love for the MPFD and the all those involved. It is the best way we can give back to the Chambal River and the NCS. The preservation, conservation and protection is best seen at the Park, and one needs to carry on with the work given to preserve the fauna of Chambal for posterity." The Eco Park is a must visit to any visitor to the Chambal as they will be able to understand the unique scenario of the river and appreciate the fact human intervention will bring about destruction, and islands of excellence like the Eco Park will become the mainstay for the future of sustainable tourism, which also provides for ample training opportunities for the foresters themselves, the students, and the local community members.

A senior officer of MPFD located in Gwalior, T. S. Suliya (personal communication, August 18, 2023), while discussing with the authors mentioned, that, "the time is ripe for ushering in community-based tourism (CBT), as this will ensure local participation, who will also get livelihood options through tourism interventions." As was pointed out by a researcher, who wanted to remain anonymous, the best option, if the forest department intervenes is the creation of EDCs where the community, and the watchful eyes of the forest personnel and other stakeholders like the Gram Sabha (local Administration) and even the tourism department can join hands to promote experiential tourism, which will in the preservation, conservation, and protection of the Chambal River and the Sanctuary, which once was engulfed in the 'dark' era of the dacoits, can see the light of day, with tourism in the forefront.

Yadav et al. (2016), in deep-dive research paper, had provided a framework (Fig. 12.16) that one could elaborate and work on further, which will support and add value to the NCS system in a way forward to move from a dark era to an era

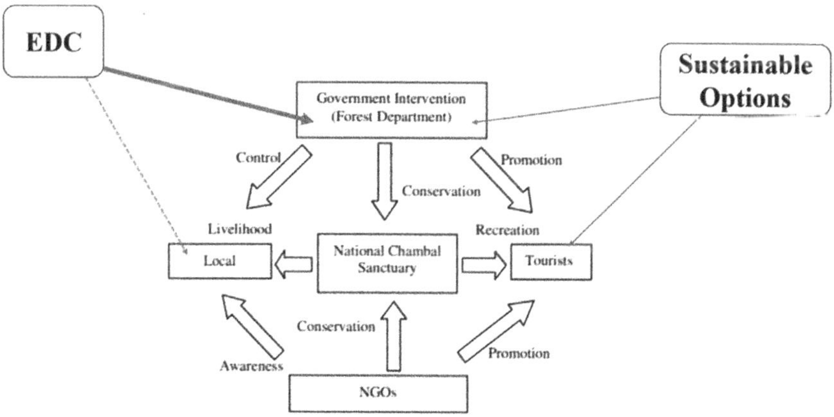

Fig. 12.16. Representative Framework of NCS Function System [the Authors Have Added Two Elements That Will Support the Community and the Issues of Sustainability]. *Source:* Yadav et al. (2016).

where tourism can be promoted with elan and one can ensure the visitors come to the Chambal River ecosystem, and get an experience that can be considered as truly incredible. The authors in their discussion with T. S. Suliya (personal communication, February 9, 2024), the Chief Conservator of Forests (CCF) for Gwalior Division, stated, "NCS is a phenomenal special purpose vehicle that the Government of the day ensured for the Chambal River. We will have to work with the local community members of all hues and ensure for the region. Chambal for long had been the fiefdom of the *Baghis*, and reasonably well protected, to evolve a good ecotourism practice, the local community should participate along with the MPFD and move ahead to build a product that blends with the Dark era of the past and juxtapose itself with the present."

References

Banerjee, P. (2017, December 7). Why Chambal's dacoits didn't ride into the sunset. https://www.hindustantimes.com/india-news/we-wouldn-t-come-charging-on-horses-like-they-show-in-bollywood-films-former-chambal-dacoits-remember-time-spent-as-bandits/story-1V8dXgX7BR50weQbN1VorJ.html

Bhot, A. M. (2022, November 6). How a dacoit and a Muslim archaeologist restored a forgotten Hindu temple complex. https://www.moneycontrol.com/news/trends/travel-trends/how-a-dreaded-dacoit-and-a-muslim-archaeologist-restored-a-forgotten-hindu-temple-complex-9458221.html

Borealis, C. (2021, August 16). Foreign language film submission to the 67th Academy Awards held in 1995 to get an Oscar nomination in the best non-English movie category. https://www.imdb.com/list/ls077719856/

Britannica. (2009, February 6). Gurjara-Pratihara dynasty. https://www.britannica.com/topic/Gurjara-Pratihara-dynasty

Business Standard. (2019, February 14). Chambal has moved on from dacoit culture but discrimination still exists: Abhishek Chaubey. https://www.business-standard.com/article/pti-stories/chambal-has-moved-on-from-dacoit-culture-but-discrimination-still-exists-abhishek-chaubey-119021400482_1.html

Chambal Safari. (n.d.). History of Chambal. https://www.chambalsafari.com/history.php#:~:text=An%20enraged%20Draupadi%2C%20on%20finding,an%20unquenchable%20thirst%20for%20vengeance

Choudhury, B. C., Singh, L. A. K., Rao, R. J., Basu, D., Sharma, R. K., Hussain, S. A., Andrews, H. V., Whitaker, N., Whitaker, R., Lenin, J., Maskey, T., Cadi, A., Rashid, S. M. A., Choudhury, A. A., Dahal, B., Ko, U. W. K., Thorbjarnarson, J., & Ross, J. P. (2007). Gavialis gangeticus. IUCN red list of threatened species, 2007. https://doi.org/10.2305/IUCN.UK.2007.RLTS.T8966A12939997.en

Dabas, M. (2016, September 10). 5 of the most dangerous dacoits who once ruled the Beehads of Chambal. https://www.indiatimes.com/culture/who-we-are/5-of-the-most-dangerous-dacoits-who-once-ruled-the-beehads-of-chambal-257186.html

Das, K. (2016, December 13). One Chambal, two tales: Why MP is hostile to river. https://www.hindustantimes.com/bhopal/one-chambal-two-tales-why-mp-is-hostile-to-river/story-HWQlLCfK4rPfvrfu54aqCJ.html

Deni, I. (2022, September 11). Top 10 most notorious dacoits of India. https://www.thedailyepisode.com/post/top-10-most-notorious-dacoits-of-india#:~:text=Daaku%20Maan%20Singh,killing%20of%2032%20police%20officers

District Bhind. (n.d.). Fort of Ater. https://bhind.nic.in/en/tourist-place/fort-of-ater/

District Morena. (n.d.). Chausath Yogini Temple, Morena. https://morena.nic.in/en/tourist-place/mitawali/

Fitzgerald, L. (2004). *The Mahabharata, the book of women, and the book of peace.* University of Chicago Press. ISBN: 9780226252506.

Ghose, S. (2018, November 14). Close encounters with dacoits of Chambal: In Chambal's badlands, former dakus say today's netas are the real 'high-tech dakus'. https://timesofindia.indiatimes.com/india/close-encounters-with-dacoits-of-chambal-in-chambals-badlands-former-dakus-say-todays-netas-are-the-real-high-tech-dakus/articleshow/66615788.cms

Gupta, R. (2022, February 6). Chausath Yogini Temple, Mitaoli: A temple that celebrates the cult of powerful women. https://www.cntraveller.in/story/chausath-yogini-temple-mitaoli-a-temple-that-celebrates-the-cult-of-powerful-women/

Hussain, S. A. (1993). Aspects of the ecology of smooth-coated otter (*Lutra perspicillata*) in the National Chambal Sanctuary, India. PhD thesis. Centre for Wildlife and Ornithology, Aligarh Muslim University. http://hdl.handle.net/10603/58828

Hussain, S. A., & Badola, R. (2001). Integrated conservation planning for Chambal River basin. In *Proceedings of the workshop on regional planning for wildlife protected areas. 2001 August 6–8*. India Habitat Centre.

Ibex Expeditions. (n.d.). Team Ibex. https://www.ibexexpeditions.com/ibex-our-story/mandip-singh-soin-frgs/

Ibex Expeditions. (n.d.A). Embracing the sustainable tourism criteria of India for the tour operator segment. https://webunwto.s3-eu-west-1.amazonaws.com/imported_images/38173/mandip_singh_soin_ecotourism_society_of_india.pdf

Kumar, P. K. (2023, April 23). Championing the cause of ecotourism in UP's Bundelkhand. https://hospitality.economictimes.indiatimes.com/news/hotels/championing-the-cause-of-eco-tourism-in-ups-bundelkhand/99773331

Madhya Pradesh Tourism. (n.d.). Ater Fort-bhind. https://www.madhya-pradesh-tourism.com/destination/Ater-Fort—Bhind-32#:~:text=Derived%20from%20the%20name%20of,visit%20this%20hill%20top%20fort

Maheshwari, K. (2022). *Resurrecting Chambal, a microscopic study of the heritage tourism.* White Falcon Publishing. ISBN: 9781636406176.

Management Plan. (n.d.). *National Chambal Sanctuary Madhya Pradesh, Management Plan (2018-19 to 2027-28)*. Madhya Pradesh Forest Department. Document available in the Office of District Forest Officer, Morena, Madhya Pradesh.

Mazoomdar, J. (2022, June 21). To end illegal mining, Madhya Pradesh plans to make it legal in parts of Chambal Sanctuary. https://indianexpress.com/article/india/end-illegal-mining-mp-legal-parts-of-chambal-sanctuary-7980795/

MoEF. (2011, January 24). *Ministry of environment and forests, wildlife division, summary records of the 21st meeting of the standing committee of National Board for Wildlife (NBWL) held on January 24th 2011.* https://forestclearance.nic.in/writereaddata/Order_and_Release/21.pdf

Mohan, L. (2010, April 18). Charms of Chambal. https://www.tribuneindia.com/2010/20100418/spectrum/main3.htm

MPEDB. (2011). *International Chambal Expedition – 2011*. Madhya Pradesh Ecotourism Development Board (MPEDB).

Nair, T. (2017, January 10). Of Government, God, and Gharial: The ecological pogrom in Chambal's Bandlands. https://thewire.in/environment/chambal-son-gharial-bansagar

Nair, T., & Krishna, Y. (2013). Vertebrate fauna of the Chambal River basin, with emphasis on the National Chambal Sanctuary, India. *Journal of Threatened Taxa*, *5*(2), 3620–3641. https://doi.org/10.11609/JoTT.o3238.3620-41Chaitanya.

Native Planet. (n.d.). Chambal Sanctuary photos. https://www.nativeplanet.com/chambal-santuary/photos/9573/

NCS Conservation Plan. (2023). *National Chambal Sanctuary Madhya Pradesh Conservation and Protection Plan (2023-24 to 2027-28)*. https://greentribunal.gov.in/sites/default/files/news_updates/Action%20Taken%20Report%20by%20State%20of%20M.P.%20in%20OA%20No.%20248%20of%202022%20(IN%20RE%20NEWS%20ITEM%20PUBLISHED%20IN%20THE%20HINDU%20DATED%2027.03.2022%20TITLED%20AS%20DIGGING%20UP%20THE%20CHAMBAL).pdf

NGT. (2022, April 5). Honorable National Green Tribunal Principal Bench. https://greentribunal.gov.in/sites/default/files/news_updates/Revised%20Factual%20and%20Action%20Taken%20Report%20by%20MPPCB%20in%20OA%20No.%20248%20of%202022%20(In%20re%20News%20item%20published%20in%20The%20Hindu%20dated%2027.03.2022%20titled%20Digging%20up%20the%20Chambal).pdf

NGT. (2023, July 20). Honorable National Green Tribunal Principal Bench. https://greentribunal.gov.in/sites/default/files/news_updates/Action%20Taken%20Report%20by%20State%20of%20M.P.%20in%20OA%20No.%20248%20of%202022%20(IN%20RE%20NEWS%20ITEM%20PUBLISHED%20IN%20THE%20HINDU%20DATED%2027.03.2022%20TITLED%20AS%20DIGGING%20UP%20THE%20CHAMBAL).pdf

Parameswaran, A. (n.d.). Chambal valley-protected by infamy, curses, and crimes. https://beyonder.travel/india/west-central-india/chambal-valley-legends/

Pezarkar, L. (2018, June 14). Chambal's haunting temples. https://www.peepultree.world/livehistoryindia/story/monuments/chambals-haunting-temples

Rafferty, P. J. (2018). Regolith. *Encyclopedia Britannica*. https://www.britannica.com/science/regolith

Raghuveer. (2023, June 10). Personal Communication. [Interview].

Rajras. (n.d.). Chambal River: Origin, tributatries, basin, dams, and concerns. https://www.rajras.in/rajasthan/rivers/chambal/#:~:text=Ancient%20name%20of%20Chambal%20is,bank%20of%20the%20Charmanwati%20river

Rai, A. (2022, March 12). Personal Communication. [Telephonic Conservation].

Reeves, R. R. (2009) Ganges dolphin declared national aquatic animal by Government of India: A great step forward. https://zoosprint.zooreach.org/index.php/zp/article/view/2300

Rongmei, P. (2023, May 14). Exploring the incredible 64 Yogini Temples in India. https://timesofindia.indiatimes.com/travel/destinations/exploring-the-incredible-64-yogini-temples-in-india/articleshow/100226156.cms#:~:text=At%20present%2C%20there%20are%20only,20%20km%20southeast%20of%20Bhubaneswar

RTSOI. (2021, March 29). Responsible tourism society of India. *Facebook*. https://m.facebook.com/ResponsibleTourismSocietyOfIndia/photos/a.1446687582293860/2553459801616627/?type=3&_rdr

Satellites Pro Morena. (n.d.). Morena map. https://satellites.pro/Morena_map.India#26.672349,78.090992,15

Sharath, L. (2018). Restoration of Bateshwar group of temples, KK Muhammed Sir interview. https://lakshmisharath.com/bateshwar-group-temples-archaeologist-kkmuhammed-interview/#:~:text=Restoration%20of%20Bateshwar%20Group%20of%20Temples%20-KK%20Muhammed%20Sir%20Interview&text=It%20was%20an%20"inner%20calling,the%209th-century%20temple%20complex

Singh, H., & Rao, R. J. (2017). Status, threats, and conservation challenges to key aquatic fauna (Crocodile and Dolphin) in National Chambal Sanctuary, India. *Aquatic Ecosystem Health and Management*, *20*(1–2), 59–70. https://doi.org/10.1080/14634988.2017.1298964

Standage, K. (2019, April 23). Kakanmath temple-defying the laws of gravity. https://kevinstandagephotography.wordpress.com/2019/04/23/kakanmath-temple-defying-the-laws-of-gravity/

Sutikshan. (2020, October 15). Chambal Ravines, a badland topography in the heart of Chambal River, take a geo-tourism around the unknown terrain. https://thejerker.com/chambal-ravines-bandland-topography-in-the-heart-of-chambal-river-take-a-geotourism-around-the-unknown-terrain/

Temples of India. (n.d.). Kakanmath temple. https://templesofindia.org/temple-view/kakanmath-temple-morena-madhya-pradesh-594kra

The Lallantop. (2021, March 28). Beehad 2.0 documentary, Chambal-A Bygone Era, Saurabh Dwivedi, Rajat San and Roohani [video]. https://www.youtube.com/watch?v=oCIoVRK0YDQ 20:16

Travel Trends. (2021, September 6). The many lives of Chambal River. https://timesofindia.indiatimes.com/travel/destinations/the-many-lives-of-chambal-river/articleshow/85976378.cms

Water Database. (n.d.). Chambal River. https://www.waterdatabase.com/rivers/Chambal-river/

WII Envis. (n.d.). Important bird areas (IBAs) in India, Wildlife Institute of India - Envis. https://wwwenvis.nic.in/database/iba_8463.aspx

Yadav, N. (2019, April 29). This temple in Chambal ravines stands out as ancient replica of Parliament House. https://www.businesstoday.in/pti-feed/story/this-temple-in-chambal-ravines-stands-out-as-ancient-replica-of-parliament-house-189171-2019-04-29

Yadav, N., Sahu, N. C., & Sahoo, D. (2016). Sustainable tourism management in the National Chambal Sanctuary, India: Impediment and opportunities. *International Journal of Innovation, Management and Technology*, *7*(5), 177–182.

Chapter 13

Marketing Strategies of Dark Tourism: Social Media

Burhan Kılıç[a] *and Hande Akyurt Kurnaz*[b]

[a]Muğla Sıtkı Koçman University, Turkey
[b]Bolu Abant Izzet Baysal University, Turkey

Abstract

Dark tourism is one of the tourism types that have developed in line with the expectations of tourists in recent years. One of the special interest tourism types, dark tourism is based on fears such as pain, fear, and sadness. From this point of view, dark tourism destinations include battlefields, execution scenes, areas where death is experienced firsthand, centers with horror themes, and regions where natural and technological disasters occur. The centers where dark tourism takes place are quite different from the centers where other types of tourism take place. Thus, there is also diversity in the tourist type. Among the travel motivations of Dark tourists are reasons such as experiencing the moment of death, feeling fear, respecting the past, and obtaining information. With the advancement of technology, the reach of dark tourism to tourists has also accelerated. Today, dark tourism destinations reach large masses by using various marketing strategies. With this information, the aim of this study is to examine marketing strategies within the scope of dark tourism. Within the scope of this examination, social media will be emphasized. Social media accounts of destinations based on dark tourism will be examined and data will be analyzed through content analysis. Thus, it is aimed to have a rich section within the scope of dark tourism and social media.

Keywords: Dark tourism; social media; marketing; marketing strategies; Türkiye

Dark Tourism, 175–183

doi:10.1108/978-1-83797-336-120241013

Introduction

Dark tourism, which has become a popular type of tourism and travel preference in recent years, is a tourism concept with a very deep meaning. Dark tourism is a popular field of study due to the different elements in its content, diversification of tourist types, and differentiation in terms of travel motivations. Dark tourism, one of the special interest tourism types, is a niche market area. From this point of view, it is among the main tourism activities in terms of marketing. Marketing activities are very important in the tourism sector. Thanks to the appropriate planning of marketing activities, success in promotional processes is inevitable. Thus, the expected tourist flow can be achieved. This chapter is important in terms of examining both tourism marketing and dark tourism, which is one of the special interest tourism types. Within the scope of this chapter, firstly, the concept of dark tourism is tried to be explained and the concept of marketing is mentioned. The main theme of this chapter is dark tourism examples within the scope of social media marketing. The topics were tried to be reinforced with the examples researched. In order to reveal the difference of this study, the study named Dark Tourism as a Social Media Marketing Product was included. In line with the results obtained from the study, the contents of dark tourism attractions within the scope of social media are mentioned.

Dark Tourism

Dark tourism is a type of tourism fed by history. Cultural products that occurred in the historical process are among the subjects of dark tourism (Akyurt Kurnaz & Kılıç, 2020). As a word meaning, dark tourism comes from Thanatourism, which is derived from the ancient Greek word thanatos or death personality (Cristiana, 2013). Dark tourism was introduced to the international literature by Foley and Lennon in the 1990s. It literally refers to tourism attractions that are directly related to death or indirectly related to death (Cohen, 2010; Dale & Robinson, 2011; Foley & Lennon, 1996; Lennon & Foley, 1999). Dark tourism has become one of the increasing areas of study in the research literature. According to the studies, it has been observed that male tourists are more interested in dark tourism than female tourists. Dark tourism is a very complex type of tourism. This is because it covers areas and topics such as culture, disaster, education, natural disasters, executions, massacres, genocides, accidents, or supernatural events. It is stated that tourists who prefer dark tourism are thrill hunters. The concept of dark tourism has a special place among tourism types. Dark tourism destinations or attractions are needed in the historical process. Because the events that took place in the historical process should be known by tourists, it is thanks to dark tourism that people think about the events that took place (Kulcsár & Sımon, 2015) (Table 13.1).

Cemetery and graveyard visits (Seaton, 2002), genocide sites (Cole, 1999), atrocity sites (Ashworth & Hartmann, 2005), prisons (Strange & Kempa, 2003; Wilson, 2008), and slavery heritage attractions (Dann & Seaton, 2001; Rice, 2009) have formed the content of this special interest tourism (Table 13.2).

Table 13.1. Definitions of Dark Tourism.

Person	Description
Seaton (1996)	Travel to a place motivated by the desire for real or symbolic encounters with death
Foley and Lennon (1997)	A visit to any site associated with death, disaster, and tragedy
Dann and Seaton (2001)	Inheritance and death-related attractions
Tarlow (2005)	Visiting places where tragedies or deaths have occurred and continue to affect life
Preece and Price (2005)	Journey to places associated with death, acts of violence, tragedy, death scenes, and crimes
Stone (2006)	Death, suffering, and the movement to travel to seemingly eerie sites
Robb (2009)	Places to visit where violence is the main attraction
Goulding and Pressey (2023)	Includes destinations featuring sites of disaster and death

Source: Akyurt Kurnaz & Kılıç (2020).

Table 13.2. Examples of Dark Tourism Destinations.

Destinations	Destinations	Destinations
Alamo Castle	Graceland	Nagasaki
Aliya Izzetbegovic Tomb	Ground Zero	Norfolk Island
Alma Bridge	Hiroşima	Normandy Landings
Anne Frank House	Hoa Lo Prison	Kovachi Martyrs' Cemetery
Anıtkabir	Hocalı Massacre	Paris Catacombs
Arlington National Cemetery	Hothell	Pearl Harbor
Auschwitz-Birkenau	Island of the Dolls	Père – Lachaise
Barents Denizi	Jack the Ripper Tours	Point Hicks Lighthouse
Batepá Massacre	Kennedy Museum	Pompeii
Belchite	Kigali Genocide Memorial Center	The Tomb of Princess Diana

(Continued)

Table 13.2. *(Continued)*

Destinations	Destinations	Destinations
Bran Castle	Kudüs via Dolorosa	Dolorosa SaoTome and Principe Island
Changi Beach	Latvian Prison Hotel	Sarıkamış
Chapel of Bones	Malta Catacombs	Sedlec Ossuary
Chernobyl	Museum of the House of Terror-Hungarian	Sinop Prison
Çığlık Tunnels	Leap Castle	Sonora-Meksika Witch Market
Coronado Hotel	Tower of London	Squatter Camp Soweto
Cu Chi (Ku Çi) Tunnels	Madımak Müzesi	Srebrenica Massacre
Dachau	Swamp of Mancha	Chile Mine Accident
De La Guillotine	Mary King Church	The Suicide Forest
Dharavi Slum	Massada Castle	The Winchester House
Eastern State Reformatory	Monte Cristo	Titanic Belfast
Dolmabahçe Palace	Morro Castle	Truk Lagoon
Easter Island	Museum of Genocide Victims	Tuol Sleng Museum
Edinburgh Castle	Mutler Museum	Ulucanlar Prison
Gallipoli Campaign	Myrtles Plantation	Cyprus Maras

Source: Akyurt Kurnaz and Kılıç (2020)

According to the study of Dunkley et al. (2011), it is possible to diversify the travel motivations of cruise tourists. These are

- visiting centers such as cemeteries and cemeteries;
- experiencing excitement and risk;
- events such as crime, disaster, or murder;
- identify the real presence of a person or place;
- self-discovery and learning;
- visit sites with immortalized historical events;
- recent visits to overcome curiosity in disaster-stricken areas; and
- religious reasons.

When travel motivations are analyzed, the reasons for traveling of dark tourists are different from other tourists. Intrinsic or spiritual reasons are more common in this type of tourists. In line with the information given, it is understood that dark tourism areas are increasing day by day, travel motivations are differentiated, and it is an interesting type of tourism. It will be possible to increase the number of attractions within the scope of future research.

Marketing Strategy and Digital Marketing

Marketing is defined by Kotler as the science and art of discovering, creating and delivering value to meet the needs of a target market at a profit. As mentioned in the definition, marketing is an art. In recent years, the marketing discipline has changed considerably. Marketing has become one of the most frequently discussed topics (Kotler, 1983). The most basic definitions of the concept of marketing over the years are given in the table. Tourism is one of the important sectors in the marketing concept. It is necessary to pay attention to marketing activities in tourism and to give importance to tourist demand.

Competition is the most fundamental element of marketing. Therefore, businesses need to accept competition in order to survive (White et al., 2019). In this respect, it is seen that new connotations have emerged in the concept of marketing. These new connotations have been the concept of digital marketing, which emerged with the internet and technology and spread rapidly. In recent years, with the emergence and use of social media networks with the development of technology (Shi et al., 2021; Yoon et al., 2022), marketing strategies have also changed. Marketing strategy has moved from offline (e.g., magazines, TV programs, etc.) to online (e.g., social media platform marketing). Thus, digital marketing has started to be used frequently (Kumar et al., 2022; Vaishnav & Ray, 2023).

As the marketing process has changed, digital marketing has become a field used by almost every destination and business. The concept of digital marketing is used to refer to all marketing activities that take place online. The important concept in this context is that it is online. In recent years, businesses have been using digital marketing to collaborate with their customers. In addition, digital marketing is also very important in terms of promotional activities. Thanks to digital marketing, promotion is becoming widespread (Dimitrios et al., 2023).

Another name for digital marketing is online marketing. It is a common marketing tool with technology in the contemporary marketing process. It is an internet-based marketing activity. Search engines, social media, e-mail, and websites used in this type of marketing are used in the digital marketing process. These systems are used to communicate with consumers (Alexander, 2022).

Digital marketing has completely changed the marketing sector. With this type of marketing, a consumer-oriented system has emerged. With more options, the demands of consumers have increased even more. Unlike the traditional marketing system, digital marketing has become an experience-oriented strategy (Dimitrios et al., 2023). Digital marketing is seen as a system that consumers trust. It is an important concept in terms of business and brand awareness (Alexander, 2022).

Tourism is a set of financial, social, and societal activities created by individuals who travel. With the advancement of technology, tourism and technology have started to move together. Digital applications establish commercial links over the internet to procure tourism-related items/products such as flights, hotel reservations, and car rentals, allowing businesses to increase their overall efficiency and effectiveness (Li et al., 2023).

Social media, one of the digital marketing tools, is one of the important mediators of tourist experience. In recent years, social media has increased the recognizability of tourism destinations, influenced tourists' travel decisions, and contributed to promotional activities. With the development of information technologies, social media has become an important means for tourists to access destination information and interact closely with tourism destinations (Kim et al., 2017). Social media has influenced not only the way travelers search, read, and process travel information, but also their attitudes, decision-making processes, and evaluation of products and services (Liu, 2024).

Dark Tourism Tripadvisor-Turkey Case in Social Media Analysis

Objective

It is planned to conduct a research that supports the literature within this chapter. In this context, the aim of the study is to identify popular sadness tourism centers in Turkey and to analyze their comments. Thus, popular sadness tourism destinations in Turkey were revealed. In addition, the purpose of the research was completed through Tripadvisor, a social media network.

Method

Tripadvisor is an online commenting site with national and international tourist reviews. It is an important qualitative study network due to the inclusion of tourist experiences. Within the scope of this study, Tripadvisor portal was searched with the keyword sadness tourism. The two results that emerged were stated as touristic attractions that are considered as sadness tourism in terms of Turkey. In the study conducted by Güzel and Kurnaz (2022), Tripadvisor comment site was also used and important research results were obtained.

Sinop Prison was the selected tourist attraction. The 423 comments on Sinop Prison are organized in date order. In this respect, Sinop Prison, which has a high number of comments, was taken into consideration. As in Güzel and Kurnaz (2022), the date order was made from old to new. The comment years started from 2021 for Sinop Prison and comments covering the year 2019 were taken.

Descriptive analysis was used in the analysis phase. Direct quotations were included with descriptive analysis. As a result of the coding and direct quotations, Sinop Prison Museum, which is Turkey's attractions within the scope of sadness tourism, was evaluated.

Findings

Important information was obtained through Tripadvisor, an online experience sharing network. When the information obtained was evaluated, especially the structure of sadness was emphasized for Sinop Prison.

> Sinop among historical cities. The city we wondered about for years, the city we wanted to visit and see. It was very sad here.

Sinop Prison, which is a center of sadness tourism, is also a very valuable element in terms of Turkey's history. When the analyzed comments are examined, there are 423 total comments; 185 of these comments (43%) were characterized as excellent and 138 comments were evaluated as very good (32%).

> Take your time when you visit this place. Imagine how people lived, what they suffered, how they were tormented.
>
> It was a very sad trip to Sinop Prison...The more you think about what happened here, the more it makes you feel...Everywhere there is a different sadness, a different story.
>
> The cold walls get under your skin. You wonder how people lived here. It leaves you with a bitter ache...

The common features obtained from the comments have been the subject of history. Other features mentioned in the comments are;

- Sad;
- Sabahattin Ali;
- Gloomy;
- Important Dark Tourism Center;
- Sadness;
- It is stated as a must-see center.

Conclusion

In recent years, sadness tourism has become a popular type of tourism. In this respect, the number of sadness tourism centers has also increased considerably. Promotional activities continue to accelerate with technology. It is useful to promote tourism attractions with social media elements. Social media elements where tourists' experiences are described are followed by people. Thus, the decision-making stages of potential tourists are also accelerated. As a result, social media is important in terms of promoting tourism products and announcing them to large masses. Sad tourism destinations consist of centers that affect people. It reveals a different aspect of tourism. As a result of the emergence of people's feelings of curiosity, the promotion of sad tourism centers in social media elements is a very popular topic.

References

Akyurt Kurnaz, H., & Kılıç, B. (2020). *Hüzün turizmi*. Nobel Akademik Yayıncılık.

Alexander, L. (2022). The who, what, why, & how of digital marketing. https://blog.hubspot.com/marketing/what-is-digital-marketing#whatis. Accessed on September 19, 2023.

Ashworth, G. J., & Hartmann, R. (2005). *Horror and human tragedy revisited: The management of sites of atrocities for tourism*. Cognizant.

Cohen, E. H. (2010). Educational dark tourism at an in populo site: The Holocaoust Museum in Jerusalem. *Annals of Tourism Research*, *38*(1), 193–209.

Cole, T. (1999). Constructing the Jew, writing the Holocaust: Hungary 1920-45. *Patterns of Prejudice*, *33*(3), 19–27.

Cristiana, P. I. (2013). Dark tourism – A niche segment for the Romanian tourism. *Romanian Economic and Business Review*, *8*, 351–358.

Dale, C., & Robinson, N. (2011). Dark tourism. In P. Robinson & S. Heitmann (Eds.), *Research themes for tourism* (pp. 205–218). CABI.

Dann, G., & Seaton, A. (2001). *Slavery, contested heritage and thanatourism*. Haworth Hospitality Press.

Dimitrios, B., Ioannis, R., Angelos, N., & Nikolaos, T. (2023). Digital marketing: The case of digital marketing strategies on luxurious hotels. *Procedia Computer Science*, *219*, 688–696.

Dunkley, R., Morgan, N., & Westwood, S. (2011). Visiting the trenches: Exploring meanings and motivations in battlefield tourism. *Tourism Management*, *32*(4), 860–868.

Foley, M., & Lennon, J. J. (1996). JKF and dark tourism: A fascination with assassination. *International Journal of Heritage Studies*, *2*(2), 198–211.

Foley, M., & Lennon, J. (1997). Dark tourism: A ethical dilemma. In M. Foley, J. J. Lennon, & G. A. Maxwell (Eds.), *Hospitality, tourism and leisure management: Issues in strategy and culture* (pp. 153–164). Cassell.

Goulding, C., & Pressey, A. (2023, July). A palimpsestic analysis of atmospheres at dark tourism. *Annals of Tourism Research*, *101*, 103577.

Güzel, Ö., & Kurnaz, H. A. (2022). Gastronomi rehberliği. In A. Kurnaz & H. Akyurt Kurnaz (Eds.), *Gastronomi Rehberliği*. Detay Yayıncılık.

Kim, S. E., Lee, K. Y., Shin, S. I., & Yang, S. B. (2017). Effects of tourism information quality in social media on destination image formation: The case of Sina Weibo. *Information & Management*, *54*(6), 687–702.

Kotler, P. (1983). *Principles of marketing* (2nd ed.). Prentice-Hall, Inc.

Kulcsár, E., & Sımon, R. Z. (2015). The magic of dark tourism. *Management and Marketing Journal, University of Craiova, Faculty of Economics and Business Administration*, *0*(1), 124–136.

Kumar, A., Paul, J., & Unnithan, A. B. (2022). 'Masstige' marketing: A review, synthesis and research agenda. *Journal of Business Research*, *113*, 384–398.

Lennon, J. J., & Foley, M. (1999). Interpretation of the unimaginable: The U.S. Holocaust Memorial Museum, Washington, D.C. and dark tourism. *Journal of Travel Research*, *1*, 38–46.

Li, P., Zhou, Y., & Huang, S. (2023). Role of information technology in the development of e-tourism marketing: A contextual suggestion. *Economic Analysis and Policy*, *78*, 307–318.

Liu, S. (2024). The impact of social media on communication and popularity in the fashion industry. *Lecture Notes in Education Psychology and Public Media*, *41*, 84–93.

Preece, T., & Price, G. (2005). Motivations of participants in dark tourism: A case study of Port Arthur. In T. C. Ryan, S. Page, & M. Aitken (Eds.), *Taking tourism to the limits: Issues, concepts and managerial perspectives* (pp. 191–197). Elsevier.

Rice, A. (2009). Museums, memorials and plantation houses in the Black Atlantic: Slavery and the development of dark tourism. In R. Sharpley & P. R. Stone (Eds.), *The darker side of travel: The theory and practice of dark tourism* (pp. 224–246). Channel View Publications.

Robb, E. (2009). Violence and recreation: Vacationing in the realm of dark tourism. *Anthropology and Humanism*, *34*(1), 51–60.

Seaton, A. V. (1996). Guided by the dark: From thanatopsis to thanatourism. *International Journal of Heritage Studies*, *2*(2), 234–244.

Seaton, A. V. (2002). Thanatourism's final frontiers? Visits to cemeteries, churchyards and funerary sites as sacred and secular pilgrimage. *Tourism Recreation Research*, *27*, 73–82.

Shi, X., Yao, S., & Luo, S. (2021). Innovative platform operations with the use of technologies in the blockchain era. *International Journal of Production Research*, *61*(11), 3651–3669.

Stone, P. R. (2006). A dark tourism spectrum: Towards a typology of death and macabre related tourist sites, attractions and exhibitions. *Tourism*, *54*(2), 145–160.

Strange, C., & Kempa, M. (2003). Shades of dark tourism Alcatraz and Robben Island. *Annals of Tourism Research*, *30*(2), 386–405.

Tarlow, P. (2005). Dark tourism: The appealing dark side of tourism and more. In M. Novelli (Ed.), *Niche tourism: Contemporary issues, trends and cases* (pp. 47–57). Elsevier.

Vaishnav, B., & Ray, S. (2023). A thematic exploration of the evolution of research in multichannel marketing. *Journal of Business Research*, *157*(2023), 113564.

White, K., Habib, R., & Hardisty, D. J. (2019). How to SHIFT consumer behaviors to be more sustainable: A literature review and guiding framework. *Journal of Marketing*, *83*(3), 22–49.

Wilson, J. Z. (2008). *Prison: Cultural memory and dark tourism*. Peter Lang Publishing.

Yoon, J. Y., Lee, C., Choi, J., Chang, S. R., & Kim, J. (2022). The effect of social media apps on shopping apps. *Journal of Business Research*, *148*, 23–32.

Chapter 14

Enlightening the "Dark" in Dark Tourism: Re-Conceptualising Dark Tourism in the Era of Late Capitalism

Max Hart, Craig Kelly and Adam Lynes

Birmingham City University, UK

Abstract

Over the past 50 years, dark tourism has seen exponential growth in terms of both physical and digital contexts. Dark tourism is primarily a concentration around documented accounts of physical violence, and theorizations centered on dark tourism studies have generally fallen within either behavioral or interpretivist perspectives. Such perspectives are indicative of the continually evolving nature of dark tourism and its receptiveness to new definitions, conceptual frameworks, and theorizations. Taking this into consideration, this chapter seeks to develop and broaden the notion of "dark tourism" within the era of late capitalism by presenting fresh theoretical perspectives stemming from critical criminological frameworks. Specifically, in drawing upon critical notions of violence and the emerging deviant leisure framework, this chapter will aim to instigate fresh academic enquiry into the nature of dark tourism, expand its theoretical underpinnings, and subsequently provide a means in which to examine how banal forms of tourism play an integral part in the proliferation of some of the most serious harms that populate the contemporary neoliberal landscape.

Keywords: Dark tourism; late-stage capitalism; violence; harm; deviant leisure

Introduction

Dark tourism primarily centers on well-documented and often widely publicized instances of physical violence that disrupt the ordinary course of day-to-day life (Žižek, 2008). When delving into the realm of dark tourism, one encounters a

Dark Tourism, 185–204

Published under exclusive licence by Emerald Publishing Limited
doi:10.1108/978-1-83797-336-120241014

strong emphasis on this form of bodily or, as Galtung (1969) defines, direct violence. This emphasis can be observed, for instance, in renowned attractions associated with Victorian-era serial killer "Jack the Ripper" in London, the so-called "murder castle" attributed to H.H. Holmes in Chicago. Additionally, there are locations dedicated to showcasing bodily violence, such as the Alcatraz East Crime Museum in Tennessee, the Museum of Death in Los Angeles, and the Museum of Death in New Orleans. Also, within this remit, the Auschwitz concentration camps in Poland, the Choeung Ek killing field in Cambodia, and the Jeju 4:3 memorial in South Korea have been established as commemorations of violent historical occurrences. Furthermore, within this sector, there are well-known destinations related to prison tourism, exemplified by the Karosta prison hotel in Latvia and Dartmoor prison in Devon. Academic discussions commonly characterize dark tourism as encompassing locations associated with death and grief worldwide, attracting tourists with a wide range of motivations (Bathory, 2018), and 'those tourists who [are] interested in the memorialization of the dead, who were concerned with historical atrocity and evil' (Hooper, 2016, p. 2). Sharpley (2009) observes that individuals have long been drawn to locations or occurrences connected, in some manner, to death, calamity, and anguish. According to Sharpley (Sharpley, 2009), the expansion of general tourism over the last five decades has paralleled the substantial growth of dark tourism, both in physical destinations and the digital realm. This surge in the creation of dark tourist attractions has, naturally, sparked considerable academic examination. For example, scholarly literature typically investigates apparent ethical dilemmas (Lisle, 2004), marketing-related issues (Seaton, 1996), challenges in interpretation (Cooper, 2006), and site management matters, including the need to ensure respect for the victims (Blom, 2000).

While these are all important and valid avenues of enquiry, this chapter aims to enhance the existing critique by incorporating significant viewpoints from modern criminological research to expand the examination of dark tourism. This expansion is particularly relevant as we progress toward the apocalyptic zero point, marked by growing inequality and environmental deterioration (Žižek, 2010). We have so far briefly outlined the broad conceptual threads that bring together most academic work concerned with dark tourism in which violence, death, and tragedy are central. Nonetheless, Sharpley (2009) acknowledges that the field is ready for additional conceptual exploration (p. 12). This chapter, in turn, endeavors to contribute to this advancement by examining and elaborating on the connection between dark tourism and violence at a deeper philosophical level. In doing so, this chapter argues that some of the most banal and culturally accepted forms of tourism generate far more harm than the orthodox examples outlined above. To locate and capture such harms, we will also need to look outside of conventional dark tourism theorizations and toward the emerging deviant leisure perspective. Specifically, its concentration on the adverse effects of commonplace and culturally ingrained forms of commercialized leisure offers a pathway to transcend normative accounts and theorizations related to dark tourism.

This chapter will commence by establishing the essential background in the remit of dark tourism, paying particular attention to the various conceptions and interpretations of this term. Subsequently, our focus will shift to prior theories concerning dark tourism, with specific emphasis on the two primary perspectives: the behavioral and interpretive approaches. We will then introduce critical viewpoints on violence and the deviant leisure framework to provide the requisite theoretical underpinnings for our later discussion.

Dark Tourism: The Usual Suspects

The public's fascination surrounding death has long been ingrained within our society before an 'official' definition of dark tourism was developed (Foley & Lennon, 1996). Early illustrations of death-related tourism included the Roman colosseums for gladiatorial games and the turnout crowds for public executions in medieval central London (Stone, 2006), making this type of tourism a phenomenon that has become both widespread and diverse over the last century (Ashworth & Isaac, 2015). Smith (1998), for instance, suggests that sites or destinations associated with war constitute the 'largest single category of tourist attractions in the world' (p. 205). However, war-related attractions, yet diverse, remain one of the numerous tourist destinations associated with death and suffering (Stone, 2006). From graveyard tours (Raine, 2013), the holocaust (Heřmanová & Abrhám, 2015), sites of devastation (Sharpley, 2005), and the macabre (Ibrahim, 2015), the dark tourism industry continues to thrive due to its ability to connect people to the past through objects, exhibits, spaces, and dramatic recreation (McDaniel, 2018). Therefore, there has been a growing recognition that it is important to consider the social, cultural and geographical contexts in which people consume dark tourism and its theoretical underpinnings (Hohenhaus, 2013).

Despite growing awareness, academic enquiry into dark tourism tends to be 'theoretically fragile' (Hooper, 2016, p. 4), a fact with which we would concur. Studies into dark tourism are highly 'eclectic' (Hooper, 2016), with terms such as 'thanatourism' (Seaton, 1996), 'morbid tourism' (Blom, 2000), 'black spot tourism' (Rojek, 1993), and 'fright tourism' (Bristow & Newman, 2004) all being proposed and interjected into the dark tourism vernacular. Stone (2006) perhaps provides the most succinct classification that synthesizes these seemingly disparate terminologies into a coherent definition, which is the 'act of travel to sites associated with death, suffering and the seemingly macabre' (cited in Sharpley, 2009, p. 10). Dann (1998) provides some distinction to this somewhat broad definition, generating a series of categorizations that distil dark tourism to 'Perilous Places' (i.e., dangerous destinations), 'Houses of Horror' (i.e., dungeons of death), 'Fields of Fatality' (i.e., the holocaust), 'tours of torment' (i.e., locations where the murder took place), and 'Themed Thanatos' (i.e., morbid museums). While such categorizations provide some distinctions between such forms of tourism, they again provide minimal attention toward theoretical considerations and such works of largely, as stated by Sharpley (2009), descriptive in nature.

There is also confusion in dark tourism studies as to whether a location being purposefully produced for the consumption of death makes it "darker" or "less dark." Sharpley (2005) argues that the deliberate classifying of encounters intended to satisfy a wish to consume death is much darker. Conversely, Stone (2006) suggests that these types of sites are often highly mediated, and the darkest experiences are conversely more authentic in comparison to these mediated forms. As noted by Denham (2023), in online spaces, we have a meeting of these two opposing stances – with user-generated content appearing in almost real time being extremely organic but curated in a contrived and artificial 'space' and setting. Stone (2006) refers to dark tourism as 'a diverse and fragmented [...] product' (p. 157) – and this is perhaps best illustrated and evidenced within the fractured spaces of the internet. Certainly, the internet is a place that some users go to recreationally consume death – and by that definition, we can understand some internet spaces as dark tourist attractions within the traditional sense. The internet has been studied alongside dark tourism principally as a mode to improve or complement the dark tourist experience for traditional, physical sites (Bolan & Simone-Charteris, 2018), but there is scope to consider the internet's role more broadly, 'as a destination in and of itself' (Denham, 2023). However, despite a long history and evidence to support societies' eagerness to visit such places and now digital locations (Tarlow, 2007), our collective understanding of what dark tourism encompasses remains delicate and lacks cohesion.

Dark Tourism: Existing Theorizations

In the realm of dark tourism, various theoretical perspectives have been proposed, including the interpretivist and behavioral viewpoints. The interpretivist perspective suggests that these sites require an emotionally charged interpretation of a war or conflict to convey a specific event's "authentic" significance or meaning (Uzzell, 1989) more effectively. This perspective has also given rise to further discussions concerning stakeholders (Ashworth, 1996), debates about historical accuracy (Muzaini et al., 2007), and the distortion and displacement of the historical legacy associated with such locations (Tunbridge & Ashworth, 1996).

In contrast, the behavioral position focuses on tourists' motivations rather than the specific characteristics of locations. It introduces the concept of a "continuum of intensity," which is contingent on the reasons for visiting specific sites and whether those reasons are general or personal, such as visiting with a relative (Seaton, 1996). Within this framework, several motivations have been identified, including morbid curiosity and a shared sense of identity 'in response to disruptive events in collective life routines' (Rojek, 1997, p. 61).

Seaton (1996) categorizes dark tourism activities into five distinct categories, which include traveling to witness public representations of death, visiting sites where individual or mass deaths have occurred, touring memorial or internment sites, exploring symbolic representations of death and re-enactments of specific events, and observing re-enactments of death. Sharpley (2009) highlights that this array of diverse theoretical perspectives has 'facilitated the deconstruction of many overarching assumptions related to dark tourism' and has laid the groundwork for further conceptual analysis (p. 12).

Beyond the Physical: Critical Notions of Violence

We will now move attention away from previous theorizations to the importance of violence. Robb (2009) stresses that dark tourism primarily consists of 'visiting destinations at which violence is the main attraction' (p. 51). So, too, Gillen (2018) highlights the importance of violent performances and enactments of violence to dark tourist spots. We will now critically examine the pivotal role of violence. To be clear, this chapter does not aim to undermine its significance, but rather seeks to broaden our comprehension of the essence of violence within the vocabulary and conceptual framework of dark tourism studies. In doing so, it will seek to shine a light on those forms of tourism that have been, ironically, left in the dark.

Ray (2011) states that the social context in which the enactment and comprehension of violence occur holds significant relevance. Stanko's (2001) frequently referenced definition of violence, which pertains to an individual who inflicts or poses a threat of 'physical, sexual, or psychological harm to others or themselves' (p. 316), is commonly employed in scholarly discussions. Notably, it becomes evident that this conceptualization of violence forms the foundation for most conventional manifestations and, thus, academic discussions on dark tourism.

Inclusive interpretations of violence strive to encompass types of harm that may be significant without necessarily being physical in nature. Galtung's (1969) concept of 'structural violence' is particularly noteworthy in this context. According to Galtung's framework, elements such as job insecurity and the reduction of welfare entitlements can all fall within the category of structural violence. Žižek (2008) further elaborates on the notion of 'structural violence.' In his characteristic dialectical reversal, he redirects our attention not toward forms of violence that disrupt what is considered a civilized norm but rather toward the actual formation and perpetuation of this civilized norm itself. Such violence is necessary for the continuation of consumer-driven societies. Therefore, these forms of violence should be regarded as a benchmark against which we can evaluate other types of violence (in Žižek's tripartite schema, these are subjective violence and symbolic violence). Subjective violence constitutes direct violence committed by a clearly identifiable individual or entity, while symbolic violence encompasses symbols, signs, words, and narratives that signify the implicit, unacknowledged violence inherent in capitalist intersubjective domination. Žižek employs Lacan's concept of the symbolic to elucidate an implicit system of authority and oppression that transcends subjective (direct) manifestations of violence. As pointed out by Recuero (2015, p. 1), while objective violence is readily discernible within the backdrop of "normality," it is precisely within this backdrop that symbolic violence exists, upholding the current status quo through language. This Žižekian perspective holds significant potential for expanding our understanding of what constitutes dark tourism and the motivations behind tourists' visits to such destinations.

Deviant Leisure

Within the field of criminology, it is essential to utilize critical theoretical frameworks capable of addressing harms that extend beyond conventional understandings of both crime and deviance. The term "deviance" often refers to behaviors that deviate from societal norms. In contrast, beyond these normative views of deviance, the concept of "leisure" in a consumer capitalist society has evolved from being considered a 'moral good' to a 'moral right' (Raymen & Smith, 2019, p. 122). Engaging in leisure activities allows us to exercise individual agency and express internal desires on cultural, economic, and political levels (Rojek, 2010). This sentiment has also been reflected in previous criminological research, where leisure took the forefront only when leisure behaviors crossed legal boundaries or when scholars erroneously identified proto-political resistance in leisure and consumerism (Raymen & Smith, 2019, p. 115). However, these perspectives have started to face critiques.

Deviant leisure constitutes an intricate and advanced theoretical framework (Raymen & Smith, 2019, 2020a, 2020b; also see Smith, O., 2019). Although limitations in space prevent a comprehensive exploration of all its fundamental aspects, this chapter primarily focuses on its categorization of subjective, environmental, socially corrosive, and embedded harms to provide a framework for discussing this reconceptualization of dark tourism. By incorporating ultra-realist theory in conjunction with contemporary cultural criminological research and zemiology, Raymen and Smith (2019) introduce a theoretical framework that transcends the boundaries of criminology. With such theoretical foundations, deviant leisure shifted its focus toward redefining social deviance and delving into examining how individual, social, economic, and environmental harms are intricately woven into various widely accepted and normalized leisure activities. This shift emphasizes the necessity for criminologists to expand their horizons beyond the boundaries of socio-legal definitions of crime and venture into the domain of harm.

Zemiology, the study of harm, becomes essential to acknowledge the widespread presence of harms ingrained in the consumer capitalist system buoyed by an ever-evolving legal structure. An illustrative example is the profound consequences of the online gambling industry, where a culture of indebtedness, 'family instability, and detrimental patterns of consumption' has been recorded in the context of the online gambling industry (Raymen & Smith, 2020a, p. 1). This perspective has also been used to examine the rise in substance use within the night-time economy (Ayres, 2019) and the negative consequences of luxury tourism, including the environmental impacts associated with travel and the development of tourist infrastructure (Smith, O., 2019). The deviant leisure perspective sheds light on broader harms that escape the notice of other criminological viewpoints linked to leisure and the existing conceptualizations and theories related to dark tourism. Within the deviant leisure perspective, there are four main categories of harm: subjective harms, environmental harms, socially corrosive harms, and embedded harms. The following sections will now examine each of these harms as we introduce several tourist destinations not conventionally considered as forms of dark tourism.

Subjective Harms

Subjective forms of harm evoke Žižek's notion of subjective violence but are refined within the context of leisure, with Smith (2016) stating that it 'involve[s] an identifiable perpetrator visiting harm upon an identifiable victim in action related to a defined leisure activity' (p. 7). While such a definition locates clear and identifiable agents behind such actions within leisure contexts, within the remit of dark tourism, we argue that such manifestations are often opaquer. It is this very murkiness that proves to be far more damaging than conventional forms of dark tourism in which violence is its primary attraction.

For instance, certain corporations are now beginning to draw upon leisure and, more pertinently, tourism tropes to evade critical attention toward their treatment of employees. Perhaps a good starting point would be Amazon's introduction of both in-person and virtual fulfillment center tours. Amazon initially launched their warehouse tours in 2014, in which they stated that members of the public can now visit such facilities, which the company deemed "safer than a department store" (Cook, 2014). The tour, which you can either arrange to attend in-person, virtually, or watched via prerecorded videos on YouTube, provides an insight into the mechanization behind your Amazon order when you click checkout on the website or app. This tour, which follows the procedures from receiving an order to its eventual outbound journey for delivery, presents a very positive, friendly, and safe work environment. However, is this indicative of the everyday experiences of working for Amazon? Here, we need to recognize the symbolic violence that 'naturalises the discourse about things and legitimates the domination system' (Recuero, 2015, p. 1) and draw upon documented accounts of working in such environments.

The authenticity of conventional forms of dark tourism has been a prevailing theme within academic studies, with many sites having been extensively reconstructed. This 'staged authenticity' (MacCannell, 1973) of life as an Amazon worker can be witnessed in such tours but ultimately implies an authenticity that is not there, often presenting a sanitized and unrealistic representation far removed from actual working conditions. For instance, studies have shown that Amazon employees often perceive their jobs as physically demanding and alienating, imposed by strict management and surveillance (Alimahomed-Wilson and Ellen Reese, 2021). This was highlighted in the investigative podcast *Megacorp* (2022), where various individuals were interviewed, including an undercover journalist and a former inmate who likened the workplace conditions to be even more unbearable than the prison where he had been incarcerated. The journalist disclosed that employees could accrue "idle points" whenever they exceeded a certain duration for bathroom breaks. When someone amassed sufficient idle points, they could face a "disciplinary point," and accumulating too many of these marks could lead to termination. The increasing performance goals and demands have had a substantial effect on the well-being of employees, with 55% of Amazon UK workers indicating that they have experienced depression during their tenure at the company (McCarthy, 2018). Such demands have also resulted in 'backaches, knee pain, and other symptoms of constant strain [that] are common enough for Amazon to install painkiller vending machines' (Newton, 2020). Deaths have also been reported, with the National Council for Occupational Safety and Health's report (2019) citing six US Amazon worker deaths between 2018 and 2019 (p. 12). In one incident where a man died after suffering a heart attack, workers were "forced to go back to work. No time to decompress. Basically, watch a man pass away and then get told to go back to work" (Sainato, 2019).

With these points in mind, the way Amazon ultimately attempts to conceal and remove traces of such forms of harm speaks to a far more insidious relationship with violence. We are witnessing an inversion of sorts, in which instead of attempting to profit from the display of violence, companies such as Amazon are instead attempting to maintain "business as usual" and ensure we, the consumer-turned-tourist, do not ponder or reconsider our use of their services. Amazon is also not alone in this attempt to conceal the violent realities of their working conditions, with companies such as Boohoo launching in-person and virtual tours in 2021 (Rodgers, 2021) and Shein using prerecorded videos of warehouses on TikTok (Shein_Official, 2022) – indicating a growth in such practices in order to conceal the harsh realities of many employees within the era of late capitalism. Here, we see how the process of green-washing (Vollero, 2022) has been co-opted by corporations who increasingly utilize what we would term "leisure-washing" to propel sublimation of recognition within the context of the subjective harms their business models necessitate.

Environmental Harms

Drawing upon the increasing recognition of harms upon the environment as a necessity of criminological attention (Beirne & South, 2007), Smith (2016) observes the utility of such perspectives for understanding the harmful nature of contemporary leisure pursuits. He notes that, from the deviant leisure perspective, the harms of the engagement of leisure should not be the limitation of this green perspective. Rather, the lens allows us to analyze how the wider culture, engulfed in an entrenched mode of competitive individualism, is spurned on by consumer culture. This culture creates, maintains, and necessitates desire for ornamental consumerism (Hall et al., 2008) in the form of Instagram-able moments (Woods, 2023) and unique, exotic experiences. Through the deviant leisure lens though, the environmental impact of such desires is laid bare. Within this context, Smith (2016) highlights the challenges of destinations such as the Maldives. Increasingly, the luxury destination is synonymous with honeymoons, and the idealized image of a postwedding luxury destination is synonymous with honeymoons and the idealized image of a postwedding paradise (Smith, O., 2019). Employing special liberty (Hall, 2012), brides and grooms disregard the litany of harms perpetrated against the local ecology of the Maldives archipelago, with the government creating the world's largest "garbage island" called "Thilafushi" to which waste is disposed of and burned to cope with the overwhelmed local infrastructure (Smith, O., 2019; Smith & Smith, 2023). Within this context, the environmental harms perpetrated for the opportunity of the perfect Instagram-able moment, taken on a phone that works due to the cobalt within it – often mined by impoverished children in the global south (Kara, 2022) – are available for the deserving individual at the cost of the local environment: increased carbon emissions and decimated ecosystems.

Away from the extremities of luxury tourism, the very nature of tourism, which is increasingly global in nature and more accessible than ever before, offers a litany of environmental harms. White (2019) notes that the leading cause of transportation CO_2 emissions is from aviation, with most of these flights being used for tourism. These staggering contributions to the process of global warming do not cease upon the arrival of the tourist, however. As White notes, upon the arrival of the average tourist, they increasingly have the expectation to eat cuisine they are accustomed to available at their hotel, further compounding the consumer's carbon footprint when we account for the carbon accumulation of moving huge amounts of food across the globe to be available at the hotel buffet the next morning (White, 2019). A plethora of environmental harms linked to tourism have been identified in recent years. These include the decimation of local habitats for tourist developments such as beaches (Smith & Smith, 2023), hiking routes (Garrihy, 2023), and mountainsides (Smith, O., 2023). Harms against animals are in abundance, ranging from overweight, gluttonous western tourists riding donkeys up the slopes of Santorini (Smith, H., 2019), defanged tigers being petted by visitors in zoos in Thailand (Alberts, 2018), and exhausted elephants tramping through the forests of Vietnam with tourists on their back under the coercion of a mahout brandishing a flail (Chen, 2023). The overt death of animals often becomes the focus of the tourists' pursuit. Prime examples of this include the slaughter of local wildlife in the form of canned hunting (Berry & Potter, 2023; Smith, O., 2019) and for the use in African Traditional Medicine as a psychopharmacological novelty for tourists (Nurse, 2023). Perhaps most stark, however, are the routine yet sublime examples of man's detrimental effect upon the environment, such as a taxidermied North American Black Bear that died from the ingestion of a large amount of cocaine it found next to the body of a deceased drug trafficker. As Linneman (2023) observes, rather than reflect on the grim nature of American hedonism and the ecological cost of the war on drugs that the bear's grim biography captures so vividly, its carcass was turned into a tourist attraction and placed in a gift shop to attract would-be consumers.

Socially Corrosive Harms

The third category of harm proposed by Raymen and Smith (2019) is the socially corrosive forms. Within this context, they draw acknowledgment of the socially corrosive nature of contemporary society. Acknowledging Baudrillard's recognition of the 'death of the social' within an increasingly consumer-oriented society, socially corrosive harms recognize the ever more atomized and cynical nature of individuality that disregards the importance of collective interests (Young, 1999; Atkinson et al., 2017). Drawing upon Lacanian psychoanalysis put forth by Žižek (2002), they observe how consumer culture underpins the economy and has thus forced a reformation of the superego toward a 'cultural injunction to enjoy'. The superego no longer acts to temper our desires; rather, it is driven by intense anxiety of missing out (Raymen & Smith, 2019, p. 23). The superego drives the individual forward through a compulsion to fill the void with enjoyment to offer a temporary escape from the culture of anxiety that drives contemporary consumer life. Within the context of modern life, however, in which identity is formulated and exhibited through consumerism, the void can only be filled for fleeting moments in a world where freedom is increasingly unfree. Freedom of choice is increasingly limited to a range of consumer products (Lasch, 1985 in Raymen & Smith, 2019). Smith (2016) recognizes the necessity for social researchers to recognize:

> ...the potential for leisure to cut individuals adrift from the social, contribute to the further erosion of social institutions such as family, class, community, and exacerbate the fragmented and individualized nature of the social under late capitalism. Smith (2016, pp. 8–9).

The range of socially corrosive harms within the context of tourism can be identified across the social strata. Smith, O. (2019) builds upon the work of Young (1999) and Atkinson (2020) in analyzing the "exclusive" society. Within this context, he highlights how leisure zones are increasingly demarcated based on both wealth and cultural capital. Within this context, he highlights ex-president Donald Trump's luxury golf courses that engulfed a huge swathe of the Aberdeenshire coastline. Destinations of permanence do not hold the monopoly on corrosive harms, however.

Hayward (2023) details such harms within the context of the Burning Man Festival in recent years. The festival, beginning in 1991, attracts over 80,000 visitors yearly to the Black Rock Desert in Nevada for nine days each August. Originally, the event and the conduct of those who attended were driven by 10 fundamental principles: radical inclusion, gifting, decommodification, radical self-reliance, radical self-expression, communal effort, civic responsibility, leaving no trace, participation, and immediacy (Hayward, 2023, p. 143). In recent years, though, many of these guiding and prosocial principles have been increasingly corroded by increasing commercialization underpinned by social media, price hikes for tickets, and a deluge of wealthy attendees who arrive on makeshift airstrips and rushed to luxury encampments in which they are attended to by paid staff.

Within the context of dark tourism within the global south, however, much more devastating examples are in abundance driven by the pursuit for exclusive and individualized experiences by primarily western tourists. The disregard for the collective interests of communities and wider society in the pursuit of enjoyment sees the disregard of human rights. For example, the hosting of the World Cup by Qatar in 2022 was underpinned by human rights violations against migrant workers contracted to rapidly build infrastructure for the games (Gallacher, 2023). Here, we see that disregard for other individual's welfare takes primacy. The 1950s saw the onset of wealthy American tourists within Nepal, shortly followed by younger American and European tourists through the 1970s across the 'hippie trail' (Smith, E., 2023). Coinciding with the influx in tourism was the arrival of Tibetan refugees escaping the persecution of the Chinese government. Smith (Smith, E., 2023) details that Kathmandu quickly became a regional hot-spot for cultural artifacts, which were routinely sold by refugees struggling to survive. Vast private collections were soon accumulated across America and Europe. Museum collections in the west continued to bloat as tourism increased, and huge swathes of artifacts from temples and sacred sites across Nepal were looted into the 1980s. The collection of objects to satisfy the compulsion to display individuality is not exclusive to physical objects. As Large (2019) notes, increasingly, individuals focus on collecting experiences to increase social capital and demark individualism. Drawing upon the example of charity tourism, much in the way that Woods (2023) explored, Large (2019) demonstrates how ornamental consumerism has transcended the physical world and grown to include the consumption of experiences to bolster social capital. Underpinning this transition within contemporary leisure spaces, however, is the othering of the locals within the tourist destination. This is perhaps most overtly demonstrated within the work of Frankis and Patel Nascimento (2023), who detail the tours of Brazilian Favelas and the commodification of violence and the lived realities of inhabitants. While much of the tourism is a key exemplar of edgework (Lyng, 2005) with tourists motivated by visiting areas they deem as risky, much of the tourism was underpinned by sex work (Frankis & Patel Nascimento, 2023).

Embedded Harms

The vast majority of leisure cultures are within the fourth harm of the deviant leisure perspective, embedded within legitimate consumer markets (Raymen & Smith, 2019). Within the context of embedded harms, Raymen and Smith propose the most ubiquitous example is gambling, noting its penetration of leisure cultures in such an extensive manner. They note that gambling has become synonymous with sports, the nighttime economy and online spaces despite the myriad of harms such activities imbue (Raymen, 2019; Brown & Osman, 2020; Raymen and Smith, 2020a, 2020b; Nicklin et al., 2021). As Raymen and Smith (2019) note, as gambling becomes more embedded within routine leisure cultures, the act of betting becomes *irretrievably entangled* with the individual's identity. Within the context of tourism, this arguably transcends the context of the individual and can be understood within the context of space and place. Take, for example, Las Vegas, the City of Sin, which is inseparable from the notions of American excess and gambling. The city is *irretrievably entangled* with the casino industry and, due to this, a leading tourist destination for hedonists across the globe. Such destinations become liminal spaces in which the typical rules of society do not apply, changing the nature of those areas entirely. The obvious example of this are the city centers, whose functions change fundamentally to cater to drinkers and hedonists in the nighttime economy (Hayward & Hobbs, 2007). Las Vegas, centered on hedonism and excess, is an obvious and extreme example of how such leisure spaces become embedded within the context of tourism. Within the global south, however, new incarnations of Las Vegas are rapidly developing. Macau has long been a go-to destination for lifestyle gamblers. In recent years though, an African doppelgänger of Las Vegas has been rapidly developing in Lagos (Business Post, 2020). In an effort to boost the Nigerian economy through the tourism industry, Lagos is rapidly being transformed into a liminal destination in which gambling, drinking, and partying are embedded and essential to the local infrastructure.

Liminal spaces, however, are no longer limited to the physical. Technology has created new digital spaces (Atkinson, 2019; Atkinson & Rodgers, 2015; Denham, 2023) in which the usual social rules do not apply. As such, another area that can be identified as culturally embedding harms can be found within social media. Just as gambling within its culturally embedded form has demonstrated an ability to become central to the shaping of identity and presentation of self (Raymen & Smith, 2019), social media has taken primacy within this context. The pervading nature of social media to frame social interaction has had an implicit influence on tourism globally. As Large (2019, p. 331) notes, 'in a society of competitive individualists, distinguishing oneself from the masses in a key cultural drive.' As social media has become a more influential and embedded tool that individuals can utilize to do this, its influence on tourism has been substantial. Volunteer tourism has been increasingly utilized to distinguish oneself (Large, 2019) with the lines between tourism and volunteering increasingly blurred (Woods, 2023). Woods (2023) recognized the growth of 'voluntourism' in Cambodia. He noted those who engage in such tourism are more concerned with the aesthetic of themselves as a 'hero humanitarian' on their social media than actually enacting change for the 'suffering subjects' next to which they pose.

Gore Capitalism, Thanatophilia, and Graze Culture

In her seminal text, Mexican philosopher Sayak Valencia (2018) proposed the term Gore Capitalism. Her thesis proposes the influence of hyper-consumerist neoliberal capitalism on the transformation of violence within Tijuana, Mexico. Within this context, Valencia argues that capitalism has sublimated production as the main accumulation of commodity as Marx famously argued. Rather, within the landscape of gore capitalism, it is the destruction of the body itself that has become the product or commodity. She states that 'death has become the most profitable business in existence' (Valencia, 2018, p. 21). Much of the thesis focuses upon the proliferation of extreme violence within the global south, underpinned by the necropolitics enforced by organized crime groups vying for control of the lucrative illicit markets that seek to quench the unquenchable thirst for pharmacopornographic -driven hedonism within the global north. Here, she recognizes that the production, trafficking, and distribution of legal and illegal drugs are central to contemporary capitalism (p. 94). The utilization of Gore Capitalism to describe Valencia's thesis was inspired by the cinematic genre of gore. It is within this context we can begin to understand how consumerism, violence, and dark tourism in the global south intersect utilizing the notion of Graze Culture (Kelly et al., 2023).

Building upon Seltzer's (1998) concept of 'Wound Culture,' Graze Culture recognizes, like Seltzer, that our awareness, understanding, and subsequent captivation of serial murder and other forms of physical violence can be attributed to "true crime" accounts and fictional representations of disturbed and unhinged criminals (Kelly et al., 2023) However, Graze Culture advances these discussions by shedding light on the impact such depictions of serial murder accounts can have. Kelly et al. (2023) explain that the prioritization of brutal violence as a commodity over factually accurate portrayals has resulted in a theorization of acts such as serial murder that disregard the structural and cultural underpinnings alongside societies consumption of such sublime violence. Ultimately, portrayals of subjective violence, that being one person causing physical damage to another, have provided us with "monsters" in which we "brush up against" to fetishistically disavow (see: Žižek, 2008) the realities of late capitalism. Such realities are fraught with social harms that, while not having a clear and identifiable perpetrator, are violent in nature and arguably much more disruptive in comparison to the aforementioned more identifiable forms of physical violence.

Valencia highlights the thanatophilic nature of contemporary capitalism, underpinned by the demarcation of the civilized global north and uncivilized global south, and the uneven development across the globe post–World War I. From this time, she argued, an epistemic break was enacted resulting in a necropolitical approach to both economic and social policies in the global south to increase capital accumulation – accumulation now often underpinned by the illicit economy and enforced with violence. Here though, Graze Culture can offer a deeper understanding of this relationship. While Valencia offers a sound and compelling narrative, she is distracted by 'instances of dismembering and disembowelment, often tied with organised crime' (Valencia, 2018, pp. 19–20) in the same way Seltzer (1998) was preoccupied by the 'the torn and open bodies' of serial killer victims. As Hall (2012) details, violence in the contemporary world has fundamentally changed. While some commentators have often argued that violence has declined (Pinker, 2011), it is more accurate to recognize violence has morphed under the influence of the economy (Hall, 2012). Violence has not been pacified, rather the nature of violence has changed to become ever more palatable to the winners of the capitalist system (the global north). While companies will no longer engage in violent and overt colonization on behalf of governments as Firestone enacted on the United States's behalf in Liberia (Harrison Church, 1969; McCoskey, 2014; Patton, 2015; Pailey, 2023), this does not mean that atrocities are no longer routinely and systematically perpetrated in the pursuit of profit (Lynes et al., 2024). Rather, the CEOs of multinational corporations can commit much more harmful acts with the stroke of a pen. Violence is, as Hall (2012) demonstrates, pseudo-pacified.

Utilizing the lens of Gore Capitalism with an awareness of the pseudo-pacified nature of contemporary violence, recognizing the full breadth of violence (including not only the actuality of violence and structural violence but also symbolic manifestations), we can utilize the four harms outlined by the deviant leisure perspective to understand how tourists engage in a prosumer-orientated Graze Culture. Tourists from the global north are increasingly disenfranchised, and the margins of capital accumulation are ever less collective (Østbø Kuldova et al., 2023). As such, individuals are increasingly aware of their precarity (Lloyd, 2018; Winlow & Hall, 2013) and the multitude of looming crises facing the world while also embedded within the democratized leisure class (Raymen & Smith, 2019; Veblen, 1965). Much as they consume violent media as a way to disavow their realities and brush up against familiar monsters (Kelly et al., 2023), they engage in seemingly de-politicized consumption within the context of tourism.

Conclusion

Often tourists are, like Seltzer (1998) and Valencia (2018), drawn to the sublime violence – thus thanatourism, as we conventionally understand dark tourism. While the recognition of thanaphilia as an underpinning element of contemporary consumer capitalism by Valencia (2018) goes some way to recognizing the realities of some harms that underpin the current global economy, it also serves to obscure the view of other harms. We focus upon sites of death and destruction and ignore death and destruction that structural and cultural violences enact in the name of tourism and development, such as the decimation of communities and ecosystems to develop vast and exclusive hotel complexes with private beaches or sports stadiums. Within this context, then, it is essential for the deviant leisure perspective and Graze Culture to be at the fore of theorization of what constitutes dark tourism. Such theoretical perspectives offer much utility for understanding dark tourism globally. The four harms offer a unique-looking glass from which we can understand how consumption and the presentation of the self culminate in deeply harmful tourist practices. In doing so, the root causes of Gore Capitalism and Graze Culture can be recognized and challenged.

Hyper-consumerism and perpetual anxiety that is disavowed through the spiral of yet more conspicuous consumption as a means of identity construction serve as the catalyst for the litany of harms perpetuated through tourism. These harms are inflicted most disproportionately and savagely upon the people and communities of the global south. In recognizing this, there is a small glimmer of hope that the worst excess can be challenged and (hopefully) curbed. The darkest nodes of tourism are not the sites of thanatourism as has often been considered. Rather, a litany of harms is lurking behind the curtain of the majority of tourist destinations globally. If we return to the work of Sharpley (2005), who argued that the deliberate classifying of encounters intended to satisfy a wish to consume death is much darker, within the context of the preceding discussion, we instead put forward the case that the inverse is, in fact, true. In order to truly understand dark tourism, it is essential to take a step back and look past the sublime acts of violence that underpin *some* attractions. In doing so, it is viable to analyze the more mundane, common, and harmful practices that underpin the industry as a whole. It is the analysis and recognition of these less tangible forms of violence normalized within the leisure economy from which substantive change can be realized. We must move past a concentration of thanatourism and transition to analysis based on what we call "zemi-tourism." To shed light on the truly dark matter within the context of dark tourism, a reconceptualization of dark tourism is needed.

References

Alberts, E. C. (2018, February 20). Popular tourist spot does the worst thing to its baby tigers. *The Dodo*. https://www.thedodo.com/in-the-wild/zoo-declaws-tiger-lion-cubs-thailand

Alimahomed-Wilson, J., & Ellen Reese, E. (2021). Surveilling Amazon's warehouse workers: Racism, retaliation and worker resistance amid the pandemic. *Work in the Global Economy*, *1*(1–2), 55–73.

Ashworth, G. J. (1996). Holocaust tourism and Jewish culture: The lessons of Krakow-Kazimierz. In M. Robinson, N. Evans, & P. Callaghan (Eds.), *Tourism and cultural change, centre for travel and tourism* (pp. 1–12). Newcastle.

Ashworth, G. J., & Isaac, R. K. (2015). Have we illuminated the dark? Shifting perspectives on 'dark' tourism. *Tourism Recreation Research*, *40*(3), 316–325.

Atkinson, R. (2019). From edge work to death drive: The pursuit of pleasure and denial of harm in a leisure society. In T. Raymen & O. Smith (Eds.), *Deviant leisure. Palgrave studies in crime, media and culture*. Palgrave Macmillan. https://doi.org/10.1007/978-3-030-17736-2_12

Atkinson, R. (2020). *Alpha city: How London was captured by the super-rich*. Verso.

Atkinson, R., McKenzie, L., & Winlow, S. (2017). *Building better societies: Promoting social justice in a world failing apart*. Policy Press.

Atkinson, R., & Rodgers, T. (2015). Pleasure zones and murder boxes: Online pornography and violent video games and zones of cultural exception. *British Journal of Criminology*. https://bjc.oxfordjournals.org/content/early/2015/10/27/bjc.azv113.full

Ayres, T. (2019). Substance use in the night-time economy: Deviant leisure? In T. Raymen & O. Smith (Eds.), *Deviant leisure: Criminological perspectives on leisure and harm* (pp. 135–161). Palgrave.

Bathory, R. (2018). *Dark tourism*. Carpet Bombing Culture.

Beirne, P., & South, N. (2007). *Issues in green criminology*. Routledge.

Berry, P., & Potter, G. (2023). Trophy hunting: Sub-Saharan Africa. In A. Lynes, C. Kelly, & J. Treadwell (Eds.), *50 dark destinations: Crime and contemporary tourism*. Policy Press.

Blom, T. (2000). Morbid tourism: A postmodern market niche with an example from Althorp. *Norwegian Journal of Geography*, *54*(1), 29–36.

Bolan, P., & Simone-Charteris, M. (2018). Shining a digital light on the dark: Harnessing online digital media to improve the dark tourism experience. In P. R.

Stone, R. Hartmann, T. Seaton, R. Sharpley, & L. White (Eds.), *The Palgrave handbook of dark tourism studies* (pp. 727–746). Palgrave.

Bristow, R. S., & Newman, M. (2004). Myth vs. fact: An exploration of fright tourism. In K. Bricker (comp., Ed.), *Proceedings of the 2004 Northeastern recreation research symposium* (pp. 215–221), Gen. Tech. Rep. NE-326. US Department of Agriculture, Forest Service.

Brown, M., & Osman, S. (2020). Addiction, gambling and gaming: Chasing the digital dragon. In C. Kelly, A. Lynes, & K. Hoffin (Eds.), *Video games, crime and next-gen deviance: Reorienting the debate* (pp. 91–113). Emerald Publishing Limited.

Business Post. (2020, August 14). Lagos like Vegas: What are the biggest Nigerian land-based casinos? *Business Post*. https://businesspost.ng/showbiz/lagos-like-vegas-what-are-the-biggest-nigerian-land-based-casinos/

Chen, H. (2023, March 10). This is what years of tourist rides do to an elephant. *CNN*. https://edition.cnn.com/travel/article/thailand-elephant-ride-abuse-intl-hnk/index.html

Cook, J. (2014, April 26). Yes, you can now take a tour of one of Amazon's massive warehouse facilities. *GeekWire*. https://www.geekwire.com/2014/yes-can-now-take-tour-massive-amazon-warehouse-facility/

Cooper, M. (2006). The Pacific War battlefields: Tourist attractions or war memorials?. *International Journal of Tourism Research*, *8*(3), 213–222.

Dann, G. M. S. (1998). The dark side of tourism. *International Center for Research and Studies in Tourism, Series J, Planning/Development, 14*.

Denham, J. (2023). The staycation: Home. In A. Lynes, C. Kelly, & J. Treadwell (Eds.), *50 dark destinations: Crime and contemporary tourism*. Policy Press.

Foley, M., & Lennon, J. J. (1996). JFK and dark tourism: A fascination with assassination. *International Journal of Heritage Studies*, *2*(4), 198–211.

Frankis, D., & Patel Nascimento, S. (2023). Favela tour: Rio de Janeiro, Brazil. In A. Lynes, C. Kelly, & J. Treadwell (Eds.), *50 dark destinations: Crime and contemporary tourism*. Policy Press.

Gallacher, G. (2023). 'The ugly side to the beautiful game': Qatar. In A. Lynes, C. Kelly, & J. Treadwell (Eds.), *50 dark destinations: Crime and contemporary tourism*. Policy Press.

Galtung, J. (1969). Violence, peace and peace research. *Journal of Peace Research*, *6*, 167–191.

Garrihy, J. (2023). Death road: La Paz to Coroico, Bolivia. In A. Lynes, C. Kelly, & J. Treadwell (Eds.), *50 dark destinations: Crime and contemporary tourism* (pp. 276–282). Bristol University Press.

Gillen, J. (2018). It begins with the bombs: Operationalising violence at a Vietnamese dark tourism site. *Cultural Geographies*, *25*(4), 519–536.

Hall, S. (2012). *Theorizing crime and deviance: A new perspective*. Sage.

Hall, S., Winlow, S., & Ancrum, C. (2008). *Criminal identities and consumer culture: Crime, exclusion and the new culture of narcissism*. Routledge.

Hanrahan, J. (2022). Ground floor at the Amazon Warehouse [podcast]. *Megacorp*. https://www.youtube.com/watch?v=pbvmye0GD6Q

Harrison Church, R. J. (1969). The Firestone rubber plantations in Liberia. *Geography*, *54*(4), 430–437. http://www.jstor.org/stable/40567142

Hayward, K. (2023). Burning man festival: Black Rock Desert, Nev, USA. In A. Lynes, C. Kelly, & J. Treadwell (Eds.), *50 dark destinations: Crime and contemporary tourism*. Policy Press.

Hayward, K., & Hobbs, D. (2007). Beyond the binge in 'Booze Britain': Market-led liminalization and the spectacle of binge drinking. http://onlinelibrary.wiley.com/doi/10.1111/j.1468-4446.2007.00159.x/full

Heřmanová, E., & Abrhám, J. (2015). Holocaust tourism as a part of dark tourism. *Czech Journal of Social Sciences Business and Economics*, *1*, 16–34.

Hohenhaus, P. (2013). Commemorating and commodifying the Rwandan genocide. In L. White & E. Frew (Eds.), *Dark tourism and place identity: Managing and interpreting dark places* (pp. 37–142). Routledge.

Hooper, G. (2016). Introduction. In G. Hooper & J. Lennon (Eds.), *Dark tourism: Practice and interpretation* (pp. 1–12). Routledge.

Ibrahim, Y. (2015). Self-representation and the disaster event: Self-imaging, morality and immortality. *Journal of Media Practice*, *16*(3), 211–227.

Kara, S. (2022). *Cobalt red: How the blood of the Congo powers our lives*. Saint Martin's Griffin.

Kelly, C., Lynes, A., & Hart, M. (2023). Graze culture' and serial murder: Brushing up against 'familiar monsters' in the wake of 9/11. In S. E. Fanning & C. O'Callaghan (Eds.), *Serial killing on screen. Palgrave studies in crime, media and culture*. Palgrave Macmillan. https://doi.org/10.1007/978-3-031-17812-2_13

Large, J. (2019). Conspicuously 'doing' charity: Exploring the relationship between doing good and doing harm in tourism. In T. Raymen & O. Smith (Eds.), *Deviant leisure. Palgrave studies in crime, media and culture*. Palgrave Macmillan. https://doi.org/10.1007/978-3-030-17736-2_12

Lasch, C. (1985). *The minimal self: Psychic survival in troubled times*. Norton.

Linneman, T. (2023). Cocaine bear: Fun Mall, Lexington, Kentucky, USA. In A. Lynes, C. Kelly, & J. Treadwell (Eds.), *50 dark destinations: Crime and contemporary tourism*. Policy Press.

Lisle, D. (2004). Gazing at ground zero: Tourism, voyeurism and spectacle. *Journal for Cultural Research*, *8*(1), 3–21.

Lloyd, A. (2018). *The Harms of work: An ultra-realist account of the service economy*. Policy Press.

Lynes, A., Treadwell, J., & Bavin, K. (2024). *Crimes of the powerful and the contemporary condition: The democratic republic of capitalism*. Policy Press.

Lyng, S. (2005). *Edgework: The sociology of risk-taking*. Routledge.

MacCannell, D. (1973). Staged authenticity: Arrangements of social space in tourist settings. *American Journal of Sociology*, *79*, 589–603.

McCarthy, N. (2018, April 17). *Report: Amazon workers skip toilet breaks to keep their jobs*. Statista. https://www.statista.com/chart/13554/report_-amazon-workers-skip-toilet-breaks-to-keep-their-jobs/

McCoskey, S. K. (2014). The Firestone Tire & Rubber company and Liberia's Civil War: Evaluating Firestone's intent to operate during chaos. *Business & Professional Ethics Journal*, *33*(2/3), 253–280. http://www.jstor.org/stable/44074816

McDaniel, K. (2018). *Virtual dark tourism*. Palgrave Macmillan.

Muzaini, H., Teo, P., & Yeoh, B. S. A. (2007). Intimations of postmodernity in dark tourism: The fate of history at Fort Siloso. *Singapore Journal of Tourism and Cultural Change*, *5*(1), 28–45.

Newton, C. (2020, April 1). Amazon's poor treatment of workers is catching up to it during the coronavirus crisis. *Verge*. https://www.theverge.com/interface/2020/4/1/21201162/amazon-delivery-delays-coronavirus-worker-strikes

Nicklin, L. L., Spicer, G., Close, J., Parke, J., Smith, O., Raymen, T., Lloyd, H., & Lloyd, J. (2021). "It's the attraction of winning that draws you in"—A qualitative investigation of reasons and facilitators for videogame loot box engagement in UK gamers. *Journal of Clinical Medicine*, *10*(10), 2103. https://doi.org/10.3390/jcm10102103

Nurse, A. (2023). Vulture brains and *muthi* markets: Johannesburg, South Africa. In A. Lynes, C. Kelly, & J. Treadwell (Eds.), *50 dark destinations: Crime and contemporary tourism*. Policy Press.

Østbø Kuldova, T., Østbø, J., & Raymen, T. (2023). *Luxury and corruption: Challenging the anti-corruption consensus*. Bristol University Press.

Pailey, R. N. (2023). Stopping Firestone and starting a citizen 'revolution from below': Reflections on the enduring exploitation of Liberian land and labour. *Third World Quarterly*. https://doi.org/10.1080/01436597.2023.2240729

Patton, A. (2015). Civil rights in America's African diaspora: Firestone rubber and segregation in Liberia. *Canadian Journal of African Studies/Revue Canadienne des Études Africaines*, *49*(2), 319–338. https://doi.org/10.1080/00083968.2015.1024438

Pinker, S. (2011). *The better angels of our nature*. Penguin Books.

Raine, R. (2013). A dark tourist spectrum. *International Journal of Culture, Tourism and Hospitality Research*, *7*(3), 242–256.

Ray, L. (2011). *Violence and society*. Sage.

Raymen, T. (2019). Lifestyle gambling in accelerated culture. In T. Raymen & O. Smith (Eds.), *Deviant leisure. Palgrave studies in crime, media and culture*. Palgrave Macmillan. https://doi.org/10.1007/978-3-030-17736-2_12

Raymen, T., & Smith, O. (2019). *Deviant leisure. Palgrave studies in crime, media and culture*. Palgrave Macmillan. https://doi.org/10.1007/978-3-030-17736-2_12

Raymen, T., & Smith, O. (2020a). Gambling and harm in 24/7 capitalism: Reflections from the post-disciplinary present. In S. Hall, T. Kuldova, & M. Horsley (Eds.), *Crime, harm and consumerism*. Taylor & Francis.

Raymen, T., & Smith, O. (2020b). Lifestyle gambling, indebtedness and anxiety: A deviant leisure perspective. *Journal of Consumer Culture*, *20*(4), 381–399. https://doi.org/10.1177/1469540517736559

Recuero, R. (2015). Social media and symbolic violence. *Social Media & Society*, *1*(1), 1–3.

Robb, E. M. (2009). Violence and recreation: Vacationing in the realm of dark tourism. *Anthropology and Humanism*, *34*(1), 51–60.

Rodgers, D. (2021, August 9). Boohoo is now giving walking tours of its allegedly exploitative factories. *Dazed*. https://www.dazeddigital.com/fashion/article/53783/1/boohoo-tours-factories-fast-fashion-exploitation-leicester-supply-chain-plt

Rojek, C. (1993). *Ways of escape*. MacMillian.

Rojek, C. (1997). Indexing, dragging and the social construction of tourist sites. In C. Rojek & J. Urry (Eds.), *Touring cultures: Transformations of travel and theory* (pp. 52–74). Routledge.

Rojek, C. (2010). *The labour of leisure*. SAGE.

Sainato, M. (2019, October 18). 'Go back to work': Outcry over deaths on Amazon's warehouse floor. *The Guardian*. https://www.theguardian.com/technology/2019/oct/17/amazon-warehouse-worker-deaths

Seaton, A. (1996). Guided by the dark: From thanatopsis to thanatourism. *International Journal of Heritage Studies*, *2*(4), 234–244.

Seltzer, M. (1998). *Serial killers: Death and life in America's 'wound culture'*. Routledge.

Sharpley, R. (2005). Travels to the edge of darkness: Towards a typology of "dark tourism". In M. Aicken, S. J. Page, & C. Ryan (Eds.), *Taking tourism to the limits* (pp. 239–250). Routledge.

Sharpley, R. (2009). Shedding light on dark tourism: An introduction. In R. Sharpley & P. R. Stone (Eds.), *The darker side of travel: The theory and practice of dark tourism* (pp. 3–23). Channel View Publications.

Shein_Official. (2022, December 30). https://www.tiktok.com/shein_official/video/7182745739720756481?lang=en

Smith, V. (1998). War and tourism: An Amercian ethnography. *Annals of Tourism Research*, *25*(1), 202–227.

Smith, O. (2016). *Deviant leisure: Emerging perspectives on leisure, consumerism, and crime*. British Society of Criminology. http://www.britsoccrim.org/wp-content/uploads/2016/04/Deviant-Leisure-Oliver-Smith.pdf

Smith, H. (2019, April 2). Tourists urged to avoid riding donkeys up Santorini's steep steps. *The Guardian*. https://www.theguardian.com/world/2019/apr/02/tourists-avoid-riding-donkeys-santorini-steps

Smith, O. (2019). Luxury, tourism and harm: A deviant leisure perspective. In T. Raymen & O. Smith (Eds.), *Deviant leisure: Criminological perspectives on leisure and harm* (pp. 305–325). Palgrave.

Smith, E. (2023). The hippie trail: Nepal, south Asia. In A. Lynes, C. Kelly, & J. Treadwell (Eds.), *50 dark destinations: Crime and contemporary tourism*. Policy Press.

Smith, O. (2023). Dark tourism, ecoside and Alpine ski resorts: The Alps, Europe. In A. Lynes, C. Kelly, & J. Treadwell (Eds.), *50 dark destinations: Crime and contemporary tourism* (pp. 288–293). Bristol University Press.

Smith, E., & Smith, O. (2023). The Maldives: Republic of Maldives, Indian Ocean. In A. Lynes, C. Kelly, & J. Treadwell (Eds.), *50 dark destinations: Crime and contemporary tourism* (pp. 271–276). Bristol University Press.

Stanko, E. A. (2001). The day to count: Reflections on a methodology to Raise awareness about the impact of domestic violence in the UK. *Criminology and Criminal Justice*, *1*, 215–226.

Stone, P. R. (2006). A dark tourism spectrum: Towards a typology of death and macabre related tourist sites, attractions and exhibitions. *Tourism: An International Interdisciplinary Journal*, *54*(2), 145–160.

Tarlow, P. (2007). Dark tourism–the appealing 'dark' side of tourism and more. In M. Novelli (Ed.), *Niche tourism* (pp. 61–72). Routledge.

The Dirty Dozen. (2019). National council for occupational safety and health. https://nationalcosh.org/sites/default/files/uploads/2019_Dirty_Dozen.pdf

Tunbridge, J. E., & Ashworth, G. J. (1996). *Dissonant heritage: The management of the past as a resource in conflict*. John Wiley and Sons.

Uzzell, D. (1989). The hot interpretation of war and conflict. *Heritage Interpretation, 1*, 33–47.

Valencia, S. (2018). *Gore capitalism*. Semiotext (e) Intervention Series 24. MIT Press.

Veblen, T. (1965). *The theory of the leisure class*. Sentry Press.

Vollero, A. (2022). *Greenwashing: Foundations and emerging research on corporate sustainability*. Emerald.

White, R. (2019). Loving the planet to death: Tourism and ecocide. In T. Raymen & O. Smith (Eds.), *Deviant leisure: Criminological perspectives on leisure and harm* (pp. 305–325). Palgrave.

Winlow, S., & Hall, S. (2013). *Rethinking social exclusion: The end of the social?* Sage.

Woods, O. (2023). Volunteer tourism-'doing it for the 'gram': Cambodia, Southeast Asia. In A. Lynes, C. Kelly, & J. Treadwell (Eds.), *50 dark destinations: Crime and contemporary tourism*. Policy Press.

Young, J. (1999). *The exclusive society: Social exclusion, crime and difference in late modernity*. Sage.

Žižek, S. (2002). *For they know not what they do: Enjoyment as a political factor*. Verso.

Žižek, S. (2008). *Violence*. Profile Books.

Žižek, S. (2010). *Living in the end times*. Verso.

Chapter 15

"Bawaal" Movie Is an Experience of Dark Tourism for the Indian Audience

Surbhi Sharma

Banasthali Vidyapith, India

Abstract

Dark tourism is a place for tourists to travel to understand the deaths, suffering, and tragedy. Nowadays, movie makers have positioned dark tourism as a source of edutainment for society, contributing to the marketing of dark places. This type of education helps children to make associations between museum artifacts, memories, battles, and visitors' lives. This study explores how movies have put forth on dark tourism and edutainment by involving audiences, especially students, through storytelling techniques. This research analyses the content of the Bollywood movie *Bawaal* (2023), which was based on education from the sites of World War II, a holocaust in Europe. Movies can improve students' and children's interest in history. There is a need to understand how movies are closely linked to dark tourism. The result shows that explaining dark tourist places is essential for the upcoming generation to maintain their connection with the world's history. This study will spread consciousness about World War II or Holocaust places and help indirectly market dark tourism.

Keywords: Bawaal movie; bollywood; edutainment; dark tourism; World War II

Introduction

Dark tourism is a term associated with the study of tourism of those locations and sites of assassination, crime, Holocaust, genocide, or imprisonment that have served to fascinate visitors. These sites act as a common place for history and visitation where questions of validity, factuality, and morality are sometimes mixed with the action of tourism facilities. Dark tourism attractions indicate the

Dark Tourism, 205–212

Published under exclusive licence by Emerald Publishing Limited
doi:10.1108/978-1-83797-336-120241015

demand but also comprise historical references, obituaries, heritage attractions, and history legacies (Lennon, 2017). Dark sites are more concerned with visitors' experiences related to disaster, death, and crime that flustered public consciousness, whereby actual and recreated locations of horror, death, depravity, or atrocity emerge as a dark obsession for visitors (Stone, 2013).

In preliterate societies, family members, or religious groups told stories about their ancestors and incidents; it was an oral tradition for mediation. However, in existential modernity, two key institutions mediate between the dead and the living, i.e., *the mass media and tourism*. When there is a disaster, priests, doctors, or family elders are not there for the information, but we turn to the mass media (newspaper, TV) for what happened. There is a close link between media and dark tourism because we do not need to dive into the bottom of the Atlantic; instead, we can watch the "Titanic" movie. However, tourism is growing to feel the waves lapping at their feet, the sun on their skin, and photographing the monument to show they have been there. The media and dark tourism are in the same business of explaining and presenting the death and hardships of millions of people and sometimes experiencing execution (Walter, 2009).

The rise in the popularity of dark tourism notably boosted the film and mass media industry, like the Television film *The Diary of Anne Frank* by Robert Dornhelma and the Film *World Trade Center* by Oliver Stone, based on the terrorist attack of September 11, 2001. Now, dark tourism is seen as controversial in the relationship between piety and commercialization. In the present situation, the commercial condition of interest is in the controversial practice of tabloid media (paparazzi) and news bringing about life-death (terrorist attacks, live coverage of wars, live executions, train crashes, etc.) (Heřmanová, 2015). Commercialization takes place with different motives, i.e., value orientation, education, and perception of time, depending on culture and past tragic events (Heřmanová, 2015). These sites can provide indispensable educational opportunities for providing different learning paths to support classroom learning, thus helping children to feel connected between museums, artifacts, lives, and memories. Learning through dark sites is about forming multilayered experiences between emotions, identities, and social relationships between people (Dresler, 2023). With the evolution of technology, there is a predominance of media that develop ideas and have the potential to build and destroy certain beliefs.

Therefore, there is a need to investigate how movies provide educational information with entertainment regarding dark tourist spots. This study explores how movies have put forth on dark tourism and edutainment by involving audiences, especially students, through storytelling techniques. For this research, we analyze the *Bawaal* Movie (director Nitesh Tiwari, 2023) based on World War II by conducting history subject classes from the Holocaust in Europe. Analysis of this film revealed the importance of site visits to learn with personal experience, and outdoor teaching is also essential for children to make things more interesting. This article attempts to teach the importance of edutainment movies to enhance the audience's knowledge.

Literature Review

Dark Tourism

Dark Tourism is the trend of visiting places or destinations described by mourning, pain, mass death, and zones of disasters. These places bury the history of violence, cruelty, or traumatic events that shocked society. It speaks of "unspeakable violence" to represent tourism's changeable nature, which conceptually draws the imaginary borders of nationhood (Korstanje, 2020). Aside from the original sites, people visit dark tourist spots as entertainment centers of horror that should offer tourists fun but at the same time also highlight the death of the actual or imaginary events (Heřmanová, 2015).

Dark tourism was first introduced by Foley and Lennon (1996) in an article in the International Journal of Science of Culture and Historical Heritage highlighting death and disaster. The Thana tourism and black tourism terms appeared in 2002, which had a connection with an English town, Soham, that became popular after the murder of two little kids. English literature has other synonyms related to this term, i.e., *Horror tourism, dark heritage, grief tourism, tragedy tourism, atrocity tourism, morbid tourism, disaster tourism,* and many others (Lennon, 2017). In 1993, Rojek first explained the concept of "Black spots" as a visitor's attraction or the commercial growth of grave sites and sites where famous fighters or mass of people have lost their lives. Rojek explained black spots in three categories: first explanation about the visit to the accident sites of James Din in 1995; second explanation about the annual lighting of candles in the remembrance of Elvis Presley at Graceland in Tennessee; and third explanation about the anniversary of the murder of Douglas Kennedy in Texas. The modern media has announced the postmodern spectacle of these three events (Minic, 2012).

Media and Dark Tourist Spots

Media successfully induced audience emotions through storytelling and images. Media viewers accept the idea of media-induced tourism, such as tourism agencies and communities appearing much later to support the unexpected rush of tourists. Media also have a peculiar relationship with dark tourism, which induces tourism and illustrates cultural representation among tourists (Tham, 2013). The nation-state and localities associate the role of dark tourism with national heritage as crucial, and the movies give it a "lighter" tone. The contribution of technologies like movies, visuals, audio, and embodied mobilities in dark tourism creates new tourist practices and imaginaries. However, watching movies can induce physical movement, actual visits to represented places, virtual travel, and embodied tourism. Therefore, the media plays a notable role in popularizing places where movies and TV series have been set by purely circulating unscientific evidence of tourists flocking there (Korstanje & Handayani, 2017).

A film and TV series have been produced to blur the boundaries between the imaginary and the real. The vanishing of "physical distance," known as "Deathstinations," has outstretched to involve both physical sites and virtual fictional suffering and death. In dark tourism, violent films and TV shows work to form popular conceptions of specific locales or eras. Film watching has long been discussed as a means of escape from society and a channel through which viewers can be transported virtually to that place. The tourist who still wants to visit the site of bloodshed and political violence for leisure indicates they have not outmatched the aptitude for a so-called Assassination vacation (Mathews, 2018). "Technology and especially multimedia have been used to connect the tourist experience to the individual."

Edutainment and Dark Tourism

Dark tourism is a critical lens to encourage academic discussions about decoding and blending entertainment and leisure with tragedy and commemoration. The interpretation of dark tourism offers tourists an emotional, entertaining, or educational experience where they can integrate their knowledge, interest, and memories with the heritage and history on the screen. Using edutainment along with dark tourism is controversial, but it can increase visitor retention, motivation, and active learning. Combining education and cultural activities with the touch of technology of entertainment and commercialization can create a notion that learning could be "Fun." It can be described as "education that entertains and informs" and "entertainment that educates" (Wyatt et al., 2019).

Educating the visitors of dark sites on historical events is essential to deliver a meaningful experience, which implies that learning positively impacts generating meaning (Zheng et al., 2020). There is a motivation among tourists to visit a dark tourist spot after watching a film or TV show; in most cases, "the attraction" is already there. Edutainment is another way of making the audience aware and marketing what has happened in that area (Busby & Devereux, 2015). A tourist wants to visit dark tourism spots to compare the reality behind the media or to acquire an emotional attachment to celebrities. It has been noticed that when visitors find a site to be a part of their heritage, they will show more interest in learning and be more emotionally involved (Lewis et al., 2022).

An Emotional Tale of Romance in the Shadows of War: Bawaal Movie 2023

Bollywood set a bar of expectations from romantic travel movies, i.e., postcard-perfect sites, lead pairs falling in love, and adventure. In this Movie, the director tries to teach life lessons and weave a love engagement with the World War II tragedy story.

Ajay Dixit, or Ajju (Varun Dhawan), is a famous middle school teacher in Lucknow who has tried to be fake about his job and appearance. He showcased that he is a perfect cookie with a perfect life. His students are in his control, and they allow him to do so during the lecture rather than teach them. He married a bright and full-of-life girl, Nisha (Janhvi Kapoor) because she is good for his reputation or image in society. However, on the wedding day, he witnesses that

Nisha has epilepsy, which might tarnish his perfect image. So, Ajay bars her from leaving the house and mistreats her at home.

After this, Ajay slaps one of the students, unaware that he is the son of a legislator, leading to Ajay's suspension with pending disciplinary action. To avoid this situation, he decided to distract everyone's attention and complete his history syllabus from the sites of the Holocaust in Europe. He sends his students videos of himself giving history lessons in key places where World War II unfolded. His parents sponsor this trip, so Nisha joins him too, and slowly, they fall in love. The rest of the movie is about reliving the war victims' pain and grief. In this process, they experience weakness, failure, and the truth of life.

This film takes the audience to some heart-wrenching sites of human tragedy, i.e., Anne Frank's house, Omaha Beach, War memorials in Berlin, and Auschwitz. In the end, when they go to Auschwitz, a survivor (an actor) assures them that "every relationship" goes through its own Auschwitz. The lead couple fights with their doubts and demons in their external and internal journey.

Explanation of Dark Tourist Places in Movie

Auschwitz – During World War II and the Holocaust, Nazi Germany operated over 40 extermination and concentration camps in Germany in 1939. Auschwitz I was the main camp in Oświęcim, Auschwitz II was the extermination and concentration camp (the crudest violation of humankind and related to the darkest period in history) with the gas chamber, Auschwitz III was a labor camp for the chemical mixture IG Farben, and many more camps were there. When Germany started World War II by conquering Poland, Auschwitz I was converted from an army barracks to a prisoner of war camp. In the two years of camp establishment, most Polish inmates were brought, but after two years, criminals from Germany were also brought there. The first "Gassing" technique was used around August 1941 for Soviet and Polish prisoners in block 11 of Auschwitz I.

Auschwitz II was started in 1942 to bring Jews from all over the places where Germany occupied Europe to the gas chambers. Those who were not murdered via Gas chambers were killed by disease, starvation, beating, exhaustion, and medical experiments. On January 27, 1945, the Soviet Red Army entered Auschwitz to end the war and sent most of the camp's population to the camps in Germany and Austria. Since 2005, January 27 has been memorialized as an International Holocaust Remembrance Day (Heřmanová, 2015).

Anne Frank House was a house and biographical museum of the writer Anne Frank, devoted to the Jewish wartime. This building is situated on a Prinsengracht canal in central Amsterdam, the Netherlands. During World War II, Anne Frank, her family, and four other people were hidden in the room of this house from Nazi execution. She started writing a diary from June 12, 1942, to August 1, 1944. She was 13-years old when she began writing and died at 15 in Belsen Concentration Camp. She was the most famous victim of the Holocaust, the most famous child of the 20th century, and, She, with her sad, shy smile, was the icon of that time. In the starting phase, she wrote for herself, but one day, in 1944, she listened to a radio broadcast where a member of the Dutch government in exile announced that after the war, he would collect accounts (especially letters and diaries) of eyewitness or sufferers which could be made available for the public. After this announcement, she was inspired to write a diary with amendments to suit the post-war public.

She did not survive the war, but her diary during the war was published in 1947. Ten years later, a foundation was established in Anne Frank's name to protect her house from the developers who wanted to destroy the block. This house was opened as a museum on May 3, 1960, and provided a space to preserve the life of Anne Frank, her family, and other victims of the Holocaust (Busby & Devereux, 2015).

Omaha Beach – Omaha Beach was one of the five beaches from landing sectors selected for the maritime assault during World War II. On June 6, 1944, two US infantry divisions, i.e., 1st and 29th· landed at Omaha beach in the morning. The troops fight through the waist-deep water and are fired upon by the Germans. The defender invaded Germany and occupied France with the Normandy landings. It was known as D-Day, which cost the loss of thousands of soldiers on the seashore of Omaha Beach when Germany's cruel defense caught them off-guard.

War Memorials in Berlin – This memorial is located in a large public park, i.e., Großer Tiergarten, which is situated west of the city center in the Berlin capital of Germany. This memorial was constructed in 1945 by the Soviet Union to honor their war dead after a few months of city capture. This place was associated mainly with the 80,000 soldiers of the Soviet Armed Forces who died during the Battle of Berlin in 1945. On May 8, the anniversary of Victory in Europe is celebrated as a ceremony at the memorial.

Conclusion

Bawaal Movie offers a virtual tour of a World War II tragedy by apparently operative assassination, capable of engaging hearing and sight. It also, entirely in a realistic way, involves the physiological sense, kinesthesis sense, and responsive sense beyond anything that would be possible on site today. According to literary and media scholar Gary Weissman, "When there is a chance to display the Holocaust in its terrors and cruelty, there is an actual problem that 'Holocaust is not unrepresentable, but it is not only representable'" (Kaelber, 2007). Put another way, those who are not sufferers or testimonies of that tragedy can neither be relived nor sympathetically experience the trauma but only experience it in its imagined and mediated forms. This research presents an understanding of promoting edutainment agendas; the movies must design their interpretation to educate their audiences and provide historically and academically accurate grounded information. Presenting raw and real formation of history in the movies will arouse audience learning and engagement. Although the aim of providing fact-based data was questioned by commercial concerns to support business, there would be a shared understanding among the filmmakers that history will be the priority and finding the right fairness between entertainment and education.

This research reveals that an agenda of edutainment is not only to entertain audiences but also to educate the audience about difficult histories. Anne Frank's house, Omaha Beach, War memorials in Berlin, and Auschwitz are dark film location tours with the twist of entertainment. These experiences taught the audience how to visualize things that are not there. Rebuilding the imaginary scenes from the films will create a different space that overwrites histories and real landscapes. After analyzing the film, it reflected that visiting dark tourist places is important to enhance children's education. Through visits to dark sites, children develop a dynamic structure of moral knowledge, action tendencies, and emotional experiences. Dark tourism could be an alternative teaching style that can be advantageous for understanding how outdoor teaching may develop children's emotional and moral selves. In addition, this study has contributed to the underdeveloped topic of dark tourism, i.e., movies and dark tourist places, edutainment, and alternative teaching methods.

References

Busby, G., & Devereux, M. H. (2015). Dark tourism in context: The diary of Anne Frank. *European Journal of Tourism, Hospitality and Recreation*, *6*, 27–38.

Dresler, E. (2023). Multiplicity of moral emotions in educational dark tourism. *Tourism Management Perspectives*, *46*, 101094. https://doi.org/10.1016/j.tmp.2023.101094

Heřmanová, E. (2015). Holocaust tourism as a part of the dark tourism. *Czech Journal of Social Sciences, Business and Economics*, *4*(1), 16–33. https://doi.org/10.24984/cjssbe.2015.4.1.2

Kaelber, L. (2007). A memorial as virtual traumascape: Darkest tourism in 3D and cyber-space to the gas chambers of Auschwitz. *E-Review of Tourism Research (ERTR)*, *5*(2).

Korstanje, M. E. (2020). The dark tourist: Consuming dark spaces in the periphery. In M. E. Korstanje & H. Seraphin (Eds.), *Tourism, terrorism and security* (pp. 135–150). Emerald Publishing Limited. https://doi.org/10.1108/978-1-83867-905-720201009

Korstanje, M. E. & Handayani, B. (Eds.). (2017). *Gazing at death: Dark tourism as an emergent horizon of research.* Nova Science Publisher's.

Lennon, J. (2017). Dark tourism. In J. Lennon (Ed.), *Oxford research encyclopedia of criminology and criminal justice.* Oxford University Press. https://doi.org/10.1093/acrefore/9780190264079.013.212

Lewis, H., Schrier, T., & Xu, S. (2022). Dark tourism: Motivations and visit intentions of tourists. *International Hospitality Review*, *36*(1), 107–123. https://doi.org/10.1108/IHR-01-2021-0004

Mathews, J. (2018). Cinematic thanatourism and the purloined past: The "game of thrones effect" and the effect of game of thrones on history. In K. N. McDaniel (Ed.), *Virtual dark tourism* (pp. 89–112). Springer International Publishing. https://doi.org/10.1007/978-3-319-74687-6_5

Minic, N. (2012). Development of "dark" tourism in the contemporary society. *Journal of the Geographical Institute Jovan Cvijic, SASA*, *62*(3), 81–103. https://doi.org/10.2298/IJGI1203081M

Stone, P. (2013). Dark tourism scholarship: A critical review. *International Journal of Culture, Tourism and Hospitality Research*, *7*(3), 307–318. https://doi.org/10.1108/IJCTHR-06-2013-0039

Tham, A. (2013). Film tourism: The original 3D movies: A review of: Places of the imagination: Media, tourism, culture. Stijn Reijnders Surrey, Ashgate, 2011, ISBN 978-1-4094-1977-8. *Tourism Geographies*, *15*(1), 173–175. https://doi.org/10.1080/14616688.2012.633224

Walter, T. (2009). Dark tourism: Mediating between the dead and the living. In R. Sharpley & P. R. Stone (Eds.), *The darker side of travel* (pp. 39–55). Multilingual Matters. https://doi.org/10.21832/9781845411169-004

Wyatt, B., Leask, A., & Barron, P. (2019). *Designing dark tourism experiences: An exploration of edutainment interpretation at lighter dark visitor attractions*. Edinburgh Napier University.

Zheng, C., Zhang, J., Qiu, M., Guo, Y., & Zhang, H. (2020). From mixed emotional experience to spiritual meaning: Learning in dark tourism places. *Tourism Geographies*, *22*(1), 105–126. https://doi.org/10.1080/14616688.2019.1618903

Chapter 16

Unexplored Avenues of Dark Tourism in Bangladesh

Md. Wasiul Islam, Md. Mahfuz Ur Rahman and Shakil Ahmed

Khulna University, Bangladesh

Abstract

Visits to locations connected to historical atrocities, tragedy, suffering, or presumably dreadful events are referred to as "dark tourism". While Bangladesh may not be widely known for dark tourism, several unexplored avenues may be of interest to those who engage in this type of unique and unconventional tourism experience. In addition to creating job opportunities and income generation in Bangladesh, it can achieve educational objectives, communicate with a broad audience, raise awareness of events of the past, and quench people's thirst for information, which can help them to comprehend a society. Though Bangladesh has several somber locations connected to tragic events including World War II, liberation war sites, mass killing sites, shipbreaking yards, Rohingya refugee camps, riots, and the mother language movement, traditional tourism predominates there. If managed responsibly, these varied resources, somber locations, and histories, some of which date back to 1800, could make Bangladesh a dark tourism destination. Although dark tourism in Bangladesh has the potential to contribute to historical awareness, preservation, educational opportunities, and socioeconomic development, it is yet unexplored due to a lack of knowledge, experience, policy, effective marketing, and some controversies. By approaching responsibly, Bangladesh can leverage its history to attract visitors' interests in exploring the darker aspects of the country's past. Hence, this chapter is designed to explore the status and potential significance, prospects, and challenges of dark tourism in Bangladesh. The findings will help policymakers, tourists, and other stakeholders to explore and enjoy enormous benefits from Bangladesh's untapped dark tourism opportunities.

Dark Tourism, 213–240

Published under exclusive licence by Emerald Publishing Limited
doi:10.1108/978-1-83797-336-120241016

Keywords: Dark tourism; historical awareness; educational opportunities; challenges; marketing; development; Bangladesh

Introduction

Tourism is a complicated phenomenon encompassing a diverse spectrum of people who are progressively seeking new and unique experiences that meet their demands, which also has an impact on the business as well as the host's socio-cultural, economic, and physical environments (Fonseca et al., 2015; Leiper, 1979). Nowadays, it has expanded beyond the conventional sun, beach, and picturesque landscapes and is always changing (Fonseca et al., 2015). Along with conventional tourist destinations, people have a fascination for places, things, and occasions that are somehow associated with death, tragedy, brutality, and catastrophic events, whether intentionally or not (Stone, 2005; Tanaya & Garry, 2006). The term "dark tourism," which was invented by Foley and Lennon (1996), is a term used often to characterize this kind of travel that breaks up the dullness of typical vacations. The roots of dark tourism can be traced back to ancient times, with Roman gladiatorial contests and public executions serving as early examples (Lennon, 2017; Stone, 2006; Stone & Sharpley, 2013). Having roots so deep in the civilization's history, it has gained popularity among tourists and researchers in recent years (Chang, 2014; Light, 2017; Stone & Sharpley, 2008). For a variety of reasons, people may be drawn to these very emotional environments, including personal connection, appetite for knowledge, education, learning, and understanding of the events that took place at that location (Magano et al., 2022). Depending on these reasons and site typologies, dark tourism is classified by experts worldwide that spans from battlegrounds to holocausts and is further measured by the extent of darkness (Fonseca et al., 2015; Stone, 2006). Tourism is well-known for the many benefits that it can provide to its host communities (Archer et al., 2005; ILO, 2016; Jamin et al., 2020; Kim et al., 2013; Liu & Var, 1986; Song et al., 2022; Zaei & Zaei, 2013). It is also one of the most important sectors of national development contributing to job creation (Hwang & Lee, 2019; Lea, 2006; Manzoor et al., 2019; Neto, 2003; Prasad & Kulshrestha, 2015; Wakimin et al., 2018) and also regarded as the third-largest export sector in the world after chemicals and petroleum (Rasool et al., 2021). If maintained responsibly, dark tourism as a segment of more conventional tourism (heritage tourism) (Chang, 2014) can contribute to local community development, cultural exchange, mutual understanding, and income generation through its ability to attract millions of tourists globally (Fonseca et al., 2015). Besides, it aims to prevent historical (dark) repetition by recalling, reflecting upon, and commemorating the most sinister parts of human history (Stone, 2005; Tanaya & Garry, 2006). Dark tourism may also be a useful tool for fostering peace, compassion, tolerance and understanding, education, and constructive social changes when managed responsibly (Friedrich et al., 2018; Light, 2017; Magano et al., 2023). Despite these significant contributions to positive societal changes, dark tourism is controversial in many aspects (Grebenar, 2018; McKenzie, 2018;

Stone, 2011); however, being connected to civilization's remarkable pasts, it aids in comprehending the societies we live in as the American historian David McCullough said, "*History is who we are and why we are the way we are*" (Salager-Meyer et al., 2012). Therefore, research on dark tourism is essential for ensuring this sector's sustainable growth and responsible and ethical management to minimize controversy in a way that benefits society.

Despite its enormous potential to be a major driver of societal progress, dark tourism is still unexpectedly underexplored in underdeveloped nations (Friedrich & Johnston, 2013). Bangladesh is a least developed country (Byron & Mridha, 2021; World Bank, 2023), and dark tourism is not frequently studied and practiced here, although numerous projects are being undertaken and implemented to grow various forms of traditional tourism (Business Standard, 2022; Rubaiyat & Shabuj, 2022; Siddiqui & Rayhan, 2022), and Bangladesh's gross domestic product (GDP) and labor market have benefited significantly from tourism in recent decades (Business Insider Media Ltd., 2022; Hafsa, 2020; Hossain & Khanal, 2020; Islam et al., 2021; Uddin et al., 2023). Bangladesh can expedite its anticipated graduation to a developed country (Rahman, 2023) by capitalizing on tourism's contribution to Sustainable Development Goals (SDGs) (Pratt, 2022; UNWTO, n.d.) and investing in this untapped potential of dark tourism. To further enhance the economic benefits of this industry, the government of Bangladesh (GoB) has undertaken a project to create a national tourism master plan for the next 20 years (Islam et al., 2021). However, the final strategic plan draft contains slight hints of dark tourism promotion rather is more focused on eco, cultural, heritage, historical, and other forms of tourism (MoCAT, 2022; Solimar International, 2023), while a distinct classification could have emphasized dark tourism in Bangladesh. Moreover, responsible management of this overlooked tourism can offer a new dimension to this development mission. Bangladesh can also steer freshers into the dark tourism industry, leveraging on their curiosity (Camp et al., 1984) which will also help to alleviate the graduate job crisis (Sarker et al., 2021; Shahriar et al., 2021) alongside contributing to other socioeconomic developments (Hoque et al., 2022; Wang et al., 2020).

A community's interactions with their departed also take on a variety of shapes with changes in geography (Light, 2017; Walter, 2012). Due to this distinction, it is no longer wise to analyze the tourism and death relationship in other regions of the world using Western frameworks (Lee et al., 2012; Yoshida et al., 2016), which demands the creation of new terminologies, and frameworks that describe the concept of local death and dead. Additionally, there has been very little research done on dark tourism in Asia (Lee et al., 2012; Yan et al., 2016; Yoshida et al., 2016; Zheng et al., 2017). Likewise, despite being home to a good number of historically profound dark tourist attractions (see Table 16.1), its significance is also underestimated in Bangladesh. These sites stand as solemn testaments to World War II, the struggle for independence, genocide, political unrest, mutiny, shipwrecking, ethnic conflict, 228 years old prison, children correction centers, and displacement (Ahmed, 2020; Amin, 2021; Dunn, 2015; Gandhi Ashram Trust, n.d.; Gwin, 2014; Karin et al., 2020; Mithu, 2019; Molla et al., 2022; Momen, 2010; Roy, 2018). Therefore, by filling in the significant gap in the

Table 16.1. Some Significant and Potential Dark Tourism Spots and Tragic Events in Bangladesh.

Spot Type/Incident	Name of the Events
Cemetery	Chittagong War Cemetery (WW2), Maynamati War Cemetery
Mass Killing Sites and Memorials	Shaheed Minar, Mujibnagar Monument, Ryerbazar Badhyabhumi, Different 1971 Mass Killing Sites, Pilkhana Mutiny, other premises
Shipyards	Chottogram Ship Wrecking Yard
Refugee Camps	Kutupalong Refugee Camp
Abandoned Houses and Prisons	Old Dhaka Central Jail, Delduar and Sakhipur Prison, and other abandoned buildings
Museums	Various Liberation War Museums (Bangladesh Police, National and regional museums, Bangabandhu Military Museum, etc.), Sheikh Mujib – Agartala Case Memorial Museum, and so on
Riots/Ethnic Conflicts	Dhaka Riot 1930, Direct Action Day 1946, Noakhali Massacres 1950, East Bengal Massacres1962, Rajshahi Massacres 1964, Hathazari Riot 2012, Fatehpur Ethnic Conflict 2012, Chirirbandar Riot 2012, and so on

Source: Authors' compilation.

literature, this chapter attempts to showcase the prospects of dark tourism in Bangladesh. In this regard, this chapter is divided into several sections: global perspectives of dark tourism, importance, status, challenges, potential, and marketing of dark tourism in Bangladesh, as well as made some recommendations to improve its status.

Global Perspectives of Dark Tourism

Since the beginning of humanity and even in the 20th century, this world has witnessed countless events of brutality, deaths, tragedies, and so on. Concentration/death camps, nuclear explosion sites, mass killing grounds, and sites of mass destruction and disasters have attracted people for different reasons including personal connection, commemoration, reflection, and learning. It also serves as a potent tool for fostering historical understanding, reconciliation, and healing processes, allowing communities to engage with difficult pasts and stimulate economic growth. Regardless of these benefits, concerns regarding ethical considerations, commodification of suffering, and potential negative impacts on local communities remain prevalent (Grebenar, 2018; McKenzie, 2018; Stone, 2011). To enhance the scope of dark tourism beyond heritage tourism, promote it as a tool for global peace, and understand its intricate nature, a multidisciplinary novel approach is a must. Further research on dark tourism can focus on examining commodification, product development, marketing, cultural conflicts, and uses of electronic media and devices to improve the tourist experience and minimize associated negative impacts.

Dark tourism which was once unexplored in many countries has experienced significant growth since 2013 (Pradhan et al., 2023). By 2030, the global dark tourism market is projected to reach between US$ 35 and 45 billion (Future Market Insights, 2022; Reports and Insights, 2023). The United Kingdom, the United States, and China are the most successful countries in dark tourism studies (Pradhan et al., 2023), and Romania has firmly established itself as a world trademark in this industry (Minić, 2012). However, the scope of research on this area has not grown as anticipated (Shekhar & Valeri, 2022).

The Western world has led the dark tourism study, resulting in a Eurocentric bias in comprehending the phenomenon, while it is less studied in Asian countries. These offer more potential for exploring dark tourism in an Eastern light since the local cultural, social, and behavioral factors are highly unique to Western civilizations. The anticipated differences between frameworks and terminologies that describe Eastern dark tourism may help to understand the interconnection, similarities/dissimilarities, and motives that caused those tragedies in the first place. To achieve these objectives, further intensive studies are required to understand the relationship between living and dead, and how it is perceived locally and why.

Dark Tourism in Bangladesh: A Case

Dark tourism is not well practiced and studied in Bangladesh resulting in a substantial lack of literature, even though it recalls, reflects, and honors the most horrific moments in human history, boosts the community, economy, education, and aids in comprehending society. There is barely any literature on dark tourism in Bangladesh that can aid in spearheading all the upcoming endeavors to improve the situation. To our knowledge, this research is the first of its kind on this specific form of tourism in Bangladesh.

Bangladesh was brutally born in 1971 during the liberation war between the East Pakistan (now Bangladesh) and West Pakistan (now Pakistan) which has its roots in 1952's language movement (Alexander, 2013; Karim & Akter, 2021). The martyrs of this war were interred in various places – known as mass graves – across the country, and yet some of them are being discovered (Ghosh, 2016; Salim, 1971). There are also cemeteries, monuments, and other spots that are connected directly or indirectly to the language movement, the liberation war, and other tragic events. Bangladesh preserving such places should have prioritized dark tourism parallelly to nature-based and other forms of tourism. The following sections will discuss various aspects of the potential of dark tourism in Bangladesh.

Importance of Dark Tourism in Bangladesh

For a country formed out of genocide, there are no adequate words to express the significance of dark events and associated tourism. It is a way to recognize, a gateway to connect to and honor the sacrifices made by almost 3 million Bengali citizens and hundreds of thousands of women who were sexually abused and killed during the independence war in 1971, (Ahmed, 2022; Jahan, 2012; National Archives, n.d.) and acknowledge the "Genocide of 1971." However, inadequate implementation of tourism policies and poor governance poses significant challenges to its development for periods unknown (Islam et al., 2021). Many historically significant and potential dark tourism sites have been built over and lost due to inadequate preservation efforts and a lack of foreseeing capabilities of the concerned bodies (Bangladesh Post, 2019b; Daily Star, 2009; Dhaka Tribune, 2019; Financial Express, 2019; Jahid & Ronie, 2017; Mithu, 2019; Seraj, 2016). By recognizing the potential of dark tourism and these spots (see Table 16.1) to raise awareness, reduce violence, and generate revenue, governments and private organizations may ensure the protection and conservation of these sites while fostering community engagement.

Visiting the above-mentioned dark sites can inspire tourists to learn more about the country's history and foster tolerance, understanding, and empathy, thereby preventing similar atrocities in the future. Such visits are also known to stimulate empathy and sentiment among adults and a sense of patriotism in children (Israfilova & Khoo-Lattimore, 2019). In Bangladesh, where corruption is a challenging issue (Monem & Baniamin, 2017; Siddique, 2015), fostering patriotism through civic duties can be a crucial weapon in fighting against corruption.

Tourism has the potential to reinforce a nation's backbone if managed responsibly (Fayissa et al., 2008; Ivanov & Webster, 2007), and due to its wide-reaching impact across SDGs, tourism presents a valuable opportunity for developing nations like Bangladesh. Dark tourism, attracting millions of visitors (Fonseca et al., 2015), offers a potential avenue for new job opportunities and revenue generation through visiting these sites and creating a popular service industry in the country. Bangladesh's numerous historical dark sites, coupled with improved marketing and service management, could attract tourists seeking unique experiences. Besides, fresh graduate engagement in this tourism sector holds promise for reducing unemployment and managing this complex and sensitive tourism that benefits local communities and the national economy as well. Therefore, prioritizing this type of tourism along with other forms of traditional tourism will help to achieve the strategic tourism goal for Bangladesh (Islam et al., 2021).

Status of Dark Tourism in Bangladesh

Management

Even after 52 years of independence, Bangladesh's dark tourism is still in its infancy. The vast majority of Bangladesh's "dark sites" are occupied by the mass graves of the liberation war, but the exact number of these graves is still not specified. The War Crimes Facts Finding Committee, Bangladesh, anticipated that there are about 5,000 mass graves in Bangladesh (Dhaka Tribune, 2019). Since dark tourism is currently not widely practiced and studied in Bangladesh, it has insufficient management, which makes it difficult to maintain most of the sites. The preservation of these historical places is further impeded by a law that was passed over a century ago (Hossain, 2019). In 2018, the Executive Committee of the National Economic Council (ECNEC) approved a BDT 442 crore (approximately US$ 40.17 million) project for preserving 271 killing grounds and construction of monuments for the mass killing sites (Dhaka Tribune, 2019). According to the Officials of the Ministry of Liberation War Affairs, 204 mass killing fields and mass graveyards were identified till 2019; however, private organizations and war researchers said otherwise (Dhaka Tribune, 2019). While there is ongoing confusion about the number of mass graves, new sites are being discovered lately (Dhaka Tribune, 2022; Salim, 1971). In addition, 200-year-old jails are being dismantled to make way for brand-new parks, retail centers, and movie theaters (Mithu, 2019). Unfortunately, many of such important and historical landmarks of dark tourism in Bangladesh are transformed into new structures rather to protect those which reduce the potential of developing this unique form of nontraditional tourism.

Training and Education

Training management for tour guides and tourism and hospitality management (THM) studies for students in Bangladesh are also underdeveloped, failing to provide them with the latest theoretical and practical knowledge and required skills to be developed as competent and confident tourism workforce. While some initiatives have been taken by private organizations and the government to address this need, they remain inadequate. The THM studies in Bangladesh are often struggling with problematic curriculum design, inadequate facilities, lack of skilled instructors and access to current, high-quality journals/resources, and a decrease in the caliber of graduates (Alam, 2022; Kabir & Rahman, 2020). Tourism being a human-intensive service sector relies on efficient human resource development. Dark tourism being controversial in many aspects (Grebenar, 2018; McKenzie, 2018; Stone, 2011), and having an educational value, calls for quality interpretation of the incidents and places (Light, 2017) to deliver a clear message about the product (incident) and place to the visitors. This requires more qualified personnel to maintain the moral, emotional, and learning standards. In fact, Bangladesh lacks building quality guides (Alauddin et al., 2021; Kamruzzaman, 2018) who may facilitate the interpretation to the guests by keeping historical integrity intact. In summary, current THM education and training are inadequate to meet the increased human resource needs for dark tourism's growth. Moreover, despite its rich history, dark tourism research and management are not prioritized in Bangladesh.

Market Development and Marketing

Tourism marketing is one of the main areas where the Bangladeshi government needs to make significant improvements, and government expenditure will be the main factor guiding the industry forward (MoCAT, 2022). The Bangladesh government established organizations like the Bangladesh Parjatan Corporation (BPC) and the Ministry of Civil Aviation and Tourism (MoCAT) in 1972 and 1986, respectively, to promote tourism (BPC, 2023; MoCAT, 2014; Roy & Roy, 2015). Subsequently, the Bangladesh Tourism Board (BTB) was established in 2010 under the Bangladesh Tourism Policy, 2010, to address the rising demand from the private sector and tourism industry professionals (Roy & Roy, 2015). These two bodies under the MoCAT lead Bangladesh's tourism marketing, strategic planning, spot identification, and preservation, while MoCAT oversees the industry's overall development (Ananya et al., 2019). Despite managing different tourism websites by these organizations, Bangladesh provides no explicit mention of dark tourism management or promotion.[1–4] Notably, some websites classify dark tourism sites under "heritage tourism," while room exists for distinct categorization. Both Beautiful Bangladesh and BPC acknowledge the landmarks associated with the Independence War but refrain from classifying them as dedicated dark tourism destinations. Bangladesh, despite its diverse cultural, natural, and historical heritage, remains a largely undiscovered destination for foreign tourists (Nahar et al., 2015; Rahman, 2021). This lack of visibility is attributed to shortcomings in marketing and management, hindering the country's tourism potential. To address this, the GoB has embarked on a collaborative project with international organizations to enhance the competitiveness of tourist destinations, reform policies, and develop a 20-year national master plan for tourism sector growth (Islam et al., 2021).

Overall, while the commodification of dark tourism necessitates careful handling distinct from traditional tourism, scant evidence exists on government websites or the strategic master plan (MoCAT, 2022) regarding such considerations. As a less studied and less popular form of tourism, literature on shaping a dark tourism market in Bangladesh remains scarce.

Challenges of Dark Tourism in Bangladesh

Despite its long and complex history of tourism, dark tourism is relatively a new concept in Bangladesh (Ashworth & Isaac, 2015; Hartmann et al., 2018). It is also not typically seen as a viable tourist attraction in Bangladesh due to ignorance, lack of information, funding, lack of skilled manpower, and cultural barriers (Hassan et al., 2013; Molla et al., 2022). To address these challenges, it is imperative to conduct more research on this topic. Alongside research on,

[1]Ministry of Civil Aviation and Tourism.
[2]Beautiful Bangladesh.
[3]Bangladesh Tourism Board.
[4]Bangladesh Parjatan Corporation.

conservative initiatives for, and familiarizing dark tourism, it is compulsory to address the fundamental constraints for developing the destinations such as the need for significant financial resources, skilled personnel, and effective marketing strategies (Šerić et al., 2020). Some of the challenges of dark tourism in Bangladesh are discussed briefly:

Public–Private Partnerships (PPPs)

The tourism industry in Bangladesh has been suffering from ineffective and absence of public–private collaboration and other associated reasons (Islam et al., 2021; Rakib & Hassan, 2020), which may threaten its future if not addressed timely. One of the prime examples is the Chottogram Ship Wrecking Yard which was once open for tourism but currently does not allow outsiders anymore. Having the risks of stigmatization, the commodification of dark tourism necessitates sophisticatedly skilled manpower and precautionary and revised integrative development initiatives to approach it holistically. PPPs have the potential to greatly contribute to these issues through multifaceted approaches.

Marketing Obstacles

Globally, tourism marketing is becoming more and more important, especially in the third-world countries like Bangladesh where the fastest-growing tourist attractions are turning into hubs for the creation and promotion of new goods and services (Ahmmed, 2013). Social media platforms, including Facebook, Instagram, YouTube, and personal blogs, exert a profound influence on consumers' perceptions of destinations and the unique experiences they offer. When utilized effectively, these platforms can make a substantial contribution to the growth of the tourism industry (Tarannum, 2020). However, Bangladesh lacks an appropriate and advanced curriculum and materials for the THM studies program and training of sales and marketing personnel to cope with advanced level marketing on social media and other virtual platforms, which is a must task to achieve for promoting Bangladesh as a tourist destination to both domestic and international tourists (Shamsuddoha, 2005).

Political Instability and Corruption

Tourism is directly and adversely affected by political instability (Ryu, 2005; Saha & Yap, 2014). It amplifies the perceived risk associated with the destination, leading to negative publicity and a resultant decline in tourist arrivals (Lepp & Gibson, 2003; Thapa, 2012). This instability also tarnishes the country's reputation, potentially discouraging potential tourists who prioritize safety and security when choosing their travel destinations (Abbasi et al., 2022). Political unrest is a common occurrence in Bangladesh and gets worse whenever the government changes (Rahman & Rashid, 2018). This instability not only hinders economic development (Chawdhury, 2016) but also defames the international image of the country, deterring tourism and causing revenue losses (Ryu, 2005). Additionally, the prevalence of corruption can have a detrimental impact on tourism demand, both directly and indirectly (Maria et al., 2022). Consequently, the resolution of political chaos and related risks and safety concerns remain crucial for unlocking the full potential of Bangladesh's dark tourism industry.

Media Influence, Education, and Sustainability

It is crucial to recognize that the public's perception of landscapes particularly those linked to dark tourism locations is greatly influenced by the media portrayals. Consequently, any initiatives aimed at promoting dark tourism must be accompanied by comprehensive education and awareness-raising campaigns that emphasize the historical, cultural, and ethical significance of these sites (Ntunda, 2014). Moreover, it is essential to acknowledge the significant financial, cultural, and infrastructural hurdles that can prevent dark tourism from being self-sustaining. Therefore, any endeavors to promote dark tourism should be carefully planned and implemented, with a focus on addressing these challenges and maintaining historical authenticity (Poirier, 2000).

Ethical Dilemmas

Visitors to dark tourism sites are not passive observers. They are critical thinkers who may be concerned about sanitized historical depictions (Austin, 2002). Dark tourism presents ethical challenges in the interpretation, presentation, and management of tragic events, and despite its rise in popularity, the ethical implications of dark tourism remain a subject of contentious debate within the academic community (Potts, 2012). Concerns have been raised about the propriety of displaying such sites for tourists (Clarke et al., 2014; Light, 2017). Critics also have highlighted the misrepresentation or trivialization of death sites but have been less likely to offer alternatives. Unfortunately, a critical gap exists in the engagement between academic scholarship and practitioners managing dark sites (Light, 2017), hindering the anticipation and avoidance of potential errors. It also has implications for both visitors and local communities, encompassing concerns related to commercial development and the potential for exploitation of such sites (Verma, 2013). These ethical dilemmas are not unique and are likely to be pertinent to the context of Bangladesh as well.

Tourism Infrastructure and Services

Despite a significant tourist influx, the tourism industry of Bangladesh lags behind its Asian counterparts. This is attributed to several challenges, including outdated infrastructure (Parveen, 2013), data deficit (Islam, 2015), inadequate health care, and insufficient transportation infrastructure and services (Biswas & Mamun-Or-Rashid, 2019; Kamruzzaman, 2018; Kamruzzaman & Fariha, 2018; Karim, 2014). This infrastructural deficit (Khadaroo & Seetanah, 2007; Masson & Petiot, 2009) poses significant safety hazards to residents and visitors alike, potentially leading to traffic accidents, injuries, and even fatalities (Mohiuddin, 2023; Mondal & Haque, 2017). Moreover, inadequate infrastructure impedes efficient law enforcement patrols, potentially increasing crime rates and discouraging visitors' extended stays (Chowdhury, 2020; Majumder, 2015). Unpleasant experiences at transit points can discourage tourists, who are frequently regarded as one of the major sources of generating foreign reserves (Matthew et al., 2021; Rasool et al., 2021). These challenges, if unaddressed, can significantly threaten both the tourism industry and the national economy, possibly hindering the growth of the dark tourism business before it can fully develop.

Site Preservation and Management

There are thousands of killing grounds dispersed across Bangladesh where heinous activities took place. Despite having several sites with profound historical, cultural, and racial values, the management and preservation initiatives revolving around these sites are questionable which threaten their historical and cultural significance (Bangladesh Post, 2019b; Dhaka Tribune, 2019; Jahid & Ronie, 2017; Mithu, 2019). This negligence is resulting in the constant degradation and disappearance of these crucial resources.

Lack of Awareness

Despite 52 years having passed since Bangladesh's liberation, a significant portion of the population remains unaware of the concept of dark tourism and its potential within the country. This lack of awareness extends to the negligence of historical events like the 1971 Chuknagar massacre and the demolition of a 200-year-old prison in Dhaka (Dutta, 2023; Mithu, 2019). The fading memory of such atrocities is further exacerbated by the disappearance of mass graves due to human activities, such as construction. This highlights a critical gap in public and institutional knowledge regarding dark tourism, hindering its potential for historical understanding, and national significance.

Security Concerns

Bangladesh's burgeoning dark tourism industry also faces significant security challenges, potentially deterring visitors, and jeopardizing safety. For example, inadequate infrastructure can leave tourists stranded in isolated locations and vulnerable to attack and theft (Mohiuddin, 2023). There are a lot of risky tourist spots in impoverished countries (Ghaderi et al., 2017; Kim, 2003), and Bangladesh has not been an exception to it (Kamruzzaman, 2018). Since safety and security are fundamental infrastructures, their development is directly tied to the growth of tourism of a nation. The GoB formed a special unit named "Tourist Police" in 2009 (Hossain & Islam, 2019) to ameliorate the situation, but still much improvement is needed to be made in this aspect (Alauddin et al., 2021).

Negative Perceptions

Dark tourism, while controversial due to its sensitive nature and potential for exploiting tragedies, can also serve as a powerful tool for historical education and victim remembrance (Friedman, 2024). It is a complex phenomenon, influenced by both positive and negative factors, challenging the notion that it is solely a fascination with death and disaster (Seaton & Lennon, 2004). Bangladesh faces significant tourism challenges such as overcrowding, disasters, unemployment, and an inadequate economic framework, although its true image is not as negative as projected abroad (Zahra, 2012). However, dark tourism in Bangladesh may pose concerns over historical authenticity, as it can lead to sanitized and distorted narratives, potentially deceiving visitors and complicating complex historical narratives. History can also be misused, exaggerated to some degree, and presented frequently in a way that is biased by political, national, and commercial interests (Chen, 2022). Therefore, it is important to ensure that dark tourism sites are developed and managed respectfully and responsibly, with input from local communities (Masanti, 2016). In addition to promoting Bangladesh's tourism industry, it is also crucial to address the negative public perception of some landscapes, particularly those associated with dark tourism (Ntunda, 2014).

Development (Market), Management, and Marketing

Tourism management encompasses functional areas such as finance, human resources, marketing and promotion, tourist servicing and satisfaction, and sustainable development (Starck, 2006; Swarbrooke & Page, 2012). Unlike traditional tourism, dark tourism emphasizes the unique experiences and emotions associated with the sites, while also incorporating educational and transformative components (Davchev, 2017). This necessitates a deep understanding of visitor motivations, which can vary depending on an individual's connection to the site or a general interest in its historical significance (Bird, 2013; Dunkley et al., 2011). Therefore, effective marketing for dark tourism destinations requires a delicate balance between acknowledging the sensitive nature of these sites and meeting the expectations of tourists.

The rise of dark tourism necessitates a nuanced understanding of its unique marketing challenges and opportunities (Veiga et al., 2022). Scholars have also advocated for a dedicated field of "dark marketing," that would aid in applying marketing principles to domains encompassing death, destruction, and ethically sensitive topics (Stone et al., 2018). Its success depends on its acceptance by local communities; therefore, the tour operators should also be aware of the potential ethical issues associated with this type of tourism besides meeting individual needs. They should avoid exploiting victims or their families, as well as sensationalizing or trivializing death and suffering (Sharpley, 2009; Sharpley & Stone, 2009).

Dark tourism development involves some key steps: site development and management, ethical considerations, partnership building, strengths, weaknesses, opportunities, and threats (SWOT) analysis, and tourism marketing plan

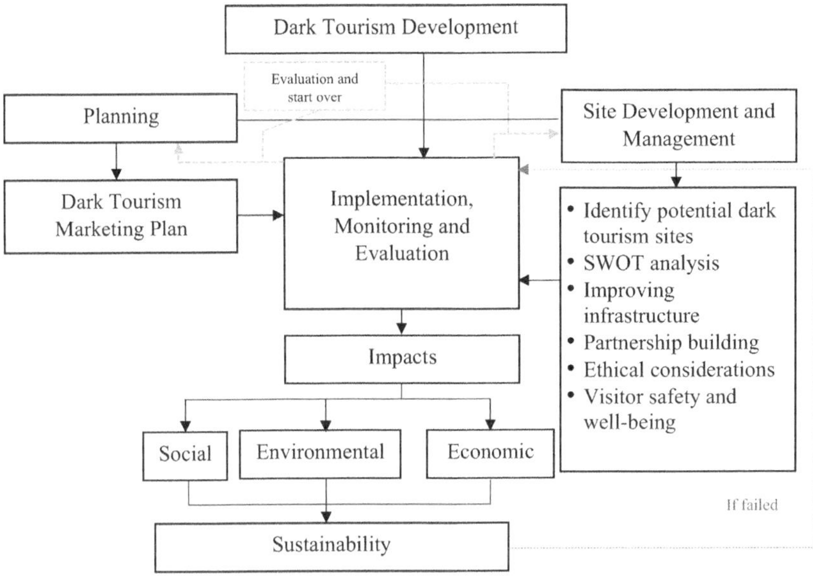

Fig. 16.1. Framework for Developing Dark Tourism in Bangladesh. *Source:* Authors' compilation.

(see Fig. 16.1). This framework for responsible dark tourism development emphasizes ethical practices, avoiding exploitation, and educating visitors about the human cost of tragedies, while sustainability, visitor safety, and well-being are paramount considerations throughout the development process. Partnership building involves collaboration with local communities and public and private stakeholders, while a SWOT analysis identifies strengths, weaknesses, opportunities, and threats (see Table 16.2). A tourism marketing plan promotes dark tourism destinations to target markets, using specific strategies like advertising, public relations, and social media. By adhering to this framework, dark tourism destinations can offer meaningful visitor experiences while ensuring responsible and respectful management.

A dark tourism product development begins with the identification of a potential site or attraction which demands contextual and successful marketing strategies and platforms (see Fig. 16.2). Careful evaluation of the historical foundations, locations, darkness of the spot, local heritages, community perception, stories, legends, and other sociocultural factors is crucial for these product developments. The technological enhancements, skilled manpower (tour operators, guides, and others), and a good integration of context analysis can elevate the visitor experience and business sustainability. As the commodification of death is highly sensitive and controversial, businesses should be developed with careful implementation of regarded policy, rules, and regulations. Alike any other

Table 16.2. A SWOT Analysis for Dark Tourism Development.

Strengths	*Weaknesses*
• Numerous dark and historical attractions • Complex history of conflict and death • Widely accessible accommodation facilities throughout the country • Low cost of products and services • Hospitable local people	• Lack of qualified personnel in the tourism sector • Absence of adequate frameworks for travelers • Poor exploration and management of dark tourism sites • Lack of safety and security • Lack of entertainment facilities • Lack of infrastructural development • Shortage of professional guides • Lack of proper promotional and marketing activities • Absence of a proper tourism policy for dark tourism • Conversion of dark tourism sites to other commercial land uses
Opportunities	*Threats*
• Designing innovative marketing strategies • Dark tourism sites protection • Supporting local communities and businesses • Promotion of peace and understanding	• Inadequate tourism management • No effective plan to attract international tourists • Political instability • Lack of investment • Poor integrity among the stakeholders for working together

Source: Authors' compilation.

tourism, tourists' experience is a vital indicator to assess the business sustainability, while it should also align with local sociocultural perceptions, historical significance, and be economically viable. The alignment with historical significance is crucial to maintain historical integrity. On the contrary, further assessments on tourist product and marketing strategies and platform can facilitate the business to sustain maintaining social, cultural, and historical significance. Ultimately, successful dark tourism products should be seamlessly integrated into the destination's broader tourism strategy, aligning with its overall goals and objectives.

Potentials of Dark Tourism in Bangladesh

Dark tourism presents a unique opportunity for education, community support, and grief processing (Seaton & Lennon, 2004). It fosters public understanding of

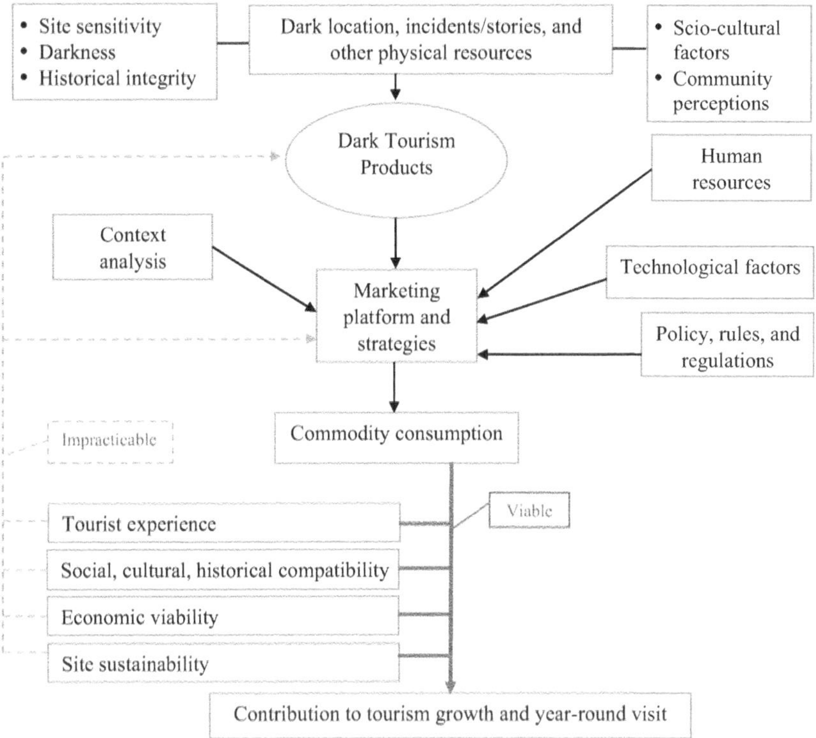

Fig. 16.2. A Model of Dark Tourism Product Development. *Source:* Authors' compilation.

historical tragedies by facilitating exploration and experience of relevant sites, while simultaneously preserving cultural heritage (Shiming, 2009). Driven by a thirst for the past and an interest in history, dark tourism attracts visitors who seek to understand the foundation of contemporary societies. Furthermore, dark tourism enhances awareness of historical events and their interconnectedness, ultimately contributing to a more comprehensive understanding of the world. Unlike conventional tourism, dark tourism offers a multifaceted and impactful experience, fostering long-term awareness and promoting personal and societal development through engagement with past events (Daams, 2007; Karsai & Morina, 2021). Although dark tourism holds promise for educational and social progress in Bangladesh, its potential remains largely unrealized due to limited exploration and implementation. It can contribute to national and individual development in the following ways:

Economic Impacts

Tourism is key to the national economy since it provides local businesses, governments, and citizens with financial assistance (Incera & Fernández, 2015; Wang et al., 2022) and contributes to the country's economy boosters – small businesses – that directly participate in the community well-being (Jamin et al., 2020). It is also the only sector that contributes to all 17 SDGs directly or indirectly and particularly addresses SDGs 8, 12, and 14 (Pratt, 2022; UNWTO, n.d.). In underdeveloped nations, it is also well known for generating a large amount of tax income and jobs. Bangladesh can utilize its unique history and past events to propel economic development, alleviate the graduate job crisis, empower local businesses, and elevate its overall living standard.

SocioCultural Benefits

Tourism plays a significant role in global employment and preventing rural migration. Dark tourism offers additional advantages, allowing locals to improve their livelihoods through employment and enhancing communication skills. Moreover, it facilitates cultural understanding by enabling visitors to engage with the local past (Niemelä, 2010). Consequently, communities can leverage their history to showcase their cultures and educate the international community about their nation's unique characteristics and traditions (Jamin et al., 2020). It can help with intergroup understanding, which aids in the processes of healing and reconciliation following tragedies or conflicts. Additionally, visiting dark tourism places can provide special insights into regional customs and civilizations, encouraging respect for diversity and cultural heritage.

Educational Value

Visiting historical sites can help children understand history and their heritage by enjoying the experience and learning about historical tragedies, sacrifices, and old equipment (McIntosh, 1999). Dark tourism can showcase real stories of triumph and tragedy, passed down from generation to generation, inspiring young people's love, and responsibility for their country, and aware them of the consequences of horrendous events. Visitors can also share their stories, experiences, and curiosity about dark tourism sites which may foster solidarity and understanding. This stimulated communication can reduce barriers and promote global understanding.

Psychological Benefits

Dark tourism can enhance visitors' environmental attitudes toward dark relics, which can further increase the perceived benefits of the experience, and conservation efficiency through raising awareness. Additionally, stronger dark tourism motivation can directly influence the psychological/health benefits of the experience, such as emotional and stress relief (Chang, 2014). For example, a visit to the

Bangabandhu Memorial Museum – the site of the planned and brutal assassination of the founder of Bangladesh Sheikh Mujibur Rahman and his family members on August 15, 1975 – can help in learning about the life and legacy of Sheikh Mujibur Rahman, as well as the tragic events that led to his assassination. Additionally, exhibits on Bangladesh's war and history offer self-reflection, commemorating war heroes and victims. As dark tourism comes with opportunities to release stress, strengthen emotional attachment, positive changes in environmental and conservation behavior, dark sites in Bangladesh with proper management can be transformed into hubs for human well-being.

Likewise, there are some other remarkable incidents in the history of Bangladesh like the brutal assassination (on May 30, 1981, in Chattogram) of former President Major General Ziaur Rahman. Moreover, the murder of four Awami League and national leaders viz., former Vice-President Syed Nazrul Islam, former Prime Minister Tajuddin Ahmed and Captain (Rtd.) Mansur Ali, and former Home Minister A H M Quamruzzaman in the jail on November 3, 1975, teaches us the life and politics of these leaders and creates a psychological attachment.

Conservation of Historical Sites

The 100-year-old policy on heritage conservation in Bangladesh (Hossain, 2019) hampers the protection of historical sites, leading to the loss of significant cultural assets, including mass killing sites and other relics. Tourism can play a bigger role in preserving these sites and expediting the process through increased awareness and familiarity with these sites, while a new and age-suited policy formulation is a time-consuming task. The associated organization or the GoB can also generate revenue from tourist influx and can reinvest in the maintenance and conservation of these locations.

Some Recommendations to Improve Dark Tourism in Bangladesh

- The Antiquities Act, 1968, must be revised to facilitate the preservation of the dark tourist sites, historical locations, and other important and historical relics in Bangladesh.
- The government should prioritize the development of dark tourism sector by supporting (education, training, and assessment) private tour operators, investing in new infrastructure, and launching marketing initiatives.
- Community knowledge and social tolerance should be enhanced, and access to tourism information must be made easier using virtual platforms and markets, traditional and social media, etc.
- Guides must be well-trained to interpret the history correctly, relay a good and sensible message to the visitors, ensure service quality, and minimize distortion/trivialization of histories.

- Demolition, reconstruction, and renovation of historically profound locations and establishments must be criticized from the touristic point of view.
- Research followed by a complete survey is imperative to develop a central database of dark tourism sites that are yet explored, unexplored, and dispersed across the country.
- Having unique attributes, further research is required on dark tourism marketing, consumption, and promotion as even the presence of souvenir shops is debated in dark tourism spots.
- Enlisting with a district-wise classification of countrywide spots of dark relics on websites like Beautiful Bangladesh for public access should also be prioritized.
- The academic curriculum on tourism and hospitality management studies should be revised by individual higher education institutes to ensure extensive study on dark tourism.
- Seasonal visit to dark tourism sites is advised for school and college students, specifically for elementary and secondary school students.
- Uses of the latest visualization technologies such as virtual reality (VR) can be introduced to provide quality and near real-life experiences to the visitors.
- In recent years, customer harassment at the Hazrat Shahjalal International Airport has increased (Bangladesh Post, 2019a; Business Standard, 2023; Daily Star, 2022); therefore, along with infrastructural development, it should be resolved to sustain the tourism industry in Bangladesh.
- The current political situation, corruption, and a long history of political instability must be addressed to attract foreign tourists and increase inbound tourism.
- Well-renowned and widely followed sites such as dark-tourism.com can be contacted to promote dark tourism in Bangladesh and control their content by providing exact information about these resources and associated situations.
- Tourism brings both positive and negative impacts on the culture, society, economy, and environment (Chong, 2020; Laakso, 2011). Therefore, any form of tourism should be promoted responsibly and sustainably to keep the negative effects minimum. Tourism in Bangladesh typically being challenged from the aspects of governance, management, and policy may go in vain.
- Public–private combined strengths and resources should be manipulated effectively to overcome obstacles and expedite the establishment of tourism infrastructure boosting destination marketing and promotion efforts and encouraging tourism practices.

Conclusion

Dark tourism is a rapidly growing industry that has the potential to improve communities, promote cross-cultural understanding, and boost the economy. Even if dark tourism may be debatable, good management can promote peace, understanding, and constructive societal progress. In nations such as Bangladesh, which have unexplored historical tourist resources, investing in and promoting dark tourism offers a chance to generate significant economic returns and accelerate national development. But even with all potential, Bangladesh's dark tourism sector growth is beset with various challenges. Lack of collaboration between the public and private sectors, marketing issues, political unrest and corruption, media prejudices, and moral dilemmas are some of the major obstacles. The growth of the sector is further hampered by inadequate infrastructure, a lack of understanding, and security concerns. To attain the desired success, embrace dark tourism, and unleash its revolutionary potential, these obstacles must be dealt with first, while a pragmatic policy and honest implementation can lead the way through.

References

Abbasi, A. R., Raza, A., & Shaikh, H. (2022). A nexus between political instability & international tourism demand. *International Journal of Social Science & Entrepreneurship*, *2*(2), 297–312.

Ahmed, J. (2020). Children's voices to ring in abandoned jails. *The Business Standard*. https://www.tbsnews.net/bangladesh/childrens-voices-ring-abandoned-jails-47181. Accessed on December 10, 2023.

Ahmed, I. (2022). *Recognising the 1971 Bangladesh genocide: An appeal for rendering justice*. https://www.ucl.ac.uk/risk-disaster-reduction/sites/risk_disaster_reduction/files/bd_genocide_booklet.pdf. Accessed on October 6, 2023.

Ahmmed, M. M. (2013). An analysis on tourism marketing in Bangladesh. *International Proceedings of Economics Development and Research*, *67*(8), 35–39.

Alam, S. S. (2022). Tourism and hospitality management education in Bangladesh: Problems and prospects. *Journal of Business Studies*, *3*(1), 145–154.

Alauddin, M., Dipu, S. M. A., & Uddin, M. R. (2021). Tourism products and services in Bangladesh: Challenges and suggestions. In *Tourism products and services in Bangladesh: Concept analysis and development suggestions* (pp. 445–459). Springer.

Alexander, C. (2013). Contested memories: The Shahid Minar and the struggle for diasporic space. *Ethnic and Racial Studies*, *36*(4), 590–610.

Amin, S. B. (2021). A brief history and overview of tourism in Bangladesh. In *The economy of tourism in Bangladesh: Prospects, constraints, and policies* (pp. 39–82). Springer.

Ananya, S. A., Islam, S., & Kuri, B. C. (2019). An exploratory study on the role of major tourism organizations in flourishing the tourism industry in Bangladesh. *IOSR Journal of Business and Management*, *21*(12), 35–43.

Archer, B., Cooper, C., & Ruhanen, L. (2005). The positive and negative impacts of tourism. In W. F. Theobald (Ed.), *Global tourism* (3rd ed., pp. 79–102). Butterworth-Heinemann.

Ashworth, G. J., & Isaac, R. K. (2015). Have we illuminated the dark? Shifting perspectives on 'dark'tourism. *Tourism Recreation Research*, *40*(3), 316–325.

Austin, N. K. (2002). Managing heritage attractions: Marketing challenges at sensitive historical sites. *International Journal of Tourism Research*, *4*(6), 447–457.

Bangladesh Post. (2019a). Harassment at Dhaka airport continues. *Bangladesh Post*. https://bangladeshpost.net/posts/no-let-up-in-passenger-harassment-at-dhaka-airport-13724. Accessed on November 16, 2023.

Bangladesh Post. (2019b). *Sylhet mass graves remains uncared.* https://bangladeshpost.net/posts/sylhet-mass-graves-remains-uncared-for-11890. Accessed on October 8, 2023.

Bird, G. R. (2013). 12 Place identities in the Normandy landscape of war. In *Dark tourism and place identity: Managing and interpreting dark places* (Vol. 37, p. 167). Routledge.

Biswas, C., & Mamun-Or-Rashid, M. (2019). Investigating barriers in tourism development: The case of Bangladesh. *International Journal of Economics, Commerce and Management*, *7*, 654–663.

BPC. (2023). *Bangladesh Parjatan Corporation Government of the People's Republic of Bangladesh.* http://www.parjatan.gov.bd/site/page/68c2d54a-112f-4804-8624-9180bf59df98/-. Accessed on October 19, 2023.

Business Insider Media Ltd. (2022). *Bangladesh to employ 17 lakh workers in tourism sector by 2023: WTTC.* https://www.businessinsiderbd.com/economy/news/21797/bangladesh-to-employ-17-lakh-workers-in-tourism-sector-by-2023-wttc#google_vignette. Accessed on September 09, 2023.

Business Standard. (2022). *12 projects to galvanise Bangladesh's tourism sector.* The Business Standard. https://thefinancialexpress.com.bd/trade/12-projects-to-galvanise-bangladeshs-tourism-sector-1641095402. Accessed on September 09, 2023.

Business Standard. (2023). *Passengers slam Dhaka airport authorities for gross mismanagement.* The Business Standard. https://thefinancialexpress.com.bd/trade/12-projects-to-galvanise-bangladeshs-tourism-sector-1641095402. https://www.tbsnews.net/economy/aviation/passengers-slam-dhaka-airport-authorities-gross-mismanagement-656190. Accessed on November 16, 2023.

Byron, R. K., & Mridha, R. U. (2021). *Becoming A developing nation: Bangladesh reaches A Milestone.* https://www.thedailystar.net/frontpage/news/becoming-developing-nation-bangladesh-reaches-milestone-2052161. Accessed on September 11, 2023.

Camp, C. J., Rodrigue, J. R., & Olson, K. R. (1984). Curiosity in young, middle-aged, and older adults. *Educational Gerontology*, *10*(4–5), 387–400.

Chang, T.-Y. (2014). Dark tourism the effects of motivation and environmental attitudes on the benefits of experience. *Revista Internacional de Sociología*, *72*(2), 69–86.

Chawdhury, J. (2016). *Political instability a major obstacle to economic growth in Bangladesh.* Thesis. Business Management, Centria University of Applied Sciences.

Chen, K. (2022). *A comprehensive analysis of dark tourism: Understanding tourist behavior and implications for future research.* Department of Urban Planning, University College London.

Chong, K. L. (2020). The side effects of mass tourism: The voices of Bali islanders. *Asia Pacific Journal of Tourism Research*, *25*(2), 157–169.

Chowdhury, S. H. (2020). Tourism in Bangladesh: Highlighting the development trend. *International Journal of Science and Business*, *4*(5), 23–36.

Clarke, R., Dutton, J., & Johnston, A. (2014). Shadow zones: Dark travel and postcolonial cultures. *Postcolonial Studies*, *17*(3), 221–235.

Daams, E. (2007). *Dark tourism: Bearing witness or crass spectacle.* Matador Network. http://www.bravenewtraveler.com/2007/10/10/dark-tourismmore-than-a-pectacle/. Accessed on February 02, 2021.

Daily Star. (2009). Mass graves of '71 left uncared for. https://www.thedailystar.net/news-detail-118340. Accessed on October 10, 2023.

Daily Star. (2022). *Passengers at mass hearing share bitter experiences at the Dhaka airport*. https://www.thedailystar.net/life-living/travel/news/passengers-mass-hearing-share-bitter-experiences-the-dhaka-airport-3092006. Accessed on November 16, 2023.

Davchev, A. (2017). The dark tourism – An alternative form of tourism or an unrevealed social phenomen. *New knowledge Journal of science*, *6*(4), 73–77.

Dhaka Tribune. (2019). *Genocide Day: Hundreds of mass graves lie unmarked or have been constructed over*. https://www.dhakatribune.com/bangladesh/nation/172308/genocide-day-hundreds-of-mass-graves-lie-unmarked. Accessed on October 8, 2023.

Dhaka Tribune. (2022). *Mass grave from Liberation War discovered in Jhenaidah*. https://www.dhakatribune.com/bangladesh/nation/270364/mass-grave-from-liberation-war-discovered-in. Accessed on October 17, 2023.

Dunkley, R., Morgan, N., & Westwood, S. (2011). Visiting the trenches: Exploring meanings and motivations in battlefield tourism. *Tourism Management*, *32*(4), 860–868.

Dunn, J. (2015). Where boats go to die: Fascinating pictures inside a Bangladeshi ship-breaking yard where once mighty liners are reduced to scrap. *Daily Mail Online*. https://www.dailymail.co.uk/news/article-3257853/Where-boats-die-Fascinating-pictures-inside-Bangladeshi-ship-breaking-yard.html. Accessed on September 9, 2023.

Dutta, P. K. (2023). The massacre at Chuknagar: A forgotten chapter of the liberation war history. *The Business Standard*. https://www.tbsnews.net/thoughts/massacre-chuknagar-forgotten-chapter-liberation-war-history-634882. Accessed on Decemb er 12, 2023.

Fayissa, B., Nsiah, C., & Tadasse, B. (2008). Impact of tourism on economic growth and development in Africa. *Tourism Economics*, *14*(4), 807–818.

Financial Express. (2019). 75 mass grave sites of Liberation War martyrs lying. *The Financial Express*. https://thefinancialexpress.com.bd/national/country/sust-student-found-dead-in-sylhet. Accessed on October 8, 2023.

Foley, M., & Lennon, J. J. (1996). JFK and dark tourism: A fascination with assassination. *International Journal of Heritage Studies*, *2*(4), 198–211.

Fonseca, A. P., Seabra, C., & Silva, C. (2015). Dark tourism: Concepts, typologies and sites. *Journal of Tourism Research & Hospitality*, *S2*(002), 2–6.

Friedman, Z. (2024). Dark tourism ethics and criticisms. https://wheretheroadforks.com/dark-tourism-ethics-and-criticisms/. Accessed on December 10, 2023.

Friedrich, M., & Johnston, T. (2013). Beauty versus tragedy: Thanatourism and the memorialisation of the 1994 Rwandan Genocide. *Journal of Tourism and Cultural Change*, *11*(4), 302–320.

Friedrich, M., Stone, P. R., & Rukesha, P. (2018). Dark tourism, difficult heritage, and memorialisation: A case of the Rwandan genocide. In *The Palgrave handbook of dark tourism studies* (pp. 261–289). Springer.

Future Market Insights. (2022). *Dark tourism market demand, size & value 2022–2032*. https://www.futuremarketinsights.com/reports/dark-tourism-sector-overview#:~:text=The%20Dark%20Tourism%20market%20is,US%24%2036.5%20Billion%20by%202032. Accessed on September 10, 2023.

Gandhi Ashram Trust. (n.d.). *Mahatma Gandhi in Noakhali - Gandhi Ashram Trust.* https://gandhiashramtrust.org/mahatma-gandhi-in-noakhali/. Accessed on September 27, 2023.

Ghaderi, Z., Saboori, B., & Khoshkam, M. (2017). Does security matter in tourism demand? *Current Issues in Tourism*, *20*(6), 552–565.

Ghosh, S. (2016). *Most mass graves in 3 southern districts still unmarked.* https://www.thedailystar.net/footages-1971-liberation-war/killing-grounds/most-mass-graves-still-unmarked-3-southern-districts. Accessed on October 8, 2023.

Grebenar, A. (2018). *The commodification of dark tourism: Conceptualising the visitor experience*. University of Central Lancashire.

Gwin, P. (2014). *The ship-breakers.* https://www.nationalgeographic.com/magazine/article/The-Ship-Breakers. Accessed on September 27, 2023.

Hafsa, S. (2020). Economic contribution of tourism industry in Bangladesh: At a glance. *Global Journal of Management and Business Research*, *20*(1), 29–38.

Hartmann, R., Lennon, J., Reynolds, D. P., Rice, A., Rosenbaum, A. T., & Stone, P. R. (2018). The history of dark tourism. *Journal of Tourism History*, *10*(3), 269–295.

Hassan, M. R., Ullah, M. M., & Chowdhury, M. S. A. (2013). Impact of tourism in Bangladesh economy. *World*, *3*(6), 1–19.

Hoque, M. A., Lovelock, B., & Carr, A. (2022). Alleviating Indigenous poverty through tourism: The role of NGOs. *Journal of Sustainable Tourism*, *30*(10), 2333–2351.

Hossain, I. (2019). Old law prevents preserving war sites. Prothom Alo. https://en.prothomalo.com/bangladesh/Old-law-prevents-preserving-war-sites. Accessed on November 14, 2023.

Hossain, M. K., & Islam, S. (2019). An analysis of destination attributes to enhance tourism competitiveness in Bangladesh. *African Journal of Hospitality, Tourism and Leisure*, *8*(2), 1–17.

Hossain, M. I., & Khanal, B. P. (2020). SWOT analysis of forest protected area based sustainable tourism development: A case study in the Lawachara National Park, Bangladesh. *Journal of Tourism and Hospitality Education*, *10*, 71–89.

Hwang, J., & Lee, J. (2019). A strategy for enhancing senior tourists' well-being perception: Focusing on the experience economy. *Journal of Travel & Tourism Marketing*, *36*(3), 314–329.

ILO. (2016). *Sustainable tourism: A driving force of job creation, economic growth and development*. https://www.ilo.org/global/about-the-ilo/newsroom/news/WCMS_480824. Accessed on August 27, 2023.

Incera, A. C., & Fernández, M. F. (2015). Tourism and income distribution: Evidence from a developed regional economy. *Tourism Management*, *48*, 11–20.

Islam, M. M. (2015). *Development of authenticity in perspective of tourism in Bangladesh*. Bachelor's Thesis. Centria University of Applied Science. https://www.theseus.fi/handle/10024/96858. Accessed on November 15, 2023.

Islam, M. W., Chowdhury, D., & Khan, A. (2021). Tourism in Bangladesh: A future outlook. In A. Sharma & A. Hassan (Eds.), *Future of tourism: An Asian perspective* (pp. 97–122). Springer.

Israfilova, F., & Khoo-Lattimore, C. (2019). Sad and violent but I enjoy it: Children's engagement with dark tourism as an educational tool. *Tourism and Hospitality Research*, *19*(4), 478–487.

Ivanov, S., & Webster, C. (2007). Measuring the impact of tourism on economic growth. *Tourism Economics*, *13*(3), 379–388.

Jahan, R. (2012). Genocide in Bangladesh. In *Centuries of genocide* (pp. 248–276). Routledge.

Jahid, A. M., & Ronie, S. (2017). *Where are the 1971 mass graves, killing fields in Dhaka?* https://www.thedailystar.net/video-stories/watch-where-are-the-1971-mass-graves-killing-fields-dhaka-189223. Accessed on October 8, 2023.

Jamin, A., Zain, Z. M., Sakarji, S. R., Ahmad, N., & Beta, R. M. D. M. (2020). The benefits of dark tourism experience among visitors in Malaysia. *KnE Social Sciences*, 219–228.

Kabir, K., & Rahman, M. S.-U. (2020). General tourism and hospitality education and research in Bangladesh. In *Tourism policy and planning in Bangladesh* (pp. 179–192).

Kamruzzaman, M. (2018). Assessment of destination-specific factors of Bangladesh: A review of Saint Martin Island as an emerging tourist destination. *Journal of Business Studies*, *39*(1), 113–125.

Kamruzzaman, M., & Fariha, M. S. (2018). Effectiveness of proper health and hygiene practices: A study on ensuring hygienic environment in Bangladeshi tourist destinations. In *Proceedings of the international conference on industrial engineering and operations management* (pp. 6–8), Bandung, Indonesia.

Karim, M. R. (2014). Policy analysis of tourism development in Bangladesh compared with the Bhutanese policy. *Journal of Public Administration and Policy Research*, *6*(6), 132–143.

Karim, M. R., & Akter, M. S. (2021). The contribution of women in the Bengali language movement (1952): A historical analysis. *British Journal of Arts and Humanities*, *3*(5), 116–127.

Karin, S., Chowdhury, M. A., Hasnat, M. A., & Tarin, N. J. (2020). Status of Rohingya in refugee camps of Bangladesh: A review study. *Open Access Library Journal*, *7*, e6575.

Karsai, Á. S., & Morina, D. (2021). Development of dark tourism destinations in Central and Eastern Europe. https://www.academia.edu/67360141/Development_of_Dark_Tourism_Destinations_in_Central_and_Eastern_Europe. Accessed on December 10, 2023.

Khadaroo, J., & Seetanah, B. (2007). Transport infrastructure and tourism development. *Annals of Tourism Research*, *34*(4), 1021–1032.

Kim, S.-M. (2003). International tourism in Korea: Barriers and challenges. *International Journal of Tourism Sciences*, *3*(2), 151–162.

Kim, K., Uysal, M., & Sirgy, M. J. (2013). How does tourism in a community impact the quality of life of community residents? *Tourism Management*, *36*, 527–540.

Laakso, P. (2011). *Tourism: Good or bad?: Impacts of tourism in Tanzania*. Bachelor's Thesis. Häme University of Applied Sciences. https://www.theseus.fi/bitstream/handle/10024/29547/Paula_Laakso.pdf?sequence=1&isAllowed=y. Accessed on December 8, 2023.

Lea, J. (2006). *Tourism and development in the Third World*. Routledge.

Lee, C. K., Bendle, L. J., Yoon, Y. S., & Kim, M. J. (2012). Thanatourism or peace tourism: Perceived value at a North Korean resort from an indigenous perspective. *International Journal of Tourism Research*, *14*(1), 71–90.

Leiper, N. (1979). The framework of tourism: Towards a definition of tourism, tourist, and the tourist industry. *Annals of Tourism Research*, *6*(4), 390–407.

Lennon, J. (2017). Dark tourism. https://researchonline.gcu.ac.uk/ws/portalfiles/portal/39971695/Lennon_Oxford_Diction_Dark_Tourism_17_10_2016_1_pdf. Accessed on December 8, 2023.

Lepp, A., & Gibson, H. (2003). Tourist roles, perceived risk and international tourism. *Annals of Tourism Research*, *30*(3), 606–624.

Light, D. (2017). Progress in dark tourism and thanatourism research: An uneasy relationship with heritage tourism. *Tourism Management*, *61*, 275–301.

Liu, J. C., & Var, T. (1986). Resident attitudes toward tourism impacts in Hawaii. *Annals of Tourism Research*, *13*(2), 193–214.

Magano, J., Fraiz-Brea, J. A., & Leite, Â. (2022). Dark tourists: Profile, practices, motivations and wellbeing. *International Journal of Environmental Research and Public Health*, *19*(19), 12100.

Magano, J., Fraiz-Brea, J. A., & Leite, Â. (2023). Dark tourism, the holocaust, and well-being: A systematic review. *Heliyon*. https://doi.org/10.1016/j.heliyon.2023.e13064

Majumder, D. (2015). Contributions and loopholes of tourism sector in Bangladesh. *Jagannath University Journal of Social Sciences*, *3*, 1–19.

Manzoor, F., Wei, L., Asif, M., Haq, M. Z. u., & Rehman, H. U. (2019). The contribution of sustainable tourism to economic growth and employment in Pakistan. *International Journal of Environmental Research and Public Health*, *16*(19), 3785.

Maria, K.-K., Ioanna, S., & Salomi, D. (2022). Nonlinear nexus between corruption and tourism arrivals: A global analysis. *Empirical Economics*, *63*(4), 1997–2024.

Masanti, M. (2016). Understanding dark tourism acceptance in Southeast Asia: The case of WWII Sandakan–Ranau Death March, Sabah, Malaysia. *Development of Tourism and the Hospitality Industry in Southeast Asia*, 113–125.

Masson, S., & Petiot, R. (2009). Can the high speed rail reinforce tourism attractiveness? The case of the high speed rail between Perpignan (France) and Barcelona (Spain). *Technovation*, *29*(9), 611–617.

Matthew, O. A., Ede, C., Osabohien, R., Ejemeyovwi, J., Ayanda, T., & Okunbor, J. (2021). Interaction effect of tourism and foreign exchange earnings on economic growth in Nigeria. *Global Business Review*, *22*(1), 7–22.

McIntosh, A. J. (1999). Into the tourist's mind: Understanding the value of the heritage experience. *Journal of Travel & Tourism Marketing*, *8*(1), 41–64.

McKenzie, B. (2018). 'Death as a commodity': The retailing of dark tourism. In *The Palgrave handbook of dark tourism studies* (pp. 667–691). https://doi.org/10.1057/978-1-137-47566-4_27

Minić, N. (2012). Development of dark tourism in the contemporary society. *Зборник Радова ГеоГрафскоГ Института Јован Цвијић" САНУ*, *62*(3), 81–103.

Mithu, A. I. (2019). Park in the prison. *The Business Standard*. https://www.tbsnews.net/feature/park-prison. Accessed on October 12, 2023.

MoCAT. (2014). *Background: Ministry of Civil Aviation and Tourism*. Government of the People's Republic of Bangladesh. https://mocat.gov.bd/site/page/7c990d63-87d4-4558-9f5b-9f6e738ef589/Background. Accessed on October 19, 2023.

MoCAT. (2022). *Tourism Master Plan (Draft)*. https://file.portal.gov.bd/uploads/8ef2a505-c131-4a8f-b6d0-2ac4760cc936//631/6c4/06a/6316c406a9807875210649.pdf. Accessed on October 30, 2023.

Mohiuddin, M. (2023). *Effects of poor infrastructure on tourism in Bangladesh*. Bachelor's Thesis. Degree Programme in Restaurant Entrepreneurship. Laurea University of Applied Sciences.

Molla, M. R., Lima, F. S., & Tareq, S. M. H. (2022). Analyzing the use of different social media marketing techniques applied by dark tourism spots to attract the attention of dark tourists: A study on different social media marketing techniques of bangabandhu memorial museum in dhaka. *American International Journal of Multidisciplinary Scientific Research*, *12*(1), 1–24.

Momen, M. (2010). Bangladesh in 2009: The peril within. *Asian Survey*, *50*(1), 157–163.

Mondal, M., & Haque, S. (2017). SWOT analysis and strategies to develop sustainable tourism in Bangladesh. *UTMS Journal of Economics*, *8*(2), 159–167.

Monem, M., & Baniamin, H. M. (2017). Public service ethics and corruption in Bangladesh. In *Public administration in South Asia* (pp. 237–256). Routledge.

Nahar, K., Islam, S., & Rahman, M. K. (2015). Selection of a tourist attractions using AHP method: The case of Bangladesh. *World Journal of Sport Sciences*, *5*(3), 211–226.

National Archives. (n.d.). *The Independence of Bangladesh in 1971*. The National Archives. https://www.nationalarchives.gov.uk/education/resources/the-independence-of-bangladesh-in-1971/. Accessed on October 6, 2023.

Neto, F. (2003). A new approach to sustainable tourism development: Moving beyond environmental protection. *Natural Resources Forum*.

Niemelä, T. (2010). *Motivation factors in dark tourism. Case: House of terror*. Bachelor's Thesis. Lahti University of Applied Sciences. Degree Programme in Tourism and Hospitality Management. The Faculty of Tourism and Hospitality.

Ntunda, J. (2014). *Investigating the challenges of promoting dark tourism in Rwanda*. Anchor Academic Publishing.

Parveen, J. A. (2013). Current status of tourism industry in Bangladesh: An empirical evaluation. *Scientific Research Journal*, *1*(1), 41–57.

Poirier, R. (2000). Tourism in the African economic milieu: A future of mixed blessings. *The political economy of tourism development in Africa*, 29–36.

Potts, T. J. (2012). 'Dark tourism' and the 'kitschification' of 9/11. *Tourist Studies*, *12*(3), 232–249.

Pradhan, B., Singh, M., Mantry, A., El-Kilany, E., Raman, M. S., & Bhatia, A. (2023). Mapping the research trends on dark tourism: A bibliometric analysis and visualization. *Acta Universitatis Bohemiae Meridionalis*, *26*(2), 1–17.

Prasad, N., & Kulshrestha, M. (2015). Employment generation in tourism industry: An input–output analysis. *Indian Journal of Labour Economics*, *58*, 563–575.

Pratt, S. (2022). Can tourism contribute to the sustainable development goals? Evidence from Pacific Island countries & Territories. *International Marketing Journal of Culture and Tourism*, *1*(1), 45–61.

Rahman, M. M. (2021). Inbound tourism in Bangladesh: The trend of competitiveness. *Bangladesh Journal of Public Administration*, *29*(2), 64–78.

Rahman, M. (2023). Lessons for Bangladesh as a graduating LDC. *The Daily Star*. https://www.thedailystar.net/supplements/32nd-anniversary/towards-smart-bangladesh/news/lessons-bangladesh-graduating-ldc-3243446. Accessed on September 28, 2023.

Rahman, R., & Rashid, M. M. (2018). Political instability and economic growth in Bangladesh. *Innovative Issues and Approaches in Social Sciences*, *11*(2), 91–105.

Rakib, M. R. H. K., & Hassan, A. (2020). Tourism facility design standards and development in Bangladesh. *Tourism policy and planning in Bangladesh*, 71–83.

Rasool, H., Maqbool, S., & Tarique, M. (2021). The relationship between tourism and economic growth among BRICS countries: A panel cointegration analysis. *Future Business Journal*, *7*(1), 1–11.

Reports and Insights. (2023). *Dark tourism market sales & growth analysis report 2023*. https://www.reportsandinsights.com/report/dark-tourism-market. Accessed on September 10, 2023.

Roy, A. (2018). Noakhali riots, October 1946. In *Making peace, making riots: Communalism and communal violence, Bengal 1940–1947* (pp. 184–213). Cambridge University Press. https://doi.org/10.1017/9781108578790.007

Roy, S. C., & Roy, M. (2015). Tourism in Bangladesh: Present status and future prospects. *International Journal of Management Science and Business Administration*, *1*(8), 53–61.

Rubaiyat, M. Q. I., & Shabuj, M. (2022). What's the govt doing to promote tourism? *The Daily Star*. https://www.thedailystar.net/weekend-read/news/whats-the-govt-doing-promote-tourism-3166951. Accessed on September 11, 2023.

Ryu, S. (2005). Political instability and its effects on tourism. *Managing risk and crisis for sustainable tourism: Research and innovation (Best Education Network Think Tank V, Kingston, Jamaica June 16-19)*.

Saha, S., & Yap, G. (2014). The moderation effects of political instability and terrorism on tourism development: A cross-country panel analysis. *Journal of Travel Research*, *53*(4), 509–521.

Salager-Meyer, F., Ariza, M. Á. A., & Briceño, M. L. (2012). The voice of scholarly dispute in medical book reviews, 1890–2010. *Stance and voice in written academic genres*, 232–248.

Salim, A. (1971). Bangladesh: Literary responses to the tragedy of 1971. In A. G. Khattak & A. Salim (Eds.), *The land of two partitions and beyond* (pp. 166–181). Sustainable Development Policy Institute.

Sarker, M. A. R., Hashim, J. B., Haque, A., & Juhdi, N. B. (2021). Graduate employability: Perception of graduate students on soft skills towards employability in Bangladesh. *Journal of International Business and Management*, *4*(4), 1–14.

Seaton, A. V., & Lennon, J. J. (2004). Thanatourism in the early 21st century: Moral panics, ulterior motives and alterior desires. In *New horizons in tourism: Strange experiences and stranger practices* (pp. 63–82). CABI Publishing.

Seraj, H. (2016). *Liberation War: Mass graves in Magura still remain uncared*. https://observerbd.com/details.php?id=48033. Accessed on October 8, 2023.

Šerić, N., Mihanović, A., & Tolj, A. (2020). Model for the development of a specialized dark tourist product. *Transactions on Maritime Science*, *9*(02), 324–330.

Shahriar, M. S., Islam, K., Zayed, N. M., HASAN, K., & Raisa, T. S. (2021). The impact of COVID-19 on Bangladesh's economy: A focus on graduate employability. *The Journal of Asian Finance, Economics and Business*, *8*(3), 1395–1403.

Shamsuddoha, M. (2005). Opportunities of developing tourism industry in Bangladesh. In *Recent Trends in Design and Manufacturing Technologies (RTDMT-2005)*. Kumaraguru College of Technology.

Sharpley, R. (2009). Shedding light on dark tourism: An introduction. In R. Sharpley & P. R. Stone (Eds.), *The darker side of travel: The theory and practice of dark tourism* (pp. 3–22). Channel View Publications.

Sharpley, R., & Stone, P. R. (Eds.). (2009). *The darker side of travel: The theory and practice of dark tourism*. Channel View Publications.

Shekhar, S., & Valeri, M. (2022). Evolving themes in dark tourism research: A review study. *Tourism*, *70*(4), 624–641.

Shiming, L. (2009). *Research on disaster site tourism development: Based on the case of Wenchuan Earthquake* (Philosophy and Social Sciences Edition). Journal of Hebei University.

Siddique, M. A. B. (2015). Corruption in Bangladesh: Review and analysis. In R. N. Ghosh & M. A. B. Siddique (Eds.), *Corruption, good governance and economic development: Contemporary analysis and case studies* (pp. 85–100). World Scientific.

Siddiqui, K., & Rayhan, J. (2022). Tourism investments pouring in on infrastructure development. *The Business Standard*. https://www.tbsnews.net/bangladesh/tourism-investments-pouring-infrastructure-development-503882. Accessed on September 11, 2023.

Solimar International. (2023). Bangladesh ecotourism and conservation alliance. Solimar International. https://www.solimarinternational.com/project/sundarbans-bangladesh/. Accessed on September 30, 2023.

Song, Y., Su, Z.-W., Tao, R., & Umut, A. (2022). Revealing the effectiveness of tourism development on health in Asian economies. *Frontiers in Public Health*, *10*, 895221.

Starck, N. (2006). *Life after death: The art of the obituary*. Melbourne University Publishing.

Stone, P. (2005). Dark tourism: An old concept in a new world. *Tourism*, *QIV*(125), 20.

Stone, P. R. (2006). A dark tourism spectrum: Towards a typology of death and macabre related tourist sites, attractions and exhibitions. *Tourism: An International Interdisciplinary Journal*, *54*(2), 145–160.

Stone, P. R. (2011). Dark tourism and the cadaveric carnival: mediating life and death narratives at Gunther von Hagens' Body Worlds. *Current Issues in Tourism*, *14*(7), 685–701.

Stone, P. R., Hartmann, R., Seaton, A., Sharpley, R., & White, L. (2018). *The Palgrave handbook of dark tourism studies*. Springer.

Stone, P., & Sharpley, R. (2008). Consuming dark tourism: A thanatological perspective. *Annals of Tourism Research*, *35*(2), 574–595.

Stone, P. R., & Sharpley, R. (2013). Deviance, dark tourism and 'dark leisure'. In *Contemporary perspectives in leisure: Meanings, motives and lifelong learning: Meanings, motives and lifelong learning* (pp. 54–64). Routledge.

Swarbrooke, J., & Page, S. J. (2012). *Development and management of visitor attractions*. Routledge.

Tanaya, P., & Garry, P. (2006). Motivations of participants in dark tourism: A case study of Port Arthur, Tasmania, Australia. In *Taking tourism to the limits* (pp. 191–198). Routledge.

Tarannum, T. (2020). *Effectiveness of social media in promoting tourism in Bangladesh*. Master Thesis. KDI School. https://archives.kdischool.ac.kr/handle/11125/40910. Accessed on December 6, 2023.

Thapa, B. (2012). Tourism in Nepal: Shangri-La's troubled times. In *Safety and security in tourism* (pp. 124–145). Routledge.

Uddin, R., Arefin, M. A., Sima, S. A., & Arman, A. (2023). *Mapping tourists' mind to introduce and nurture data driven decision-making culture in the tourism sector for creating informal employment opportunities*. https://www.undp.org/bangladesh/blog/mapping-tourists-mind-introduce-and-nurture-data-driven-decision-making-culture-tourism-sector-creating-informal-employment#:~:text=Approximately%205%20million%20people%20are,this%20sector%20gradually%20within%202030. Accessed on September 29, 2023.

UNWTO. (n.d.). *Tourism in the 2030 Agenda*. UNWTO. https://www.unwto.org/tourism-in-2030-agenda. Accessed on October 28, 2023.

Veiga, J. M., Veloso, J. D., Pereira, S. Q., & Sousa, B. B. (2022). Market segmentation and dark tourism and the (Post) pandemic scenario. In *Handbook of research on green, circular, and digital economies as tools for recovery and sustainability* (pp. 295–310). IGI Global.

Verma, D. S. (2013). Exploiting tragedy: Dark tourism. https://doi.org/10.2139/ssrn.2306870

Wakimin, N., Azlinaa, A., & Hazman, S. (2018). Tourism demand in Asean-5 countries: Evidence from panel data analysis. *Management Science Letters*, *8*(6), 677–690.

Walter, T. (2012). Why different countries manage death differently: A comparative analysis of modern urban societies 1. *British Journal of Sociology*, *63*(1), 123–145.

Wang, K., Gan, C., Chen, L., & Voda, M. (2020). Poor residents' perceptions of the impacts of tourism on poverty alleviation: From the perspective of multidimensional poverty. *Sustainability*, *12*(18), 7515.

Wang, Y., Han, L., & Ma, X. (2022). International tourism and economic vulnerability. *Annals of Tourism Research*, *94*, 103388.

World Bank. (2023). *Bangladesh Overview: Development news, research, data*. World Bank. https://www.worldbank.org/en/country/bangladesh/overview#1. Accessed on September 11, 2023.

Yan, B.-J., Zhang, J., Zhang, H.-L., Lu, S.-J., & Guo, Y.-R. (2016). Investigating the motivation–experience relationship in a dark tourism space: A case study of the Beichuan earthquake relics, China. *Tourism Management*, *53*, 108–121.

Yoshida, K., Bui, H. T., & Lee, T. J. (2016). Does tourism illuminate the darkness of Hiroshima and Nagasaki? *Journal of Destination Marketing & Management*, *5*(4), 333–340.

Zaei, M. E., & Zaei, M. E. (2013). The impacts of tourism industry on host community. *European Journal of Tourism, Hospitality and Recreation*, *1*(2), 12–21.

Zahra, I. (2012). Destination image and tourism: A case study of Bangladesh. *European Journal of Business and Management*, *4*(6), 18–27.

Zheng, C., Zhang, J., Zhang, H., & Qian, L. (2017). Exploring sub-dimensions of intrapersonal constraints to visiting "dark tourism" sites: A comparison of participants and non-participants. *Asia Pacific Journal of Tourism Research*, *22*(1), 21–33.

Chapter 17

Reviving Forgotten Heritage as a Tourist Destination: A Case Study on Malcha Mahal Haunted Heritage Walks in Delhi

Terrance Ancheary[a,*], *Puneet Mehta*[b] *and Anish Mondal*[c]

[a]Indian Institute of Management, India
[b]Central University of Kerala, India
[c]National Council for Hotel Management - Institute of Hospitality, India

Abstract

Dark tourism, a phenomenon encompassing visits to sites associated with death, suffering, and historical atrocities, has gained increasing scholarly attention in recent years. This chapter explores the concept's historical roots, motivations for engagement, and its transformative impact on heritage sites, culminating in a case study of the Malcha Mahal Haunted Heritage Walks in Delhi. Despite being abandoned for centuries, this historic site gained prominence with the occupation by the self-proclaimed royal family of Oudh in the 1980s. The introduction of the Haunted Heritage Walk by Delhi Tourism aimed to harness the dark tourism potential of this previously neglected monument. However, initial challenges such as deteriorating conditions and safety concerns necessitated collaborative efforts with government departments for conservation and site preparation. Interviews with visitors revealed diverse motivations for attendance, ranging from thrill-seeking to appreciation of the monument's historical significance. Since the walk's inception, conservation efforts have commenced, accompanied by landscaping initiatives and amenities for visitors. While the future conservation and tranquility of Malcha Mahal remain uncertain amid increasing tourist interest, its transformation into an alternative destination within Delhi underscores the potential of dark tourism to revitalize neglected heritage sites. Sustainable tourism planning is essential to preserve the

*Corresponding author.

Dark Tourism, 241–264

doi:10.1108/978-1-83797-336-120241017

monument's unique features and manage tourism while maintaining its integrity for future generations.

Keywords: Dark tourism; Malcha Mahal; heritage; conservation; tourist motivations; sustainable tourism planning

Introduction

The term "dark tourism" was first coined by Foley and Lennon (1996), subsequently becoming the title of a book that, arguably, remains the most widely cited study of the phenomenon (Lennon & Foley, 2000). Their work was not however the first to focus upon the relationship between tourism and death, whether violent, untimely, or otherwise. Sites associated with war and atrocities have long been considered within a broader heritage tourism context, particularly from an interpretative perspective. For example, Uzell (1992) argues for the "hot" interpretation of war and conflict (interpretation that is as intense or passionate as the site/event), while Tunbridge and Ashworth's (1996) subsequent work on "dissonant heritage" develops an important conceptual framework for the management of such sites. This literature review explores the concept of dark tourism, its historical roots, motivations for engagement, and the transformative impact on heritage sites. Furthermore, it delves into the case study of Malcha Mahal Haunted Heritage Walks in Delhi to analyze how dark tourism is contributing to the revival of this forgotten heritage site.

Dark tourism, also known as thanatourism, has its roots in the historical practice of pilgrimage to sites associated with martyrdom or religious significance (Seaton, 1996). The evolution of dark tourism can be traced to the interest in visiting places linked to death and suffering, with early examples including visits to battlefields, execution sites, and graves of famous personalities (Lennon & Foley, 2000). Over time, the concept has expanded to encompass a diverse range of sites, including prisons, concentration camps, disaster zones, and haunted places. The phenomenon of dark tourism encompasses a wide array of destinations and forms, reflecting the diverse and complex nature of death-related attractions. Scholars frequently refer to specific destinations, such as the Sixth Floor in Dallas, Texas, infamous for the assassination of President Kennedy (Foley & Lennon, 1996). Additionally, various forms of dark tourism include visits to graveyards (Seaton, 2002), Holocaust memorials (Beech, 2000), sites of atrocities (Ashworth & Hartmann, 2005), prisons (Strange & Kempa, 2003; Wilson, 2004), and heritage tourism related to slavery (Dann & Seaton, 2001). The spectrum extends from unique experiences like the "Dracula Experience" in Whitby, UK, to Vienna's Funeral Museum, as well as visits to sites associated with 'famous' deaths or major disasters such as Ground Zero (Alderman, 2002).

Categorizing these death-related attractions is a complex task due to their sheer diversity, ranging from museums to sites of historical significance. Despite the long history of travel to such places, academic attention to the collective phenomenon known as "dark tourism" has only recently emerged (Foley and Lennon, 1996; Lennon & Foley, 2000). Scholars have attempted to define this tourist activity, using terms like "thanatourism" (Seaton, 1996), "morbid" (Blom, 2000), "black-spot" (Rojek, 1993), or, as Dann (1994a) creatively puts it, "milking the macabre." The academic exploration of dark tourism goes beyond defining terms and extends to the analysis of specific manifestations. Studies delve into war museums, employing both traditional and contemporary museology methods (Lennon & Wight, 2004), as well as genocide commemoration sites and the political ideologies attached to remembrance (Williams, 2004). While less explored, attention has been given to visitor motivations, ranging from morbid curiosity to a collective sense of identity or survival "in the face of violent disruptions of collective life routines" (Rojek & Urry, 1997, 61). The nuanced nature of dark tourism and the myriad motivations behind it highlight the complexity and richness of this evolving field of study.

Understanding the motivations behind engaging in dark tourism is crucial for comprehending the appeal of such experiences to tourists. MacCannell's (1976) seminal work on the "staged authenticity" of tourist attractions suggests that visitors are drawn to sites that provide an authentic and immersive experience, even if it involves confronting darker aspects of human history. Lennon and Foley (2000) identified curiosity, education, entertainment, and commemoration as key motivations for dark tourism engagement. Tourists may seek a deeper understanding of historical events, a thrilling experience, or an opportunity for contemplation and remembrance.

Dark tourism has the potential to transform overlooked and neglected heritage sites into economically viable and culturally significant tourist destinations. The increased visitor numbers can lead to economic revitalization for local communities, conservation efforts, and heightened awareness of historical and cultural significance (Stone, 2006). However, the delicate balance between preserving the integrity of the site and catering to the tourist demand poses challenges (Sharpley & Stone, 2009). Striking this balance is essential to ensure that dark tourism contributes positively to the preservation and appreciation of heritage.

Malcha Mahal Haunted Heritage Walks in Delhi

Malcha Mahal, located in the heart of Delhi, is a historical structure with a rich yet forgotten past (Smith, 2015). Once the residence of the exiled princesses of Oudh, it has long been abandoned and rumored to be haunted. In recent years, Malcha Mahal has gained attention through the organization of Haunted Heritage Walks, allowing visitors to explore the mysterious history of the site.

Malcha Mahal, also referred to as Wilayat Mahal, stands as a testament to the rich tapestry of history woven into the landscape of Delhi (Act, P. O. C. S. O., 2023). Dating back to the 14th century, this architectural gem was established as a hunting lodge during the reign of Feroz Shah Tughlaq, a pivotal era in the Tughalaq dynasty (Joshi, 2013; Sharma, 2006). Situated in the quaint Malcha village, the palace reflects the opulence and grandeur characteristic of Tughlaq architecture, offering a glimpse into the cultural diversity that has defined the region over the centuries. Constructed with meticulous detail, Malcha Mahal served as a retreat for the royal family during the Tughalaq era. The sprawling grounds and strategic location amid lush greenery provided an idyllic setting for hunting excursions, adding to the palace's allure. The architectural features of the palace, including intricate carvings and ornate design elements, showcase the skilled craftsmanship of the time and underline the palace's significance as a symbol of royal leisure and elegance.

As the pages of history turned, Malcha Mahal underwent various transformations, adapting to the changing tides of rulers and regimes. The palace, once a refuge for Tughalaq royalty, saw shifts in purpose and function under subsequent rulers. Its ability to endure these changes speaks to the resilience of its architectural design and the cultural importance bestowed upon it by successive generations.

Despite the passage of centuries, Malcha Mahal has retained its distinctiveness and historical significance, earning it the status of a protected heritage site. This designation not only underscores the architectural prowess embedded in its walls but also emphasizes the importance of preserving the cultural legacy it represents. The palace stands as a living chronicle, narrating stories of bygone eras and offering a tangible link to Delhi's historical roots. In the present day, Malcha Mahal continues to be a site of fascination for historians, archaeologists, and enthusiasts alike. Its well-preserved structure serves as a portal through which one can traverse the corridors of time and immerse themselves in the grandeur of the Tughlaq period. The palace stands as a silent witness to the ebb and flow of history, inviting contemplation on the cultural and architectural evolution that has shaped Delhi into the vibrant metropolis it is today.

As visitors explore Malcha Mahal, they are not only greeted by its architectural splendor but also invited to unravel the layers of history that have unfolded within its walls. The palace, with its intricate details and storied past, beckons those who seek to understand and appreciate the historical fabric of Delhi. In the ever-changing urban landscape, Malcha Mahal remains a steadfast sentinel, echoing the whispers of the past and standing as a testament to the enduring legacy of Tughlaq architecture in the heart of India's capital. The Haunted Heritage Walks at Malcha Mahal exemplify the integration of dark tourism with heritage preservation. Tourists are drawn to the eerie ambiance and the opportunity to engage with the site's haunted reputation. This case study explores the motivations of tourists participating in Malcha Mahal Haunted Heritage Walks, the impact on the local community, and the challenges in balancing preservation with tourism demands.

The haunted reputation surrounding Malcha Mahal can be attributed to a confluence of historical events, local legends, and contemporary anecdotes, each playing a significant role in shaping the perception of the site as a paranormal hotspot (Gupta, 2011). At the heart of this mystique lies the demise of the last occupants, Princesses Wilayat and Sakina, who asserted their lineage as descendants of the last Nawab of Oudh (Moreno-Álvarez, 2021). The secluded lifestyle led by the princesses within the walls of Malcha Mahal became a focal point in the development of its haunted reputation. The deliberate withdrawal from the public eye, coupled with the dense forest that envelops the palace, created an aura of mystery and seclusion. This intentional isolation became a breeding ground for local speculation and rumors, often attributing supernatural elements to the princesses' lives and deaths. As a result, the palace garnered a reputation as a place imbued with an otherworldly energy, setting the stage for the emergence of haunted narratives.

The mysterious circumstances surrounding the deaths of Wilayat and Sakina have further solidified the haunted reputation of Malcha Mahal. The lack of clarity and official documentation regarding their demise has allowed for the proliferation of various speculative accounts, giving rise to ghost stories and legends. These tales often revolve around unfulfilled wishes, tragic love stories, and the lingering spirits of the princesses. The ambiguity surrounding their deaths serves as a fertile ground for the construction of narratives that blur the lines between history and the supernatural, contributing to the site's haunted aura.

Moreover, the dilapidated state of Malcha Mahal and its surroundings significantly adds to the eerie atmosphere that envelops the site. The once grand hunting lodge, now in a state of disrepair, serves as a physical manifestation of the passage of time and the historical events it has witnessed. Crumbling walls, overgrown vegetation, and the general sense of neglect evoke a sense of desolation, amplifying the haunted narrative associated with the palace. Visitors, confronted with the visual decay of the site, are predisposed to interpret their surroundings through the lens of the supernatural, further perpetuating the haunted reputation.

Contemporary anecdotes and testimonials from visitors who claim to have experienced paranormal activities at Malcha Mahal contribute to the ongoing construction of its haunted reputation. These accounts, whether based on personal experiences or local hearsay, become integral components of the evolving narrative surrounding the palace. The collective sharing of such stories through word of mouth, social media, and various forms of media perpetuates the image of Malcha Mahal as a haunted site, attracting a curious audience seeking the thrill of the supernatural.

The haunted reputation of Malcha Mahal is a product of a complex interplay between historical events, local legends, and contemporary anecdotes. The deliberate seclusion of its last occupants, the mysterious circumstances surrounding their deaths, the dilapidated state of the palace, and the accounts of paranormal experiences collectively contribute to the creation and perpetuation of the haunted narrative associated with this historical site. Understanding these factors provides insight into the allure of Haunted Heritage Walks and the enduring fascination with the supernatural within the context of Malcha Mahal in Delhi.

The motivations of tourists participating in Malcha Mahal haunted heritage walks are multifaceted. Some seek the thrill of exploring a reputedly haunted location, drawn by the intrigue of supernatural stories associated with the site. Others may be interested in the historical and cultural aspects, using the opportunity to learn about the lives of the exiled princesses and the palace's significance in Delhi's history. The immersive experience offered by the Haunted Heritage Walks, combined with the authenticity of the site's historical and haunted narratives, aligns with MacCannell's (1976) concept of staged authenticity. Visitors are not passive observers but active participants in the exploration of Malcha Mahal's dark and mysterious past.

Haunted Heritage Walks at Malcha Mahal, while undeniably adding to the allure of the site, are not without their fair share of critiques and challenges. One prominent critique revolves around the argument that the emphasis on supernatural elements in these walks detracts from the genuine historical and cultural significance of Malcha

Mahal. Additionally, concerns have been raised regarding the potential exploitation of haunting narratives for commercial gain, prompting ethical questions about the commodification of the supernatural (Stone & Sharpley, 2008). One of the primary concerns voiced by critics is the potential distortion of Malcha Mahal's historical and cultural narrative due to the prominence of supernatural elements in the heritage walks. The site, with its rich historical background dating back to the 14th century, has borne witness to the ebb and flow of various dynasties, reflecting the architectural splendor of its time (Singh, 2010). The presence of the Begum Wilayat Mahal and her sons, claiming royal lineage, further adds layers of intrigue to Malcha Mahal (Smith, 2005). However, the emphasis on ghost stories and paranormal encounters during heritage walks may overshadow the genuine historical and cultural significance of the site. Critics argue that this distortion could potentially contribute to a superficial understanding of Malcha Mahal, with visitors focusing more on the supernatural narratives than on appreciating the authentic historical and cultural context (Sharpley, 2005).

Moreover, the commercialization of haunting narratives raises ethical concerns within the realm of heritage tourism. The popularity of Haunted Heritage Walks has transformed Malcha Mahal into a hotspot for tourists seeking a unique blend of history and the supernatural. While this surge in interest benefits local tourism and the surrounding community, some argue that it may come at the cost of ethical considerations. Critics question the fine line between providing an immersive experience for tourists and the potential exploitation of haunting narratives for commercial gain. The danger lies in turning the supernatural aspects of Malcha Mahal into mere commodities, overshadowing the genuine cultural and historical value of the site (Stone & Sharpley, 2008). The ethical dilemma extends to the narrative crafting by tour operators and the potential exaggeration or fabrication of ghost stories to enhance the thrill factor for visitors. The risk of perpetuating inaccurate or sensationalized tales raises questions about the responsibility of heritage tourism operators in presenting an authentic representation of Malcha Mahal. Striking a balance between entertaining narratives and maintaining the integrity of historical and cultural authenticity becomes imperative in ensuring that visitors leave with a nuanced understanding of the site.

Despite these critiques, it is essential to acknowledge that the challenges faced by Haunted Heritage Walks at Malcha Mahal are not insurmountable. Efforts can be made to address these concerns and strike a balance between the supernatural allure and the genuine historical and cultural significance of the site. This might involve collaborative initiatives between heritage conservationists, local communities, and tour operators to develop guidelines for responsible storytelling and tourism practices at Malcha Mahal.

The influx of tourists to Malcha Mahal has brought both economic opportunities and challenges to the local community. The increased footfall has stimulated the local economy through the provision of services such as guided tours, transportation, and hospitality. However, the delicate balance between catering to the tourist demand and preserving the site's integrity poses challenges. There is a need for sustainable tourism practices to ensure that the local community benefits without compromising the authenticity and historical significance of Malcha

Mahal. Preserving Malcha Mahal as a heritage site while accommodating the growing interest in Haunted Heritage Walks presents challenges that require careful consideration. The delicate nature of historical structures necessitates measures to prevent damage caused by increased foot traffic. Implementing guidelines for responsible tourism, community involvement in preservation efforts, and collaboration with heritage conservation organizations are vital steps to strike a balance between tourism and preservation.

While Haunted Heritage Walks at Malcha Mahal contribute significantly to its appeal and the local economy, they also encounter valid critiques and challenges. Striking a delicate balance between the supernatural allure and preserving the genuine historical and cultural significance of the site is crucial. Ethical concerns surrounding the potential commercialization of haunting narratives underscore the need for responsible tourism practices. By addressing these challenges, stakeholders can ensure that Malcha Mahal continues to be a captivating destination that not only thrills visitors with its haunted tales but also educates them about its rich historical and cultural heritage.

Dark tourism has emerged as a dynamic phenomenon that not only satisfies the morbid curiosity of tourists but also has the potential to revive forgotten heritage sites. The case study of Malcha Mahal and its Haunted Heritage Walks in Delhi illustrates how dark tourism can contribute to the economic revitalization of neglected sites while presenting challenges in terms of preservation and community impact. As dark tourism continues to evolve, it is imperative to adopt sustainable practices that ensure the long-term preservation of heritage sites, fostering a balance between tourism demands and historical authenticity.

Nestled within the Central Ridge of Delhi, next to the Delhi earth station of ISRO, Malcha Mahal stands as a silent witness to more than seven centuries of history. Built during the reign of Firoz Shah Tughlaq, this hunting lodge encapsulates the architectural splendor of medieval Delhi. However, Malcha Mahal is not merely an antiquated structure; it is a repository of haunting legends, an eerie testament to forgotten tales that have lingered through the decades. There is limited literature on the ownership of Malcha Mahal before it became the dwelling place of the self-claimed descendants of Oudh, except that it was referred to as Bistadari ruins mentioned in the Indian Express article on 06-05-23. Once a truly forgotten heritage, Malcha Mahal is now undergoing a revival, driven by government initiatives and nongovernment organizations. The study aims to explore the impact of the Haunted Heritage Walk, initiated by Delhi Tourism, on the revival of Malcha Mahal and its surroundings. Additionally, it seeks to assess the dark tourism potential of this secluded heritage site before and during the restoration.

History and Architectural Details

Constructed in the 14th century, the Malcha Mahal is an architectural relic from the era of the Tughlaqs, the third ruling dynasty over the Delhi Sultanate in medieval India. Malcha Mahal is one of the four hunting palaces built by Firoz Shah Tughlaq (AD 1351-88) along with the Kushk-Mahal in the south Ridge, the

Bhuli-Bhatiyari-ka-Mahal on the Ridge in Central Delhi, and the Pir Ghalib on the northern Ridge near the Delhi University. Malcha Mahal is situated in Malcha, one of the historic villages in proximity to Raisina Hill. During the 1920s construction of the new capital, New Delhi, the British relocated Malcha, along with Raisina, Todapur, Aliganj, Pillanji, Jaisinghpura, and Kushak villages (Dastidar & Khandekar, 2011).

The structure is constructed entirely from sandstone and features a main entrance on its eastern side. Access to the lodge is provided by two wide sandstone staircases on either side of the entrance. Inside, the central hall can be reached through the main entrance (Fig. 17.1).

The interior spaces are created using arches, allowing for ventilation throughout the building. The lodge comprises nine rooms, with the central one being larger and open on all four sides. The four corner rooms have arches on two sides. There are a total of six staircases within the structure, including two circular

Fig. 17.1. Malcha Mahal's Interior – Showcases the Arches of a 14th-Century Hunting Lodge. *Source:* Taken on May 13, 2023, during one of the Haunted Heritage walks at Malcha Mahal.

front staircases and L-shaped staircases located in the center of the rooms. The inner chamber openings on four sides have flat lintels, a unique feature compared to the pointed arches found in other parts of the structure. The facade of the building is adorned with repetitive brackets above the arches and a Kangura pattern with medallions on the parapet level, situated above these brackets. The parapet is only 0.6m in height and features rainwater spouts, currently existing only on the north side ("Malcha Mahal," Department of Archeology, Delhi, retrieved on October 1, 2023) (Figs. 17.2 and 17.3).

Existing Condition

The monument appears run-down, displaying extensive neglect with numerous sections suffering significant damage. Stairs and windows are overgrown with trees, and part of the complex has collapsed. Ceilings have crumbled, and floors have caved in various areas. Deep cracks mar the pillars, and haphazard construction is evident in some sections. Bats hang from the ceilings, infusing the monument with their pungent dropping on the floor (Fig. 17.4).

On one hand, the physical decay of this once-grand hunting lodge is evident, giving it an ethereal quality that aligns with its haunted reputation. On the other hand, despite guards, the lodge walls bear graffiti and scribblings from visitors. "While the interiors may seem clean, hidden nooks reveal scattered liquor bottles

Fig. 17.2. Malcha Mahal's Exterior – Showcases the Southeastern Corner Room of Malcha Mahal. *Source:* Taken on May 13, 2023.

Fig. 17.3. Main Entrance of Malcha Mahal Facing East. *Source:* Taken on May 13, 2023.

and cigarette packets throughout the complex, and surrounding the outskirts are piles of garbage" *Hindustan Times* writes in its news article published on 10-05-2023. Though the restoration is underway now, the vicinity is being cleaned out, and surrounding trees and vegetation are either pruned or cut. According to the officials from the Archaeology Department of Delhi in an interview for *Hindustan Times* Article published on July 22, 2023, they are landscaping the premises as a first step to restoring the monument (Figs. 17.5 and 17.6).

Last Inhabitants of Malcha Mahal

There are several available articles narrating the story of the final resident of Malcha Mahal, recounting the arrival of a woman with her two children in 1985, accompanied by servants and dogs. She identified herself as Wilayat Mahal, claiming to be the granddaughter of the last Nawab of Oudh (Awadh). Her elder daughter, Princess Sakina, and younger son, initially known as Prince Ali Raza and later referred to as Prince Cyrus, accompanied her. Most literature suggests that the Malcha Mahal was granted to them by then-Prime Minister Indira Gandhi after they protested for roughly a decade in the waiting room of the New Delhi railway station. However, upon their arrival in 1985, Indira Gandhi had passed away, and there exists no written evidence supporting this grant.

Fig. 17.4. Caved Floor of Malcha Mahal. *Source:* Taken on April 2023, during the survey of the monument for Haunted Heritage Walk.

For the next 30 years, the family deliberately avoided communication with Indian media and exclusively engaged with foreign media, keeping themselves isolated from the public. They surrounded the entire mahal with razor-sharp wires and posted a warning board stating "INTRUDERS SHALL BE GUNDOWN." Wilayat Mahal reportedly appealed to the Indian government to reclaim her claimed grandfather Wajid Ali Shah's property, seized by the British in the 19th century. Some accounts mention that upon moving to Malcha Mahal, Indian authorities promised to renovate the structure to royal standards, which never materialized. Wilayat Mahal's alleged suicide in 1993, as reported by her children in a foreign media interview, was attributed to this unfulfilled promise, with claims that she consumed crushed diamonds to end her life.

Information regarding the siblings' solitary life in the Mahal was scarce until Sakina's death in the early to mid-2010s, confirmed by Prince Ali Raza in a phone conversation with Ellen Barry, the South Asia Bureau Chief of *The New York Times* in New Delhi at that time. During his final period alone, Cyrus sought

Fig. 17.5. Undergoing Landscaping of the Premises of Malcha Mahal. *Source:* Taken on October 1, 2023. During one of the Haunted Heritage Walk.

interactions to alleviate his solitude. Some articles suggest Cyrus engaged in aiding Chanakyapuri residents with their dogs, to earn some income. Additionally, he had good connections with the guards at Delhi Earth Station. However, when he showed symptoms of dengue, he purportedly refused hospital treatment, saying hospitals were for "commoners" and he was a "Royal," he succumbed to the same fever in September 2017 (Figs. 17.7 and 17.8).

Legends and Folklore

Beyond its weathered sandstone walls, Malcha Mahal resonates with captivating legends that echo through the corridors of time. Whispers of spectral inhabitants, mysterious occurrences, and the macabre history of its former residents have woven an enigmatic tapestry around the palace. The legend of Begum Wilayat Mahal, her alleged suicide, and the family's reclusive existence create an aura that beckons the curious. "As per locals, the palace is haunted by the spirit of Begum Wilayat Mahal." The TOI writes in its news article on August 6, 2023.

Yet, the eeriness that envelopes the Malcha Mahal extends beyond the tragic end of the Begum Wilayat Mahal, and the dense canopy of mesquite trees resonates with frequent jackal howls.

The mysteries that revolve around the formal inhabitants of Malcha Mahal have given rise to numerous legends and conceptions about the place. Some suggest they may have been a family of djins (Barry, 2019), judging by their pale skin, peculiar physical structure, and secretive reclusive lifestyle in the isolation of

Fig. 17.6. Pruned Tree Was Growing Out of One of the L Shaped Staircases of Malcha Mahal. *Source:* Taken on May 13, 2023.

a kikar (mesquite tree) forest. Others who have encountered them and interacted with them firmly believed they indeed belonged to a royal family as mentioned in The Chakkar article 'Reclaiming the Lost House of Oudh' published on June 18, 2023. The solitude that enveloped Sakina and Cyrus, left to navigate the dilapidated halls after their mother's alleged suicide, speaks volumes. Father John, a Catholic charity worker, recalled them as "children submissive to their mother, more obedient than the dogs, reluctant even to accept a banana without her permission" (Barry, 2019). Sakina in her book mentioned asking permission from her mother in their interactions or when to sit together for meals (Kumar, 2023). In a despairing interview with Reuters in 1996, Sakina expressed her hopelessness, stating, "I do not wish to live in this world; this is a meaningless place." She hinted at the idea of following in her mother's footsteps. The loneliness and anger they harbored may have cast a spectral shadow over the place, attracting more legends and folklore. Rumors suggested that Prince Ali Reza kept his sister's corpse in Malcha Mahal for several days before a proper burial (Kumar, 2023).

Fig. 17.7. Prince Cyrus, Princess Sakina, and a Servant on the Roof of the Malcha Mahal. *Source:* Taken by Barry Bearak in 1998 for *The New York Times*.

After Cyrus's demise, in 2017, the Malcha Mahal again lay abandoned for nearly two years, during which it deteriorated further, and legends continued to accumulate around the mahal (Fig. 17.9).

The Haunted Heritage Walk

In 2019, the Delhi Government received a proposal from Indian National Trust for Art & Cultural Heritage (INTACH) for the restoration of this forgotten heritage (Smith, 2019). Subsequently, as mentioned in *Hindustan Times* in its May 10, 2023 news article, in July 2022, the Delhi government announced its intentions to give the structure a facelift. During the same year, Delhi Tourism, the state's tourism development corporation, conceived the idea of the Haunted Heritage Walk, initially focusing on multiple locations, with Malcha Mahal being the first. The inaugural walk occurred a year later on May 6, 2023, during the evening. This guided tour was exclusively available for prebooked groups ranging from a minimum of six to a maximum of 20 participants, scheduled specifically for weekends. The walk's duration was set at one and a half hours. To participate, individuals had to be at least 15 years old, considering various factors: the monument's remote location within a jungle, potential encounters with wildlife (like hornets and mosquitoes), the perception of the site as haunted, and the potential psychological impact on younger attendees. A member of the Delhi Tourism team involved in designing the walk

Fig. 17.8. Princess Wilayat at Malcha Mahal. *Source:* Image Taken via New York Times.

highlighted that "the minimum age restriction and evening schedule were strategically established to heighten intrigue among potential tourists." The walk was priced at 800 rupees, offering a package with a Delhi Tourism–branded yellow jute bag containing essential items such as a water bottle, juice, seasonal fruits, a muffin, a cap, a badge, a wristband, and a postcard.

These walks are led by a storyteller. The storyteller narrates tales and legends associated with the haunted lodge in an engaging manner. The walks begin at the entry gate of Delhi Earth Station on Sardar Patal Marg, covering approximately 750 meters on the Bistdari road to Malcha Mahal. This route passes through the dense Kikar canopy, home to abundant wildlife, including monkeys, jackals, and cattle. "While vehicles can access the jungle area, we intentionally chose to commence the walk from the Delhi Earth Station gate to allow tourists to relish the natural setting and wildlife en route to the monument," explained a Delhi Tourism staff member involved in designing the walk (Figs. 17.10–17.12).

Fig. 17.9. Malcha Mahal After Princes Cyrus' Death. *Source:* Taken by Bryan Denton for *The New York Times*.

Fig. 17.10. The Storyteller (on Right in White Kurta) is Narrating While Walking Through the Bistadri Road to the Monument. *Source:* Taken on May 13, 2023.

Fig. 17.11. Storyteller Describing the Architecture of Monument to the Tourists, Along With One of the Employees of Delhi Tourism (on Right in Plain White Shirt) Who Had a Crucial Role in the Inception of the Haunted Heritage Walk. *Source:* Taken on May 13, 2023.

One tourist expressed during an interview, 'It provided an escape from the bustling life of the city.' The walk is meticulously coordinated by official representatives of Delhi Tourism, ensuring the security of the tourists with the presence of a security guard. The walk conductor or storyteller halts at various points during the walk to showcase photographs or unveil interesting aspects of the story, maintaining the interest of the tourists. Tourists are encouraged to ask questions during the walk, and the stories narrated by different storytellers vary, encompassing legends, haunted experiences, and insights into the truth about Malcha Mahal and its inhabitants.

At the walk's conclusion, tourists reach the hunting lodge, where they have the opportunity to explore the monument independently while the storyteller continues to provide insights into the monument's history and the way its inhabitants used to live. 'I have this pendulum with me, and when I asked it if there were any ghosts, it indicated four of them,' mentioned a lady during the feedback interview with Delhi Tourism.

The culmination of the walk is a highlight, as the entire group ascends one of the two front circular stone stairs to the terrace of the lodge. From the terrace, visitors are treated to a panoramic view of Delhi. On the east side, they can see prominent buildings of Connaught Place and Raisina Hill, including landmarks

Fig. 17.12. Storyteller Showing the Photos of Inhabitants of Malcha Mahal, Assisting Him the Delhi Tourism Employee (on the Left). *Source:* Taken on May 20, 2023.

such as India Gate and Rashtrapati Bhawan. 'Right now, we're in the heart of Delhi. From here (Terrace of Malcha Mahal), I can see Rashtrapati Bhawan. Although I've been to Rashtrapati Bhawan several times, this unique view from the midst of the forest, allowing me to see its rear side, is something entirely new to me,' a tourist said (in Hindi) in a feedback interview with Delhi Tourism. Additionally, on a clear day, the Lotus Temple in the far southeast becomes visible. The west side offers tourists a chance to relish a beautiful sunset during the summertime. The walk ends at seven o'clock in the evening.

The authors of this chapter were actively involved in marketing the walk. Their involvement ranged from engaging social media influencers, print and digital media invites, to designing promotional posters and assessing potential storytellers for the walk. The inaugural walk garnered significant media coverage in both Hindi and English, sparking curiosity among Delhiites. Subsequently, Delhi Tourism received a surge in queries regarding Malcha Mahal and the Haunted

Heritage Walk, with prebookings filling up for the upcoming weeks. Interestingly, not all queries were for booking the walk; some were simply seeking information about this hidden monument.

It's noteworthy that the inquiries primarily matched the assumed target market of upper-middle-class Delhiites, inferred from the walk's pricing. However, the majority of actual visitors to the monument were school and college-going youth, as indicated in interviews. According to a guard stationed at Malcha Mahal, these young visitors typically frequent the site on weekdays and during daytime hours.

Before the introduction of the Haunted Heritage Walk, the monument saw minimal foot traffic, mainly from drug addicts and young couples who often defaced the walls with graffiti. The guard mentioned in an interview that since the walk's announcement, the Archaeological Department has been more vigilant, actively monitoring and curbing these unwanted activities. "They explicitly instructed me to be more vigilant against these activities," emphasized the guard.

After the launch of the Haunted Heritage Walk and the subsequent attention received by Malcha Mahal, the Lieutenant Governor of Delhi, accompanied by the Chief Secretary, participated in a special walk to assess the monument's condition. "This further bolstered the restoration process," mentioned one of the Delhi Tourism employees during an informal phone interview.

Within a couple of months of the walk's launch, restoration work appears to have commenced. The area surrounding the monument is undergoing cleanup, the pathway leading to it has been cleared and widened, the forest department is cleaning the Bistadari Road and planting indigenous trees, most of the cacti close to the monument have been removed, and a signboard has been placed, marking the property under the purview of the Archaeology Department of Delhi. Additionally, the Archaeology Department is planning to restrict entry without permission by installing grilles to enclose the premises.

Research Findings and Conclusion

Abandoned for centuries, Malcha Mahal was one of the monuments listed as "worthy of conservation" by Maulvi Zafar Hasan, an expert from the Archaeological Survey of India (ASI), in 1916 (Gandhiok & Akhtar, 2023). It gained prominence in 1985 when the self-proclaimed royal family of Oudh occupied its deteriorating structure, residing there until the demise of the last member, Ali Raza, in September 2017. The intriguing tales surrounding the final occupants of Malcha Mahal – Wilayat Mahal's suicide, the mysterious death of Princess Sakina, and Prince Cyrus's depression and loneliness before his demise – fueled fascination and curiosity across all demographics. The identity, these occupants bestowed upon this long-forgotten monument, cultivated potential for dark tourism. This potential remained dormant until Delhi Tourism introduced the Haunted Heritage Walk. Working with Delhi Tourism, the authors had the opportunity to observe the comprehensive process involved in making the Haunted Heritage Walk possible. In April of 2023, Delhi Tourism initiated a location survey to assess the feasibility of this walk. Despite preparatory cleaning efforts prior to Delhi officials' visit, graffiti and scribbles were all over the walls of the monument, and the vicinity was marred by the prevalent presence of litter and plastic bottles. Traversing from the entrance of Delhi Earth Station to the monument, the path was littered with various forms of waste, predominantly plastic, greatly overshadowing the natural surroundings. Notably, this was the designated route selected by Delhi Tourism for the walk. Although the Delhi government had stationed a guard to protect the monument during daylight hours, the site attracted thrill-seekers and individuals engaging in night-time activities, contributing to its degradation.

The primary objective of this walk was to introduce an alternative tourism attraction in Delhi, a city already teeming with renowned monuments by harnessing the dark tourism potential of the monument. However, the initial condition of the monument was considerably deteriorated. Hornet nests were found on one of the front stairs, and the structure had several holes from which snakes were frequently seen emerging. Given these conditions, introducing the walk was dangerous. Therefore, Delhi Tourism collaborated with the Forest Department and the Department of Archaeology to make the monument and its surroundings safe for expected tourists. The last section of the walk, covering 20–25 meters from the Bistadari Road to the monument, underwent widening and cleaning, eliminating scattered shards of glass from liquor bottles to facilitate safe and convenient passage for walking groups. Additionally, access to the left front circular stairs was restricted, and ad hoc pruning was performed on overgrown vegetation, particularly on the monument's terrace, ensuring safety without causing harm to the structure.

The authors' role as coordinators of the inaugural walk and subsequent walks provided them with a unique vantage point to observe tourists' firsthand experiences. In interviews conducted after the walks, many visitors confessed that the allure of the word 'Haunted' in the walk's promotion was the primary reason for their attendance. Among them were those drawn by the desire to experience the natural beauty and serenity missing in the city. Despite varying motivations among visitors, it was clear that the Haunted Heritage Walk's launch revitalized interest of people in the monument. "The Government of Delhi had promised to restore the monument, back in 2019, but the restoration work had somewhat stopped or never happened until this heritage walk brought the monument into the limelight," a tourist shared in the interview. Since the walk's inception, conservation efforts have commenced. Landscaping work is currently underway. In the month of July, the Archaeology Department obtained permission from the Forest and Wildlife Department to fell 38 trees and prune another 150 within the Central Ridge in Chanakyapuri, where the monument complex is situated.

"We are committed to conserving the monument and safeguarding its premises. Landscaping initiatives are in progress. Trees encroaching upon the monument have been pruned," explained an official from the Archaeology Department in an interview with Hindustan Times. [20b] "Within the 25 × 25 m area surrounding the monument, grilles and barricades are being set up. Lighting arrangements will be installed, and facilities like toilets will be provided for visitors," [20c] added another official quoted in the same article. Most respondents agreed that before the introduction of the walk, they were unaware of the monument's history or the stories of its inhabitants. In addition to the eerie experience within the monument, the walk also served as a serene nature retreat away from the city's hustle and bustle, emerging as a new and alternative attraction for both Delhiites and tourists. Importantly, there was unanimous agreement on the significance of conservation efforts for the monument.

It's premature to guarantee complete conservation as the process is ongoing. Likewise, predicting the monument's tranquility amid an increasing tourist influx remains challenging. However, what's evident is that the dark tourism potential transformed this neglected monument into an alternative destination within bustling Delhi. Dark tourism drew attention to this forgotten dilapidated structure, which sparked public interest, playing a role in its ongoing conservation efforts. To ensure the preservation of this hidden heritage gem, along with its unique selling points such as the natural escape it offers, the panoramic city views, wilderness, intriguing stories, remoteness, and eerie ambiance, it's imperative that the ongoing conservation process is accompanied by sustainable tourism planning. This approach aims to safeguard these distinct features while managing tourism in a way that maintains their integrity for future visitors to experience and appreciate.

References

Act, P. O. C. S. O. (2023). *Protection of Children from Sexual Offences (POCSO) Act 2012*.

Alderman, D. H. (2002). Street names as memorial arenas: The reputational politics of commemorating Martin Luther King Jr. in a Georgia county. *Historical Geography*, *30*, 99–120.

Ashworth, G. J., & Hartmann, R. (2005). *Horror and human tragedy revisited the management of sites of atrocities for tourism*. Cognizant Communication.

Barry, E. (2019, November 22). The jungle prince of Delhi. *The New York Times*. https://www.nytimes.com/2019/11/22/world/asia/the-jungle-prince-of-delhi.html

Beech, J. (2000). The Enigma of Holocaust sites as tourist attractions – The case of Buchenwald. *Managing Leisure*, *5*, 29–41.

Blom, T. (2000). Morbid tourism - A postmodern market niche with an example from Althorp. *Norsk Geografisk Tidsskrift - Norwegian Journal of Geography*, *54*(1), 29–36. https://doi.org/10.1080/002919500423564

Dann, G. M. S. (1994a) Tourism and nostalgia: Looking forward to going back. *Vrijetid en Samenleving*, *12*(1/2), 75–94.

Dann, G. M., & Seaton, A. V. (2001). Slavery, contested heritage and thanatourism. *International Journal of Hospitality & Tourism Administration*, *2*(3–4), 1–29. https://doi.org/10.1300/j149v02n03_01

Dastidar, A. G., & Khandekar, N. (2011, January 5). From rocks & ridges rose a New Delhi. *Hindustan Times*. https://web.archive.org/web/20140314133344/http://www.hindustantimes.com/india-news/newdelhi/from-rocks-amp-ridge-rose-a-new-delhi/article1-646426.aspx. Accessed on March 12, 2014.

Department of Archaeology, Delhi. (2023, October 1). Malcha Mahal. https://archaeology.delhi.gov.in/archaeology/malcha-mahal

Foley, M., & Lennon, J. (1996). JFK and dark tourism: A fascination with assassination. *International Journal of Heritage Studies*, *2*(4), 198–211.

Gandhiok, J., & Akhtar, S. (2023, July 22). To conserve Malcha Mahal, 14th-century lodge in Delhi forest set for a facelift. *Hindustan Times*. https://www.hindustantimes.com/cities/delhi-news/14thcentury-lodge-in-delhi-forest-set-for-a-facelift-101689963104700.html

Gupta, D. R. (2011). *Paper 11: Special Interest Tourism Module 25: Dark tourism*. Doctoral dissertation, Panjab University.

Joshi, H. (2013). Le patrimoine historique de Delhi: Réappropriation populaire, conflits d'intérêt et enjeux politiques. *Synergies Inde*, *6*, 25–71.

Kumar, A. (2023, May 18). Reclaiming the Lost House of Oudh. *The Chakkar*. https://www.thechakkar.com/home/houseofoudh

Lennon, J., & Foley, M. (2000), *Dark tourism: The attraction of death and disaster*. Cengage Learning EMEA

Lennon, J. J., & Wight, A. C. (2004). Towards and understanding of visitor perceptions of 'dark' sites: the case of the Imperial War Museum of the North, Manchester. *Journal of Hospitality & Tourism*, *2*(2), 105–22.

MacCannell, D. (1976). The tourist. In *A new theory of the leisure class*. Schocken Books.

Moreno-Álvarez, A. (2021). Believe me, do not believe me: Jhumpa Lahiri and the royal family of Oudh. *Revista Canaria de Estudios Ingleses*, 67–76. https://doi.org/10.25145/j.recaesin.2021.83.05

Rojek, C. (1993). *Ways of escape: Modern transformations in leisure and travel*. Macmillan.

Rojek, C., & Urry, J. (Eds.). (1997). *Touring cultures: Transformations of travel and theory* (1st ed.). Routledge. https://doi.org/10.4324/9780203427736

Seaton, A. V. (1996). Guided by the dark: From thanatopsis to thanatourism. *International Journal of Heritage Studies 2*:234–244.

Seaton, A. V. (2002). Thanatourism's final frontiers? Visits to cemeteries, churchyards and funerary sites as sacred and secular pilgrimage. *Tourism Recreation Research*, *27*(2), 73–82. https://doi.org/10.1080/02508281.2002.11081223

Sharma, A. K. (2006). Prehistoric Delhi and its neighbourhood: Physical features. *Delhi Ancient History* (pp. 14–25). Social Science Press.

Sharpley, R. (2005). Travels to the edge of darkness: Towards a typology of dark tourism. In C. Ryan, S. Page, & M. Aicken. (eds.), *Taking tourism to the limits: Issues, concepts and managerial perspectives* (pp. 217–228). Elsevier.

Sharpley, R., & Stone, P. R. (Eds.). (2009). The darker side of travel: The theory and practice of dark tourism. In *Aspects of tourism series*. Channel View Publications.

Singh, R. (2010, August 21). *Rajasthan's 'ghost' town Bhangarh now a popular tourist spot. Times of India*. http://timesofindia.indiatimes.com/india/Rajasthans-ghost-town-Bhangarh-now-a-popular-tourist-spot/articleshow/6386066.cms. Accessed on August 25, 2015.

Smith, R. V. (2005). *The Delhi that no-one knows*. Orient Blackswan.

Smith, R. V. (2015). *Delhi: Unknown tales of a city*. Roli Books Private Limited.

Smith, R. V. (2019, October 22). The story of Malcha Mahal. *The Hindu*. https://www.thehindu.com/society/history-and-culture/the-story-of-malcha-mahal/article29766170.ece

Stone, P. R. (2006). A dark tourism spectrum: Towards a typology of death and macabre related tourist sites, attractions and exhibitions. *Tourism: An International Interdisciplinary Journal*, *54*(2), 145–160.

Stone, P., & Sharpley, R. (2008). Consuming dark tourism: A thanatological perspective. *Annals of Tourism Research*, *35*(2), 574–595. https://doi.org/10.1016/j.annals.2008.02.003

Strange, C. & Kempa, M.. (2003). Shades of dark tourism: Alcatraz and Robben Island. *Annals of Tourism Research*, *30*, 386–405.

Tunbridge, J. E., & Ashworth, G. J. (1996). *Dissonant heritage: The management of the past as a resource in conflict*. J. Wiley.

Uzell, D. (1992). The hot interpretation of war and conflict. In D. Uzell (Ed.), *Heritage interpretation, Volume I: The natural and built environment* (pp. 33–47). Bellhaven Press.

Williams, H. (2004). Death warmed up. The agency of bodies and bones in early Anglo-Saxon cremation rites. *Journal of Material Culture*, *9*, 263–291.

Wilson, J. Z. (2004). Dark tourism and the celebrity prisoner: Front and back regions in representations of an Australian historical prison. *Journal of Australian Studies*, *28*(82), 1–13. https://doi.org/10.1080/14443050409387951

Bibliography

Screenocean, 1996 (1996, January 23). *India: Feature - India Great princely families - Paradise lost*. Screenocean.

Akhtar, S. (2023, May 10). Delhi's Malcha Mahal gets tourist boost but lies in ruins. *Hindustan Times*.

Delhi Tourism. [@delhitourism_official]. (2023, June 1). *Heritage Walk experience shared by a tourist [Video]*. Instagram.

delhitourism_official. (2023, June 5). *Visitors experiences for Haunted Heritage Walk at Malcha Mahal. Come and explore exciting Heritage Walks with Delhi tourism [Video]*. Instagram.

Denton, B. (2019, November 22). Thrill seekers tramped through Malcha Mahal after Cyrus's death [Photograph]. In *The Jungle Prince of Delhi*.

Express Web Desk. (2023, May 6). What is Malcha Mahal, the place where Delhi's first 'haunted walk' leads to? *The Indian Express*.

Gandhiok, J., & Akhtar, S. (2023, July 22). To conserve Malcha Mahal, 14th-century lodge in Delhi forest set for a facelift. *Hindustan Times*.

Kumar, A. (2023, May 18). Reclaiming the Lost House of Oudh. *The Chakkar*.

Times of India. (2023, August 6). Malcha Mahal: Discovering Delhi's haunted past. *Times of India*.

Webliography

https://www.hindustantimes.com/cities/delhi-news/mysterious-malcha-mahal-delhi-s-haunted-hunting-lodge-seeped-in-history-and-neglect-awaits-conservation-101683658208098.html

https://timesofindia.indiatimes.com/travel/destinations/malcha-mahal-discovering-delhis-haunted-past/articleshow/102465979.cms

https://indianexpress.com/article/what-is/malcha-mahal-delhi-haunted-walk-8594750/
https://reuters.screenocean.com/record/1000163
https://www.hindustantimes.com/cities/delhi-news/mysterious-malcha-mahal-delhi-s-haunted-hunting-lodge-seeped-in-history-and-neglect-awaits-conservation-101683658208098.html
https://www.thechakkar.com/home/houseofoudh
https://www.instagram.com/reel/CuKG0gLAL7O/?utm_source=ig_web_copy_link&igshid=MzRlODBiNWFlZA==
https://www.hindustantimes.com/cities/delhi-news/14thcentury-lodge-in-delhi-forest-set-for-a-facelift-101689963104700.html
https://www.instagram.com/reel/CtGtfTsrZ8o/?utm_source=ig_web_copy_link&igshid=MzRlODBiNWFlZA==
https://www.nytimes.com/2019/11/22/world/asia/the-jungle-prince-of-delhi.html

Chapter 18

Conceptual Multiplicity and Contradiction that Impede Dark Tourism Promotion

Samik Ray

IITG, Ministry of Tourism, India

Abstract

Dark tourism is a new coinage rooted in the perception of tourism activities at the sites or destinations connected to phenomena that bear varied, flexible, dynamic, diverse, and graded dark shades of life and civilization. It is now customary to subsume it within the sets of niche tourism. Some dark tourism sites attract visitors and generate sizable amounts of revenue, yet most of the world does not register much demand compared to other niches. Accordingly, promotion pursuit turned crucial to draw the market's attention, creating its competitively distinct position.

Indeed, inherent issues, such as conceptual multiplicity within nature, aspect, attribute, and product paradigm, turn dark tourism into a complex phenomenon and put a challenge toward creating its distinct market position. Additionally, contradictions in semantic and functional significances, conflicts in framing morbid memory and authentic portrayal, variances in ethical, cultural and ideological interpretations, transition of liminal space identity, and diverse focuses in stakeholder engagement in imaging impede efforts to transform dark tourism attractions into a significant driver of tourism.

This chapter will locate and address the issues that challenge the marketability of dark attractions and dark tourism promotion more directly, with attention to the Indian context.

Keywords: Challenges; conceptual multiplicity; contradiction; dark tourism; dark stories; promotion

Dark Tourism, 265–278

Published under exclusive licence by Emerald Publishing Limited
doi:10.1108/978-1-83797-336-120241018

Introduction

Dark tourism is a promotional activity, marketing and managing visitation and visits at locations bearing historical or contemporary darker stories of human civilization. Although the dark story of a destination has been a crucial travel motivator for centuries, it is difficult to determine the origin and diachronic development of the desire to visit those sites and destinations.

The earliest known travelers to a postbattle location presumably were Alexander III of Macedon and Emperor Asoka of India (Hultzsch, 1925; Wilcken, 1967). The urge to visit the graves and relics of saints, prophets, and monks was evident since the pre-Common era. The tradition of visiting relics of Buddhist monks was not uncommon in India, Tibet, and Bhot lands in both pre-Common and Common eras, e.g., Emperor Asoka (C.260 BCE) reportedly visited Kushinagar (Lord Buddha's death place). While Archbishop Thomas Becket's shrine at Canterbury began attracting travelers from 1170 CE, Europe's High Romantic Period (1770–1830) experienced an increasing trend of visiting graves and relics (Reed, 2018). Modern Europe has observed the development of travel to places related to death among Western spiritual travelers (Sallay et al., 2022). It was chiefly an individual activity until the 1840s (Ray, 2015). The initiatives to visit Waterloo (the last battle site of Napoleon Bonaparte), the buried city of Pompeii, the battle sites of Gettysburg (USA) and Boer (South Africa), and other thanatoptic destinations (Seaton, 1996) began in the mid-19th century within an organized institutional framework at the behest of Thomas Cook (Lioyd, 1998; Ray, 2015). The World War sites in the post-1950s became significant drivers of battlefield tourism, grief tourism, and Holocaust tourism. By the early 21st century, Auschwitz, Hiroshima, Pearl Harbour, Cambodian genocide sites, Dallas, Ford's Theatre of Washington, DC, Memphis of Tennessee, War cemeteries (India), Gandhi Memorial (New Delhi), Central Saheed Minar (Dhaka), and many other similar sites linked to dark stories and histories are becoming significant on the travel map.

In the recent past, the term dark tourism gained considerable attention with the development of academic pursuits and media projections. Dark tourism now encompasses a wide range of events and experiences, including the holocaust, war, battlefield, slavery, cemetery, internment, disaster, black, grief, death, and many other dark shades. Accordingly, the conceptual multiplicity embedded within its nature, aspect, attribute, and product paradigm makes the phenomenon complex. The stakeholders face challenges in translating dark attraction offers into a significant driver of tourism due to contradictions in imaging diverse shades, value development, frame of remembrances, authenticity, ethical obligations, and cultural and ideological interpretation.

Literature Review

As an important and newly emerging driver, dark tourism began to draw academic attention in the last decade of the 20th century, and a large amount of literature is present to frame the dark tourism concept. Only a few of those are worth mentioning (Table 18.1).

While dark tourism started to gain researchers' attention, agreement in using a general or generic term and developing a universal concept is still not apparent. Investigations on dark tourism promotion issues are inadequate, meager, and fragmented (Table 18.2) when the study on motivation, consumption, and experience is available on a large scale.

Table 18.1. Research Areas.

Study Area		**Researchers**
Concept and definition		Seaton (1996), Foley and Lennon (1997), Lennon and Foley (1999), Dann and Seaton (2001), Preece and Price (2005), Tarlow (2005), Best (2007), Ashworth (2008), Robb (2009), Knudsen (2011), Stone (2011a), Johnston (2015), Ray (2015), Johnston and Mandelartz (2016), and Stone (2016)
Dark tourism destination and sites	Associated to battle and war	Seaton (1999), Bigley et al. (2010), Carr (2010), Miles (2014), Murphy (2015), MacCarthy (2016), and Johnston (2016)
	Connected to genocide	Hughes (2008), Sharpley (2012), and Hohenhaus (2013)
	Linked with natural disaster	Rittichainuwat (2008), Robbie (2008), and Biran et al. (2014)
	Prison	Strange and Kempa (2003), Preece and Price (2005), Best (2007), and Kang and Lee (2013)
	Cemeteries	Seaton (2002), Raine (2013), and Seaton et al. (2015)
Ethical issues		Foley and Lennon (1996, 1997), Lennon and Foley (1999, 2000), Seaton and Lennon (2004), Lennon and Foley (2000), Lennon (2009), and Dale and Robinson (2011)

Table 18.2. Research Areas on Dark Tourism Market.

Study Area		Researchers
General spectrum		Brown et al. (2012), Farmaki (2013), Johnston (2013), MacCarthy (2016), and Johnston et al. (2016)
Commodification issues		Foley and Lennon (1996, 1997), Lennon and Foley (1999, 2000), Cole (2000), Dann and Potter (2001), Marcuse (2005), Lennon (2009). Rice (2009), Braithwaite and Leiper (2010), Carr (2010), and Dale and Robinson (2011)
Product perspective		Ashworth (2004, 2005), Isaac and Ashworth (2011), and Isaac and Çakmak (2014)
Dark tourism motives	Individual connection	Biran et al. (2011), Lelo and Jamal (2013), and Yankholmes and McKercher (2015)
	Learning	Preece and Price (2005), Best (2007), Bigley et al. (2010), Biran et al. (2011), Kang and Lee (2013), Sharpley (2012), Farmaki (2013), Lelo and Jamal (2013), Simone-Charteris et al. (2013), Isaac and Çakmak (2014), and Isaac and Çakmak (2016)
	Interest in dark stories and histories	Preece and Price (2005), Best (2007), Biran et al. (2014), Raine (2013), and Yankholmes and McKercher (2015)
	Obligation	Kang and Lee (2013), Sharpley (2012), and Tinson et al. (2015)
	Others	Preece and Price (2005), Bigley et al. (2010), Dunkley et al. (2007), Lelo and Jamal (2013), Isaac and Çakmak (2014), Raine (2013), Simone-Charteris et al. (2013), Tinson et al. (2015), and Isaac and Çakmak (2016)
Consumption perspective		Stone and Sharpley (2008, 2014), Stone (2009a, 2009b, 2009c, 2011b, 2011c, 2012a, 2012b, 2013a, 2013b), Walter (2009) Buda (2015a, 2015b), Buda and McIntosh (2013), and Buda and Shim (2015)
Branding issues		Simone-Charteris et al. (2013), Volcic et al. (2014), and Wu et al. (2014)

Dark Tourism Concept Linking Promotion Issues

Conceptual Multiplicity

The market analysts estimate the value of the dark tourism market (2022) was US$ 30 billion against the tourism market (as a whole) value of US$ 10.5 trillion and anticipate that the tourism market value will rise to US$ 17.1 trillion at a 5% compound annual growth rate (CAGR), while the dark tourism market value will be US$ 36.5 billion at a 2% CAGR by 2032 (futuremarketinsight, 2023). Though academics prefer to call the practice of traveling and making tour arrangements at a dark story and history destination "Dark Tourism," only 0.010% of inexact 4,000 tourism business operators use the term (Reports and insights, 2023; Statista, 2023). Thus, a large section is still hesitant about the term since it conceptually takes in multiple, flexible, dynamic, and varied shades of dark events, phenomena, and experiences. The conceptual multiplicity leads to emerging varieties of terminologies alongside dark tourism. They are thanatourism (Seaton, 1996), Holocaust tourism (Ashworth, 1996; Lennon & Foley, 2000; Tunbridge & Ashworth, 1996), battlefield tourism (Henderson, 2000), morbid tourism (Blom, 2000) and slavery-heritage tourism (Dann & Seaton, 2001; Seaton, 2001), cemetery and internment tourism (Seaton, 2002), celebrity death site tourism (Alderman, 2002), prison tourism (Strange & Kempa, 2003), fright tourism (Bristow & Newman, 2005), genocide tourism (Dunkley et al., 2007), war tourism, ghost tourism, disaster tourism, black tourism, grief tourism, death tourism, and many others.

Those syntactically compound terminologies combine static head lexeme "tourism" and immediate preconstituent variables, a modifier (descriptive attribute) (Ray, 2015), which is none other than a signifier, belong to a paradigmatic set of lexemes bearing a few apparent similar and diverse semantic and perceptive features at a time, thus able to alter the functional, semantic, and perceptional values of the term conforming the corresponding abstraction of the referent or signified. Indeed, apparent similitude is nothing but the tragic or negative sense embedded within the paradigm. This premise possibly prompts the academics to refer to the set as dark. On the contrary, the relation between dark tourism and its referents is causal and contextual since modifying constituents either restricting or specifying the values attach the term to an abstraction corroborating particular signified bearing diverse tangible manifestations and intangible experiences from the others. Thus, the possibility of developing separate market space for each referent becomes evident.

Contradictions

Contradiction in imaging, value determining, authenticity, ethical obligations, cultural and ideological interpretation, and frames of remembrances lie within the dichotomy between the presence and absence of tragic/dark/negative value. The cemetery would attract visitors of diverse and contradictory motivations, such as knowledge, commemoration, remembrance, and pride (set A) or death, morbidity, disaster, battle, atrocity, horror, and mourning (set B). In contrast, the main draw is either death or disaster. Although the main draw is death, the said dark or negative values are overtly absent in set A and present in set B. While the motivation and gaze of set B push a cemetery to become a dark tourism location, e.g., Wojslawice (Poland), set A promotes it as a heritage or cemetery/grave tourism site, e.g., cemeteries of Calcutta. If the main draw at the cemetery is battle or warfare stories, it could be a war tourism location, e.g., Commonwealth war cemeteries of India. Thus, a burial ground could be promoted or projected as a location to engage in varied tourism activities and experiences. Accordingly, frames of remembrances at visitation and imaging, cultural norms or ideological interpretation, and value determining of the location vary and sometimes contradict. While signifier–signified corroboration ascertains terminology's semantic value and location's image, the main draw, tourist motivation and gaze, push factors, market demand determine perceptive value, and pull factors and supply-side driven projection reveal functional value.

Dark tourism locations conceptually inhabit a liminal state (Van Gennep, 1960; Turner and Abrahams, 1969), a transition phase from an old to a new status or identity. As locations traverse among and between diverse identities, they do not hold any specific status or identity. Therefore, a graveyard travels from a traditionally accepted identity of the burial ground to a cemetery or dark tourism location and the site of a past battlefield, a forgotten or remembered battle place, to a war or dark tourism location. Hence, identity-related conceptual liminality (Van Gennep, 1960; Turner and Abrahams, 1969) prevails as locations traversing from one culturally defined and traditionally accepted state to a newly emerging state of identity. Contradiction appears when locations struggle to emerge with a new identity instead of swapping or switching between two or more identities. For example, a cemetery location swaps between heritage and burial ground or cemetery and dark tourism status. Change is progressing at the behest of tourism business stakeholders who perform rites to swap the state of identity from one to another, developing museums, monuments, memorials, and other related things around the actual site toward the formation of a recreated identity or commoditized remembrance (Ray, 2015).

The debates over ethical obligations in dark tourism applications have been ongoing. Obligations oscillate along a spectrum drawn between highly moral and highly immoral or highly ethical engagement and high disengagement. The activity of visiting or presenting for a tour at holocaust sites in Auschwitz, a poverty-stricken location in India, a postearthquake location in Nepal, and the Hindu cremation grounds of Varanasi is not highly immoral when the main draw is to acquire knowledge about cultural settings of the location away. The same practice will become morally disengaged if visitors take selfies/groupies or photos or get permission to do the same there. On the other hand, visiting and presenting a civil cemetery not in use for centuries, such as South Park Street Cemetery of Calcutta, is comparatively more ethical than visitation at the cremation sites of Varanasi. Thus, contradiction over moral obligations prevails in two different spheres. They are tourist and supply-side stakeholder spheres.

Dark Tourism Promotion: Challenges

The emergence of dark tourism as a niche market in the past few decades is crucial. Despite the above fact, the growth of dark tourism is not at par with the other niche tourism products. Its survival is heavily dependent on promotion pursuits. Primarily, people have a basic instinct to know, read, and watch movies that portray horror, suffering, death, and tragedy but are hardly keen to connect those dark phenomena, except the conditions compel them to be in it. Tourism stakeholders, including tourists, are no exception in this regard. They, then, look away from locations associated with dark stories and histories. Hence, the promotion of these locations as tourism destinations does not arise. The destinations that deal with stories and histories of darker shades do not have the mechanism, in general, to translate those stories into tourism resources. National tourism policies in most countries failed to understand the scope of dark tourism growth. They are keen to promote other niche tourism markets, such as ecotourism, adventure tourism, heritage tourism, and cultural tourism. Thus, the contribution of the dark tourism market to the economy of the destinations is yet to be understood. Although significant research on dark tourism recognized the growth opportunity of dark tourism, most stakeholders are still hesitant to accept it as a crucial driver of the tourism economy. Besides the above culture-driven hindrances, there are psycho-cultural blockades that impede the promotion of dark tourism, such as ethical obligations in projecting or dealing with contemporary or past tragedy, sufferings, death, atrocities, grief for the sake of money, profit, and tourism entertainment. Contradiction in the minds of stakeholders, including tourists, about moral engagement and disengagement in the practical field also affects their promotion and purchase decisions. Thus, the dilemma about ethical obligations has a direct bearing on the promotion pursuit of dark tourism. It makes many stakeholders uncomfortable in translating dark emotions into a tourism commodity.

Certain conceptual factors, besides cultural ones, impede the promotion pursuits. As a result, translating dark attractions into meaningful and noticeable drivers of tourism would become difficult. Conceptually, dark tourism is a generic term for the tourism business dealing with varied dark events and phenomena of human life and civilization. Since it takes in multiple, flexible, dynamic, and different but distinct shades of dark events, phenomena, and experiences, diverse types of tourism activities develop within its periphery. Those types often operate independently. Some of those did exist even before the emergence of the dark tourism concept. Indeed, dark tourism takes those previously existing types, such as cemetery tourism, battlefield tourism, disaster tourism, and many others, within its conceptual periphery. Conflict of identity creates the possibility of developing separate market space for each referent type. The contradiction between the discretely emerging separate market and the general market space (for dark tourism) is logical. The stakeholders are, as a result, in a state of confusion about whether to promote the generic type or subtype. The contradiction between an old and emerging identity impedes the stakeholders' decision about promotion pursuit, though signified is the same to them. The conceptual multiplicity and contradiction over the state of identity develop dilemmas in imaging, value determining, cultural and ideological interpretation, and frames of remembrance. It also affects tourist gazing and purchase decision-making.

India has various stories and histories associated with natural calamities, battles, wars, poverty, the rise and fall of empires and kingdoms, superstition, dark shades of the struggle for freedom, and many more apparent dark events. Despite this, the scope for dark tourism in India is not open across the country since national tourism policymakers and stakeholders fail to understand the importance of dark tourism in the growing Indian economy. Thus, skilled, knowledgeable, trained storytellers and interpreters, infrastructure to develop a connection between the dark story or history and visitors, and other human resources to care for dark-tourist demands are inadequate. Above all, the cultural and spiritual backdrop of Indian psychology hinders the stakeholders from developing dark tourism. Thus, the Golden Temple draws more visitors than the site of the Jallianwala Bagh massacre.

Conclusion

While a location is a geographical construct, a tourism destination is a commercial construct but focuses on tourism contributions to the economy. Thus, it crafts to be a location for tourists. Promotion pursuit acts as a catalyst for the economic development of the destination. Promotion is a marketing strategic tool to communicate or influence the actual and potential markets or tourist visitors and convince the tourists to visit the destination, disseminating its uniqueness and projecting the place as favorable for travel and tour. Thus, it is a tool for image creation or building and improvement. Imaging through the promotion process creates interest in the tourist's mind about the destination (Ray, 2022).

The emergence of a location associated with the stories and histories of varied dark shades and experiences as a tourism commodity is the most dramatic and striking phenomenon of the late 20th century. Some inherent cultural and conceptual factors, such as conceptual multiplicity, contradictory ethical obligations, confusion in determining the semantic, perceptional, and functional value, conceptual liminality issues, and conflicting cultural and ideological interpretations, often impede the promotion of dark tourism at different destinations. Promotion pursuit involves various agents (Ray, 2022) or stakeholders, including tourists, with distinct cultural and ethical backgrounds and diverse understandings of dark tourism. Thus, stakeholder engagement turns diverse and contradictory to their backgrounds and comprehensions. It consequently develops a dilemma in taking promotion pursuit. The states compel all the promotion initiatives taken by various tourism stakeholders to align with the created image. It often ignores the importance of dark tourism to the growth of the local economy. Hence, all those dilemmas and ignorance pose challenges to promoting any destination as a dark tourism destination and translating the dark stories and histories into a potential tourism economy driver.

References

Alderman, D. H. (2002). Writing on the Graceland on the importance of authorship in pilgrimage landscapes. *Tourism Recreation Research*, *27*(2), 27–35.

Ashworth, G. J. (1996). Holocaust tourism and Jewish culture: The lessons of Krakow- Kazimierz. In M. Robinson, N. Evans, & P. Callaghan (Eds.), *Tourism and cultural change* (pp. 1–12). Centre for Travel and Tourism.

Ashworth, G. J. (2004). Tourism and the heritage of atrocity: Managing the heritage of South African apartheid for entertainment. In T. V. Singh (Ed.), *New horizons in tourism: Strange experiences and stranger practices* (pp. 95–108). CABI International.

Ashworth, G. (2005). When time heals: The present interpretation of 18th century Acadian, planter and loyalist heritage sites in Nova Scotia, Canada. In G. Ashworth & R. Hartmann (Eds.), *Horror and human tragedy revisited: The management of sites of atrocities for tourism* (pp. 41–59). Cognizant Communication Corporation.

Ashworth, G. J. (2008). The memorialisation of violence and tragedy: Human trauma as heritage. In B. Graham & P. Howard (Eds.), *The Ashgate companion to heritage and identity* (pp. 231–244). Ashgate.

Best, M. (2007). Norfolk Island: Thanatourism, history and visitor emotions. *Shima. International Journal of Research into Island Cultures*, *1*, 30–48.

Bigley, J. D., Lee, C.-K., Chon, J., & Yoon, Y. (2010). Motivations for war-related tourism: A case of DMZ visitors in Korea. *Tourism Geographies*, *12*(3), 371–394.

Biran, A., Liu, W., Li, G., & Eichhorn, V. (2014). Consuming post-disaster destinations: The case of Sichuan, China. *Annals of Tourism Research*, *47*, 1–17.

Biran, A., Poria, Y., & Oren, G. (2011). Sought experiences at (dark) heritage sites. *Annals of Tourism Research*, *38*(3), 820–841.

Blom, T. (2000). Morbid tourism—A postmodern market niche with an example from Althorp. *Norsk Geografisk Tidsskrift*, *54*(1), 29–36.

Braithwaite, R., & Leiper, N. (2010). Contests on the River Kwai: How a wartime tragedy became a recreation, commercial and nationalistic plaything. *Current Issues in Tourism*, *13*(4), 311–332.

Bristow, R. S., & Newman, M. (2005). Myth vs. fact: An exploration of fright tourism. In K. Bricker & S. J. Millington (Eds.), *Proceedings of the 2004 Northeastern Recreation Research Symposium* (pp. 215–221). USDA Forest Service. Northeastern Research Station general technical report NE-326.

Brown, S., McDonagh, P., & Shultz, C. (2012). Dark marketing: Ghost in the machine or skeleton in the cupboard? *European Business Review*, *24*(3), 196–215.

Buda, D. M. (2015a). The death drive in tourism studies. *Annals of Tourism Research*, *50*, 39–51.

Buda, D. M. (2015b). *Affective tourism: Dark routes in conflict*. Routledge.

Buda, D. M., & McIntosh, A. J. (2013). Dark tourism and voyeurism: Tourist arrested for “spying” in Iran. *International Journal of Culture, Tourism and Hospitality Research*, *7*(3), 214–226.

Buda, D. M., & Shim, D. (2015). Desiring the dark: ‘A taste for the unusual’ in North Korean tourism? *Current Issues in Tourism*, *18*(1), 1–6.

Carr, G. (2010). Shining a light on dark tourism: German bunkers in the British Channel Islands. *Public Archaeology*, *9*(2), 64–84.

Cole, T. (2000). *Selling the Holocaust: From Auschwitz to Schindler: How history is bought, packaged and sold.* Routledge.
Dale, C., & Robinson, N. (2011). Dark tourism. In P. Robinson, S. Heitmann, & P. U. C. Dieke (Eds.), *Research themes for tourism* (pp. 205–217). CABI.
Dann, G. M. S., & Potter, R. B. (2001). Supplanting the planters: Hawking heritage in Barbados. *International Journal of Hospitality & Tourism Administration*, *2*(3/4), 51–84.
Dann, G. M. S., & Seaton, A. V. (2001). Slavery, contested heritage and thanatourism. *International Journal of Hospitality & Tourism Administration*, *2*, 1–29. https://doi.org/10.1300/J149v02n03_01
Dunkley, R. A., Morgan, N., & Westwood, S. (2007). A shot in the dark? Developing a new conceptual framework for thanatourism. *Asian Journal of Tourism and Hospitality*, *1*(1), 54–63.
Farmaki, A. (2013). Dark tourism revisited: A supply/demand conceptualisation. *International Journal of Culture, Tourism and Hospitality Research*, *7*(3), 281–292.
Foley, M., & Lennon, J. J. (1996). JFK and dark tourism: A fascination with assassination. *International Journal of Heritage Studies*, *2*, 198–211. https://doi.org/10.1080/13527259608722175
Foley, M., & Lennon, J. J. (1997). Dark tourism: A ethical dilemma. In M. Foley, J. J. Lennon, & G. A. Maxwell (Eds.), *Hospitality, tourism and leisure management: Issues in strategy and culture* (pp. 153–164). Cassell.
Henderson, J. C. (2000). War as a tourist attraction: The case of Vietnam. *International Journal of Tourism Research*, *2*, 269–280.
Hohenhaus, P. (2013). Commemorating and commodifying the Rwandan genocide: Memorial sites in a politically difficult context. In L. White & E. Frew (Eds.), *Dark tourism and place identity: Managing and interpreting dark places* (pp. 142–155). Routledge.
Hughes, R. (2008). Dutiful tourism: Encountering the Cambodian genocide. *Asian-Pacific Viewpoint*, *49*(3), 318–330.
Hultzsch, E. (1925). *Inscriptions of Asoka. Government of India.*
Isaac, R. K., & Ashworth, G. J. (2011). Moving from pilgrimage to 'dark' tourism. Leveraging tourism in Palestine. *Tourism, Culture and Communication*, *11*(3), 149–164.
Isaac, R. K., & Çakmak, E. (2014). Understanding visitor's motivations at sites of death and disaster: The case of former transit camp Westerbork, The Netherlands. *Current Issues in Tourism*, *17*(2), 164–179.
Isaac, R., & Çakmak, E. (2016). Understanding the motivations and emotions of visitors at Tuol Sleng Genocide Prison Museum (S-21) in Phnom Penh, Cambodia. *International Journal of Tourism Cities*, *2*(3), 232–247.
Johnston, T. (2013). Mark Twain and the Innocents Abroad: Illuminating the tourist gaze on death. *International Journal of Culture, Tourism and Hospitality Research*, *7*(3), 199–213.
Johnston, T. (2015). The geographies of thanatourism. *Geography*, *100*(1), 20–27.
Johnston, T. (2016). Blogging the dark side of travel: Consuming the siege of Sarajevo. In T. Johnston & P. andelartz (Eds.), *Thanatourism: Case studies in travel to the dark side* (pp. 1–22). Goodfellow.
Johnston, T., & Mandelartz, P. (2016). Introduction. In T. Johnston & P. Mandelartz (Eds.), *Thanatourism: Case studies in travel to the dark side*. Goodfellow.

Johnston, T., Tigre-Moura, F., & Mandelartz, P. (2016). Welcome to the home of Auschwitz tours': The online marketing of genecide tourism. In T. Johnston & P. Mandelartz (Eds.), *Thanatourism: Case studies in travel to the dark side* (pp. 155–170). Goodfellow.

Kang, E.-J., & Lee, T. J. (2013). War and ideological conflict: Prisoner of war camps as a tourist experience in South Korea. In L. White & E. Frew (Eds.), *Dark tourism and place identity: Managing and interpreting dark places* (pp. 236–247). Routledge.

Kang, E.-J., Scott, N., Lee, T. J., & Ballantyne, R. (2012). Benefits of visiting a 'dark tourism' site: The case of the Jeju April 3rd Peace Park, Korea. *Tourism Management*, *33*, 257–265.

Knudsen, B. T. (2011). Thanatourism: Witnessing difficult pasts. *Tourist Studies*, *11*(1), 55–72.

Lelo, L., & Jamal, T. (2013). African Americans at sites of darkness: Roots-making, diasporic identities and place making. In L. White & E. Frew (Eds.), *Dark tourism and place identity: Managing and interpreting dark places* (pp. 28–45). Routledge.

Lennon, J. (2009). Tragedy and heritage in peril: The case of Cambodia. *Tourism Recreation Research*, *34*(1), 35–44.

Lennon, J. J., & Foley, M. (1999). Interpretation of the unimaginable. *Journal of Travel Research*, *38*(1), 46–50. The United States Holocaust Memorial Museum: Washington, DC, USA, and "dark tourism".

Lennon, J. J., & Foley, M. (2000). *Dark tourism: The attraction of death and disaster*. Continuum.

Lioyd, D. (1998). *Battelfield tourism*. Berg.

MacCarthy, M. (2016). Consuming symbolism: Marketing D-Day and Normandy. *Journal of Heritage Tourism*, *12*, 191–203.

Marcuse, H. (2005). Reshaping Dachau for visitors: 1933-2000. In G. Ashworth & R. Hartmann (Eds.), *Horror and human tragedy revisited: The management of sites of atrocities for tourism* (pp. 118–148). Cognizant Communication Corporation.

Miles, S. (2014). Battlefield sites as dark tourism attractions: An analysis of experience. *Journal of Heritage Tourism*, *9*(2), 134–147.

Murphy, B. (2015). Dark tourism and the Michelin World War 1 battlefield guides. *Journal of Franco-Irish Studies*, *4*(1), 1–9.

Preece, T., & Price, G. G. (2005). Motivations of participants in dark tourism: A case study of Port Arthur, Tasmania. In C. Ryan, S. Page, & M. Aitken (Eds.), *Taking tourism to the limits: Issues, concepts and managerial perspectives* (pp. 191–197). Elsevier.

Raine, R. (2013). A dark tourism spectrum. *International Journal of Culture, Tourism and Hospitality Research*, *7*(3), 242–256.

Ray, S. (2015). Dark tourism: In search of Raison d'être for its multiplicity and nomenclature. *Tourism Theory and Practice*, *13*(1–2), 21–44.

Ray, S. (2022). Technology and destination promotion in Asian tourism: Challenges, changes, and bearing. In A. Hassan (Ed.), *Handbook of technology application in tourism in Asia*. Springer. https://doi.org/10.1007/978-981-16-2210-6_18

Reed, K. (2018). *Thanatourism to dark tourism: The transition of the religious sacred to secular sacred* (Vol. 14). Museum Studies Theses. http://digitalcommons.buffalostate.edu/museumstudies_theses/14

Rice, A. (2009). Museums, memorials and plantation houses in the Black Atlantic: Slavery and the development of dark tourism. In R. Sharpley & P. R. Stone (Eds.),

The darker side of travel: The theory and practice of dark tourism (pp. 224–246). Channel View.
Rittichainuwat, B. N. (2008). Responding to disaster: Thai and Scandinavian tourists' motivation to visit Phuket, Thailand. *Journal of Travel Research*, *46*, 422–432.
Robb, E. (2009). Violence and recreation: Vacationing in the realm of dark tourism. *Anthropology and Humanism*, *34*(1), 51–60.
Robbie, D. (2008). Touring Katrina: Authentic identities and disaster tourism in New Orleans. *Journal of Heritage Tourism*, *3*(4), 257–266.
Sallay, Á., Mikházi, Z., Gecséné Tar, I., & Takács, K. (2022). Cemeteries as a part of green infrastructure and tourism. *Sustainability*, *14*(5), 2918. https://doi.org/10.3390/su14052918
Seaton, A. V. (1996). Guided by the dark: From thanatopsis to thanatourism. *International Journal of Heritage Studies*, *2*(4), 234–244. https://doi.org/10.1080/13527259608722178
Seaton, A. V. (1999). War and thanatourism: Waterloo 1815–1914. *Annals of Tourism Research*, *26*(1), 130–158. https://doi.org/10.1016/S0160-7383(98)00057-7
Seaton, A. V. (2001). Sources of slavery e destinations of slavery: The silences and disclosures of slavery heritage in the UK and US. *International Journal of Hospitality & Tourism Administration*, *2*(3/4), 107–129. https://doi.org/10.1300/J149v02n03_05
Seaton, A. V. (2002). Thanatourism's final frontiers? Visits to cemeteries, churchyards and funerary sites as sacred and secular pilgrimage. *Tourism Recreation Research*, *27*(2), 73–82. https://doi.org/10.1080/02508281.2002.11081223
Seaton, T., North, M., & Gajda, G. (2015). Last resting places? Recreational spaces or thanatourism attractions and the future of historic cemeteries and churchyards in Europe. In S. Gammon & S. Elkington (Eds.), *Landscapes of leisure: Space, place and identities* (pp. 71–95). Palgrave.
Seaton, A. V., & Lennon, J. J. (2004). Thanatourism in the early 21st century: Moral panic, ulterior motives and alterior desires. In T. V. Singh (Ed.), *New horizons in tourism: Strange experiences and stranger practices* (pp. 63–82). CABI International.
Sharpley, R. (2012). Towards an understanding of 'genocide tourism': An analysis of visitors accounts of their experience of recent genocide sites. In R. Sharpley & P. R. Stone (Eds.), *Contemporary tourist experience: Concepts and consequences* (pp. 95–109). Routledge.
Simone-Charteris, T. M., Boyd, S. W., & Burns, A. (2013). The contribution of dark tourism to place identity in Northern Ireland. In L. White & E. Frew (Eds.), *Dark tourism and place identity: Managing and interpreting dark places* (pp. 60–78). Routledge.
Stone, P. R. (2009a). Making absent death present: Consuming dark tourism in contemporary society. In R. Sharpley & P. R. Stone (Eds.), *The darker side of travel: The theory and practice of dark tourism* (pp. 23–38). Channel View.
Stone (2009b). Dark tourism: Morality and new moral spaces. In R. Sharpley & P. R. Stone (Eds.), *The darker side of travel: The theory and practice of dark tourism* (pp. 56–72). Channel View.
Stone, P. R. (2009c). 'It's a bloody guide': Fun, fear and a lighter side of dark tourism at the Dungeon visitor attractions, UK. In R. Sharpley & P. R. Stone (Eds.), *The darker side of travel: The theory and practice of dark tourism* (pp. 167–185). Channel View.

Stone, P. (2011a). Dark tourism: Towards a new post-disciplinary research agenda. *International Journal of Tourism Anthropology*, *1*(3/4), 318–332.

Stone, P. R. (2011b). Dark tourism and the cadaveric carnival: Mediating life and death narratives at Gunther von Hagens' Body Worlds. *Current Issues in Tourism*, *14*(7), 685–701.

Stone, P. R. (2011c). Dark tourism experiences: Mediating between life and death. In R. Sharpley & P. R. Stone (Eds.), *Tourist experience: Contemporary perspectives* (pp. 21–27). Routledge.

Stone, P. R. (2012a). Dark tourism as 'mortality capital': The case of Ground Zero and the significant other dead. In R. Sharpley & P. R. Stone (Eds.), *Contemporary tourist experience: Concepts and consequences* (pp. 71–94). Routledge.

Stone, P. (2012b). Dark tourism and significant other death: Towards a model of mortality mediation. *Annals of Tourism Research*, *39*(3), 1565–1587.

Stone, P. (2013a). Dark tourism scholarship: A critical review. *International Journal of Culture, Tourism and Hospitality Research*, *7*(3), 307–318.

Stone, P. R. (2013b). Dark tourism, heterotopia and post-apocalyptic places: The case of Chernobyl. In L. White & E. Frew (Eds.), *Dark tourism and place identity: Managing and interpreting dark places* (pp. 79–93). Routledge.

Stone, P. (2016). Enlightening the 'dark' in dark tourism. *Interpretation Journal*, *21*(2), 22–24.

Stone, P., & Sharpley, R. (2008). Consuming dark tourism: A thanatological perspective. *Annals of Tourism Research*, *36*(2), 574–595.

Stone, P. R., & Sharpley, R. (2014). Deviance, dark tourism and 'dark leisure': Towards a (re)configuration of morality and the taboo in secular society. In S. Elkington & S. J. Garmon (Eds.), *Contemporary perspectives in leisure: Meanings, motives and lifelong learning* (pp. 54–64). Routledge.

Strange, C., & Kempa, M. (2003). Shades of dark tourism: Alcatraz and Robben Island. *Annals of Tourism Research*, *30*(2), 386–405.

Tarlow, P. E. (2005). Dark tourism: The appealing 'dark' side of tourism and more. In M. Novelli (Ed.), *Niche tourism: Contemporary issues, trends and cases* (pp. 47–58). Elsevier.

Tinson, J. S., Saren, M. A. J., & Roth, B. E. (2015). Exploring the role of dark tourism in the creation of national identity of young Americans. *Journal of Marketing Management*, *31*(7/8), 856–880.

Tunbridge, J. E., & Ashworth, G. J. (1996). *Dissonant heritage: The management of the past as a resource in conflict*. Wiley.

Turner, V., & Abrahams, R. D. (1969). Liminality and communitas. In V. Turner, R. D. Abrahams, & A. Harris (Eds.), *The ritual process: Structure and anti-structure* (pp. 94–113). Aldine.

Van Gennep, A. (1960). *The rites of passage*. University of Chicago Press. Original work published 1909.

Volcic, Z., Erjavec, K., & Peak, M. (2014). Branding post-war Sarajevo. *Journalism Studies*, *15*(6), 726–742.

Walter, T. (2009). Dark tourism: Mediating between the dead and the living. In R. Sharpley & P. R. Stone (Eds.), *The darker side of travel: The theory and practice of dark tourism* (pp. 39–55). Channel View.

Wilcken, U. (1967). *Alexander the great*. W. W. Norton.

Wu, C., Funck, C., & Hayashi, Y. (2014). The impact of host community on destination (re)branding: A case study of Hiroshima. *International Journal of Tourism Research*, *16*(6), 546–555.

Yankholmes, A., & McKercher, B. (2015). Understanding visitors to slavery heritage sites in Ghana. *Tourism Management*, *51*, 22–32. https://doi.org/10.1016/j.tourman.2015.04.003

Weblinks

https://www.futuremarketinsights.com/reports/dark-tourism-sector-overview

https://www.futuremarketinsights.com/reports/tourism-market

https://www.reportsandinsights.com/report/dark-tourism-market

https://www.statista.com/statistics/1179124/global-travel-agency-sector-key-data/#:~:text=The%20market%20size%20of%20the,employed%20approximately%20two%20million%20workers

https://www.grin.com/document/265137?lang=en

Chapter 19

Unearthing the Digital Dark: Enhancing Dark Tourism Through Online Platforms

Shruti Sharma and Nidhi Sharma

Institute of Management Studies, Ghaziabad (University Courses Campus), India

Abstract

Exploring historically significant and often eerie locations, dark tourism has surged in popularity lately. With advancing technology, online platforms have become influential tools that amplify the dark tourism encounter. This research delves into how online platforms contribute to enriching the dark tourism experience. It addresses the effective utilization of online media to improve exploration, understanding, and engagement with dark tourism destinations.

The research questions focus on how online platforms enhance exploration and interpretation, their impacts on visitor understanding and emotional engagement, and their facilitation of community engagement and knowledge sharing within dark tourism.

Employing a mixed-methods access, in addition to literature review, qualitative interviews, and quantitative surveys, the study adopts a theoretical framework integrating destination management, digital marketing, and visitor experience theories.

This empirical study adds to the existing literature on how online platforms influence dark tourism, providing practical insights for destination managers and platform developers to enhance visitor experiences and engagement at dark tourism destinations.

Keywords: Dark tourism; online platforms; visitor experience; exploration; understanding; cultural exchange; engagement; immersive experiences; historical context

Dark Tourism, 279–295

doi:10.1108/978-1-83797-336-120241019

Introduction

The tourism industry has experienced significant growth and advancement recently, as noted by Ivanov and Webster, 2013. Concurrently, the digital industry is rapidly progressing, permeating both daily life and work environments, as highlighted by Chaffey in 2015. Within this framework, every aspect and variety of tourism should acknowledge the crucial significance of digital platforms in marketing, interacting with visitors, and providing impactful experiences. Nowadays, tourists regularly employ digital resources to delve into destination details, investigate visitor spots, make reservations for accommodations and flights, and broadcast their experiences across social media platforms. These digital tools are fundamental in shaping and enhancing the overall tourism experience, influencing travelers at every stage of their journey. The prevalence of mobile devices among tourists is on the rise, as emphasized by Baggio and Chiappa (2014), further enriching and personalizing their overall travel experience.

This chapter delves into incorporating digital media into dark tourism to elevate and craft a more immersive and unforgettable experience for those engaged in this type of tourism. Addressing this critical element involves adopting a threefold approach. First one is an examination of social media that delves into its potential for marketing and promoting dark tourism destinations. Additionally, it explores how these platforms can be leveraged to meaningfully engage with dark tourists online, fostering a connection that elevates their overall experiences. Second one is that this chapter delves into the realm of phone applications, catering to the increasing preferences of tourist people who increasingly rely on smartphones and tablets for connectivity. Mobile gamification could also be used to make learning about dark history more engaging. The goal is to give information, engage, and deeply immerse them in an entirely newly realm of experience.

In conclusion, the studies explore idea of any blog and associated online forums, emphasizing the role of these digital platforms in nurturing information sharing and fostering collaborative research within the domain of dark tourism. Throughout each of these dimensions, instances of exemplary practices will be showcased, alongside an examination of the evolving nature of technology for the future. This chapter wrapped up with summarizing essential insights that could guide the formulation of new strategic directions for the more effective development of dark tourism in the digital landscape.

Phenomenon of Dark Tourism

Dark tourism, centered on visiting sites marked by tragedy, isn't a new trend. Throughout history, travelers, similar to pilgrims, have been drawn to locations linked to crucifixions, sacrificial deaths, or where it was a form of entertainment, such as the Colosseum in ancient Rome. Even in the medieval period, crowds gathered for public executions, evident in places like the "Maumbury Rings" in "Dorchester," UK. Lately, places like Ground Zero in New York and the tunnel of Paris linked to Princess Diana's tragic event have emerged as significant destinations on tourists' itineraries (Stone, 2006). The array of macabre-related attractions varies widely, ranging from simulated deaths at the London Dungeon to authentic locations of mass extermination like the Nazi death camps scattered across Poland and various parts of Europe. These locations also encompass sites linked to renowned celebrity deaths, like John Lennon or Elvis Presley, and notable catastrophes, such as post-Hurricane Katrina New Orleans. In the domain of tourism research, scholars have sought to classify diverse types of dark tourism, encompassing categories such as battlefield tourism, cemetery tourism, disaster tourism, holocaust tourism, political tourism, among others (Simone-Charteris & Boyd, 2010; Trotta, 2006). Efforts have been made to delineate different intensities within dark tourism. Miles (2002), for example, distinguishes between locations linked to death and places where the actual event occurred. This differentiation allows for the categorization of 'dark' and 'darker' tourism, where visits to sites of death denote a heightened level of empathetic travel, referred to as darker tourism. Likewise, Stone (2006, p. 147) presents the idea of "shades of darkness." He proposes that dark tourism encounters span a continuum of intensity that is adaptable and evolving. Within this spectrum, certain sites may be viewed as "darker" than others due to distinct characteristics, perceptions, and features of the experience. Seaton (1996, 1999) previously delineated five potential categories of dark traveling activities:

Traveling to witnessing real-time depictions of death, illustrated by tourists hurrying to witness plane crashes and ship sinking. Visiting locations of mass or individual deaths after the occurrences, as observed in trips to Ground Zero in New York following the events held on 9/11.

Touring sites of internment and memorials devoted to the departed, like monuments paying tribute to soldiers who lost their lives in both World Wars. Visiting locations indirectly associated with the original events yet displaying physical proof or symbolic representations of particular deaths, such as journeys to the Holocaust Museum in Washington, DC. Participating in travel involving re-enactments or simulations of death, akin to the staged battles by American Civil War enthusiasts in the United States.

Looking at the tourist perspective, studies have aimed to uncover the reasons motivating individuals to seek out and interact with dark sites and attractions. This type of motivation encompasses various factors like as curiosity, commemoration, nationalistic feelings, the pursuit of uniqueness and genuineness, educational aspirations, and empathy toward the victims (Ashworth, 2002; Causevic & Lynch, 2007; Stone, 2006; Seaton, 1999; Stone and Sharpley 2008; White and Frew 2013). However, irrespective of the motivations for visiting places connected to death and anguish, the attraction toward different destinations and specific attractions linked to dark tourism remains steadfast (Ntunda, 2012). Visitors' curiosity seems aligned with the scholarly focus on this specialized tourism sector, which culminated in the formation of the Institute for Dark Tourism Research (iDTR) at the University of Central Lancashire (UCL an), UK, launched by Dr Philip Stone in 2012.

The Dark Tourist and Social Media Platform

Although Stone (2006) acknowledges that "notion of dark tourism, across its diverse forms, has stimulated considerable research and media attention" (refer also to Chapter 14), there has been a noticeable lack of extensive exploration into the digital platform aspect of dark tourism and the use of online platforms where dark tourists can convene and share their experiences and sentiments. Moreover, there's been a lack of investigation into how those involved in managing and promoting this type of tourism can effectively utilize digital platforms and diverse technologies in their tourism strategies. Specifically, these digital components include social media and mobile applications (apps), and it's essential to consider Wi-Fi connectivity for effectively leveraging both aspects.

In the domain of merging digital media platform and innovative technology, the ascent of social media platform has fundamentally transformed how businesses communicate with customers. Within the tourism industry, this change has consequently reshaped our methods of engaging with tourists. The social media platform is described as "a form of media disseminated through online social interactions and encompasses diverse formats such as social networking sites, blogs, wikis, podcasts, photo and video sharing, social bookmarking, and virtual environments" (Flectcher & Lee, 2012). In the expansive digital realm of social media, a select few platforms wield significant influence. Stelzner (2015) acknowledges the "famous five" – Facebook, Twitter, LinkedIn, blogging, and YouTube. Moreover, there has been notable growth in image-focused social media platforms such as Instagram and Pinterest (Chaffey, 2015). Most frequently discussed topics on social media platforms is travel, with users sharing their vacation experiences. Consequently, the modern traveler actively employs social media before their trip to gather information and increasingly utilizes these platforms during their vacation to enrich their experience and share elements of it with others (White, 2009). The pace of change in this area has been swift. Remarkably, "it took 38 years for the radio to reach 50 million listeners, and 13 years for television to captivate 50 million viewers. The internet attracted 50 million subscribers in just 4 years, and Facebook gained 50 million subscribers in a mere one-and-a-half years" (Nair, 2011). Concerning dark tourism, also recognized as than tourism, it has firmly entered the sphere of social media platform. For instance, in the realm of more "entertainment-focused" dark tourism, consider the 'Jack the Ripper Tour' in London, which has amassed a significant following and is increasingly active on Twitter, boasting over 1,300 followers. Jim Morrison's grave in Paris has its own dedicated Facebook page. On the more somber end of Stone's (2006) taxonomy of the "dark tourism spectrum," the former German Nazi concentration camp at Auschwitz-Birkenau in Poland has several widely followed social media pages spanning across platforms like Facebook, Instagram, and Pinterest. Several social media pages linked to Auschwitz-Birkenau have gathered over 2 lakh likes, featuring subscribers' posts that have spurred discussion threads with up to 55 comments per post. Many visitors have shared photographs, and some have uploaded video clips, revealing the intent of these "dark tourists" to document their visit, convey their perspectives, articulate emotions, and recount their experiences. Their intent extends beyond mere sharing; they seek engagement, even in a virtual context, with individuals who have undergone similar feelings and experiences. One could argue that they are pursuing a sense of community among fellow dark tourists, possibly even sharing a connection based on a sense of "suffering," while also engaging with historical elements and nostalgic aspects associated with these tragic events. Furthermore, the creation of dedicated social media pages for these profoundly significant visitor sites can be seen as a form of "commemoration" – an additional means of ensuring that the tragedy remains part of collective memory. Social media is revealing fresh insights into social interaction, communication, and information sharing within the realm of dark tourism places and locations. This not only serves as a way to promote these destinations to tourists but also creates online "community hubs" where dark tourists can connect, share experiences, and engage in discussions about the impact and significance of these places. As an example, "Dark Tourism Photography" is a public group on Facebook that invites members to share photographs taken at dark tourism attractions and sites worldwide, boasting nearly a 1,000 members currently. These social media platforms can act as valuable information sources for industry professionals overseeing dark tourism destinations, providing easily accessible repositories of valuable data for researchers investigating this phenomenon. Additionally, they can operate as forums where scholars exchange information and engaged in collaborating research initiatives.

In this regards, the leading Facebook groups falling into this category are presently the "Dark Tourism Forum" and "Dark Tourism World." Dr Philip Stone (UK) oversees the "Dark Tourism Forum," which has garnered over 3,500 'likes' at present. Meanwhile, "Dark Tourism World," managed by Jorge Coelho (Portugal), operates as a bilingual group, offering posts in both English and Portuguese, currently boasting over 2,500 "likes." Both groups are committed to advancing and sharing scholarly research and updates related to dark tourism. Another notable Facebook group, "Dark Tourism Worldwide," with slightly over 400 members, stands out from the prior two by not exclusively concentrating on academic material. This community caters to a mix of academics, students, and both novice and seasoned "dark tourists." Its goal is to explore dark tourism comprehensively, spanning from its historical roots to its modern manifestations – addressing its physical, ethical, and core principles (Dark Tourism Worldwide, 2016).

Leveraging social media platform has become a necessity across all types of tourism places today, as emphasized by Munar and Jacobsen (2014) and Nusair et al. (2013), and dark tourism is no different. The fascination with the macabre has expanded its reach into the digital realm of social media, impacting various facets of dark tourism. The availability of shared online content and information, combined with the expansive reach of social media platforms (such as user counts and content volume), influences how individuals perceive the significance of sharing online (Munar & Jacobsen, 2014). Consequently, this not only involves but also motivates travelers to actively pursue the viewpoints and recommendations of fellow tourists via using social media online platform.

Digital Dark Tourism Spaces

Interactive computer displaying to create the illusion of being transported to another location are commonly referred to as virtual environment displays. Different names have been used to depict this concept, including artificial intelligence (AI), virtual reality (VR), cyberspace, virtual worlds, and virtual environment (Ellis, 2016). Crafting tourist destinations in a virtual or re-envisioned manner requires advanced technology, considering that these tours go beyond the confines of two-dimensional space. Travelers can engage their equilibrioceptive (balance, direction), proprioceptive (movement), and nociceptive (pain/wellness) senses in the domain of 3D virtual tourism, in addition to their visual and aural senses. This is commonly achieved through formats like VR, Quick Time, or Flash movies, allowing viewers to panoramically explore sites and zoom in and out (Kaelber, 2007). While some studies express legitimate worries about the efficacy of online versus offline encounters (Cowan, 2005; Dawson, 2005), others predict that visits will evolve into multimedia-induced hyper-reality as the future of travel and tourism (Ritzer, 2005; Kaelber, 2006a, 2006b). Virtual tours are offered for a variety of attractions, including Canterbury Cathedral in England, the Palace of Versailles in France, and even Drum Castle in Scotland. These virtual tours provide internet-based encounters, offering a preview for potential future visits or as independent visual experiences for those unable to visit physically (Arnold, 2005). Nevertheless, within the realm of dark tourism, there's currently a scarcity of virtual environments dedicated to dark tourism experiences.

While most dark tourism sites utilize websites for promotional objectives, they primarily exist as physical destinations necessitating in-person visits. For example, to authentically immerse oneself in the convict experience at former prison sites, visitors must encounter the cold, damp stone walls, perceive the claustrophobic ambiance within confined cells, rest on uncomfortable prison mattresses, and feel every irregularity on the basic metal beds.

Nevertheless, certain sites of dark tourism feature highly sophisticating "online gateways" designed to lead people through a virtual experience, aiming to entice them for visiting the actual physical site. Hence, to a degree, one can venture into dark sites from the comfort of home, seated in front of a laptop. The limited availability of virtual environments dedicated to dark tourism might, partly, result from the challenge of delivering suitable interpretation for such visitor destinations. Moreover, attempting to engage a broad array of online viewers within the fleeting attention span that 'surfers' allocate to a website, all while being mindful of the sensitive subject matter, poses a significant challenge. Dark tourism sites encounter the dual task of meeting the diverse requirements to educate and engage online audiences, similar to their efforts for visitors at physical locations (The Digital Panopticon, 2014). Yet, there are a few "destinations" that predominantly or solely exist as cyber-sites within the scope of dark tourism, such as Alcatraz and the gas chambers of Auschwitz-Birkenau.

From 1934 until 1963, the federal penitentiary on Alcatraz Island in California's San Francisco Bay held some of America's most notorious offenders. Al "Scarface" Capone (1899–1947), the famed mobster, and Robert "Birdman of Alcatraz" Stroud (1890–1963), renowned for murder, were among the prominent inmates in this high-security jail. Despite numerous attempts, no prisoner successfully escaped from the prison. Alcatraz Island is still a famous tourist site today, and it has also evolved into a cyber-tourism attraction (History, 2016). Virtual tours of Alcatraz are available on several websites, including www.fullscreen360.com/alcatraz, http://www.virtuar.com/alcatraz/, and http://www.digitalproperties.ca/alcatraz/. Notably, www.virtuar.com/alcatraz/ offers guests a variety of virtual tour alternatives that cover various features of the Island and San Francisco Bay. This includes specialized excursions that focus just on the penitentiary. Visitors can explore areas such as the south entrance, the visitation center, where inmates occasionally met visitors, the main block known as "Broadway," a cell within the main block, the shower room (a notorious place for inmate conflicts), the barber shop, the entrance to D block housing solitary confinement cells, the dining hall, and the inmate entrance wing on the ground floor (Virtuar, 2016). The virtual tours' high-definition quality offers a strikingly genuine experience, closely resembling being physically present inside the actual prison. The tours evoke a range of emotions akin to those felt by the authors when visiting other prisons in person, such as the Crumlin Road Gaol in Belfast, Northern Ireland, and the Napier Prison in Hawkes Bay, New Zealand. "Auschwitz-Birkenau" holds a distinctive place in collective awareness, symbolizing unparalleled moral importance. The Nazi German regime's systematic slaughter of Jews, known as the Holocaust or Shoah, has become a worldwide emblem of human suffering and ethical outrage, providing an opportunity for educational reflection (Alexander, 2002). The number of visitors to Poland's Auschwitz-Birkenau Memorial and Museum has increased significantly, reaching 1,725,700 in 2015 (Auschwitz-Birkenau State Museum, 2015b).

Still, the museum faces an enduring struggle concerning representation, given that a significant portion of the current site differs greatly from its state when liberated by Soviet forces in 1945. Due to the challenges of accessing the physical site, which contains remnants that have deteriorated, undergone significant changes, or are inaccessible physically, virtual tours of Auschwitz offer an alternative means to interact with a site marked by obliterated trauma (Kaelber, 2007). One of these virtual tours is available on the website of the State Museum of Auschwitz-Birkenau (2015a). This trip, featured on "Remember.org," allows visitors to examine the ruins and surroundings of two large Birkenau crematoria. Hotspots indicate adjacent landscape elements and are complemented by explanatory text.

The VR tours of the Auschwitz crematoria predominantly focus on visually exploration. In contrast to on-site visitors constrained by physical limitations, those taking virtual tours can freely scrutinize the facilities from various angles – exterior, interior, and aerial views. However, the immersive sensory qualities of hearing, feeling, smelling, and tasting are not engaged in the same way that the contemporary Alcatraz virtual tours are (Kaelber, 2007). Furthermore, users must download QuickTime and Flash media players on their own computers to access these virtual tours, which may discourage some people from participating. Furthermore, virtual tours of dark sites appear to have a limited duration, as evidenced by the discontinuation of certain tours highlighted by Kaelber (2007), such as the one presented by the Block Museum of Art in its online exhibit "The Last Expression" and the British Broadcasting Corporation (BBC) in its "Auschwitz" documentary. As a result, virtual settings in dark tourism may be better suited as previews for possible trips rather as alternatives for actual visits until technology progresses to enable 3D experiences engaging all senses.

Engaging in Dark Tourism Encounter

The mobile device, notably the smartphone, has captured the public's fascination. Essentially, a smartphone merges a cellular telephone with diverse applications and internet connectivity. It amalgamates digital voice services with text messaging, email, web browsing, portable media players, digital cameras, pocket video cameras, and Global Positioning System (GPS) navigation (Dickinson et al., 2014).

Smartphones have evolved into a digital Swiss army knife, effectively replacing items such as watches, cameras, books, diaries, laptops, and even televisions. In the realm of tourism, businesses, operators, and organizations must recognize this widespread adoption of mobile technology. There's no denying that the industry must fully leverage mobile technology and consistently explore its potential to enhance how products and services are marketed and promoted.

Moreover, this technology should facilitate improved information dissemination and elevate the overall visitor experience. Modern travelers routinely carry their smartphones and tablets while on the move, actively intending to make use of them. The interpretative and experiential details traditionally conveyed through signage, leaflets, guidebooks, maps, and guided tours are now readily accessible on mobile devices. This encompasses features such as augmented reality, immersive multimedia content (video and audio), personalized tour selections, elements of gamification, GPS, Location-Based Service (LBS), and Quick Response (QR) code technology. This creates opportunities for those capable of harnessing the growing dependence on mobile devices, aligning with the escalating interest in various facets of dark tourism. Take, for instance, the "Killer GPS: Crime Scene, Murder Locations, and Serial Killers" app, offering information about over 470 global locations where notorious murders occurred or bodies were disposed.

This application equips users with interactive maps, travel directions, and profiles of notorious serial killers directly accessible on their smartphones or tablets. As expounded by Tussyadiah and Zach (2012, p. 783), the escalating intrigue in geographic technologies shaping immersive tourism experiences demands an innovative approach to meet the evolving expectations of tourists. The "Killer GPS" app stands out as a striking illustration of this paradigm shift, seeking to cater to a sustained fascination with the darker realm of serial killers. It is a dynamic, location-based application meticulously tailored for dark tourism enthusiasts to explore while on the move using their mobile devices.

In 2012, the debut of an iPhone/iPad application marked the emergence of advanced digital technology recounting the Jewish history of Oświęcim, the industrial town in southern Poland that served as the location for the former Nazi German concentration and death camp Auschwitz-Birkenau. Dubbed "Oshpitzin," echoing the Yiddish name for the town, this application elevates the digital experience, complementing the previously discussed social media aspects associated with former Nazi concentration camp sites. This immersive technology revitalizes the narratives and solemn history linked to these sites in a manner that traditional interpretive information cannot replicate. The creators of the application aimed to preserve memories through a digital medium, viewing this specific application as a contemporary tool to aid travelers in unraveling the 'real stories' behind these unfolding tragedies (Gruber, 2012).

In Northern Ireland, the "Titanic in Belfast" app recounts the story of the ill-fated ship using GPS technology. As tourists navigate the "Titanic Quarter" in the city, the app provides comprehensive information, delivering interpretation at significant locations and linking visitors to transportation and dining choices. This stands as an enlightening example of an app focused on a poignant site, albeit one that provides a broader range of visitor-oriented information to meet the diverse needs of tourists. Essentially, this case suggests that inclusive concept apps can integrate elements of dark tourism without exclusively focusing on that specific type of tourism.

Self-Portraits at Dark Tourism Sites

Another aspect of mobile digital tourism is the prevalence of "selfies," a trend that extends to dark tourism. As outlined by Hodalska (2015), the act of taking selfies at sites associated with horror, like camps of concentration, Ground Zero, or areas affected by disasters, has notably surged on online social media platforms. It seems that locations marked by tragedy and loss are impacted by the contemporary practice of capturing self-portraits at close range using smartphones. The act of photographing oneself at dark tourist spots isn't a recent phenomenon.

In earlier eras, including ancient Greek and Roman times, travelers left their names, places of origin, and dates etched on stones within the Egyptian Valley of Kings (Quack, 2016; as cited in Tourism Review, 2016). However, the emergence of the "selfie generation" has transformed the manner in which these images are now captured.

As emphasized by Hodalska (2015), "in areas linked with death, adolescents often feel compelled to commemorate life in what they perceive as the most 'imaginative' manner." Nonetheless, the act of individuals taking seemingly cheerful selfies, especially those depicting smiles, laughter, and casual behavior, could be deemed highly disrespectful by numerous individuals in associated places with the anguish and demise of others.

As a result, selfies captured at locations of profound tragedy may be perceived as somewhat eerie mobile keepsakes, prompting the question of their true cost. Reading, as cited in Garde-Hansen (2011, p. 139), labels this phenomenon as "memobilia," referring to "mobile digital phone memories or memo ilia – wearable, shareable multimedia data records of events or communications... that are deeply personal yet instantly become part of a global memory cape through the World Wide Web." Undoubtedly, these selfies are frequently shared on social media platforms, eliciting engagement in the form of likes, shares, and tweets from online friends and followers. This serves as what some argue is a "real-time" portrayal of oneself directed at an audience situated elsewhere (Levine, 2014).

Quack, as cited in Tourism Review (2016), contends that individuals taking selfies are predominantly focused on themselves – "Look, I'm in a Favela," "Look, I am at the Costa Concordia sinking site" (referring to the Italian Island of Giglio, where the Costa Concordia ship ran aground and capsized in 2012). Quack, as mentioned in Tourism Review (2016), suggests that some individuals visit disaster zones like the earthquake regions of Haiti to enhance their own status. They may feign concern but are primarily there to reassure themselves, thinking, "How fortunate this didn't happen to me." Their motive could involve observing others' hardships from a safe distance, allowing them to capture distressing images and return home with memory cards filled with poignant photos as evidence of their visit (Hodalska, 2015).

Enduring questions remained – is this morally permissible? Should the aftermath of a nuclear disaster, like the one at the Chernobyl nuclear plant in Ukraine in 1986, be utilized as a personal backdrop, akin to decorative tropical scenery? (Tourism Review, 2016).

The challenge lies in the fact that camera phones and tablets, facilitating the taking of selfies, have become socially accepted devices that contribute to emotional detachment, allowing individuals to cope with underlying fears. Additionally, while capturing photos validates an experience, it simultaneously acts as a means of creating distance from it (Hodalska, 2015), a concept expressed by Sontag (1977, p. 9) as "reducing the experience to something photogenic, transforming it into an image, a keepsake." Essentially, through kitsch, we divert our attention from tragedy (Harris, 2002). Durkin (2003, p. 47) shares a similar perspective, asserting that "by turning death into amusement and entertainment, we effectively neutralize it, making it innocuous and thus less threatening."

Perhaps Phoebe McWilliams had a similar intention back in August 2012 when she captured a selfie at Ground Zero and shared it on Twitter with the hashtag #hangover (Hodalska, 2015). Consequently, while smartphones and mobile devices contribute positive elements, especially through social media and specialized apps utilizing augmented reality, gamification, GPS positioning, and QR code technology, they also carry a potentially negative aspect, particularly concerning the phenomenon of selfies. Whether unintentional or deliberate, it is evident that tourists taking casual self-portraits at dark tourism sites can elicit feelings of disrespect in some and even strong disapproval, as argued by authors like Hodalska (2015).

Blogosphere of "Dark Tourism"

Introduced by Tim Berners-Lee in 1992, blogs have undergone a rapid expansion in recent years, achieving substantial popularity. As of 2013, WP Virtuoso reported the existence of approximately 152 million blogs on the internet, with a new blog emerging roughly every half second, resulting in a daily addition of 172,800 new blogs (WP Virtuoso, 2013). Current estimates indicate the presence of around 450 million active blogs (Bolan, 2015). This surge in blogging carries noteworthy implications for the travel and tourism industry.

The challenge facing the tourism sector lies in the inherent difficulty of appraising the quality and merits of a tourism offering prior to consumption, owing to the absence of direct experiences. Blogs emerge as a viable solution, providing a platform for prospective travelers to engage in a virtual community that shares firsthand tourism encounters online (Wang et al., 2002). Historically, word of mouth (WOM) has held a superior reputation as a source of information compared to official tourist channels. In the contemporary landscape, blogs have evolved into a novel digital rendition of WOM. The tourism blogosphere encompasses diverse categories, including consumer-to-consumer (C2C), business-to-business (B2B), business-to-consumer (B2C), and government-to-consumer (G2C) blogs, each bearing significant ramifications for tourism and destination marketing (Schmallegger & Carson, 2008).

The bulk of travel and tourism blogs prevalent on the internet fall under the personal consumer-to-consumer (C2C) category. In this domain, these blogs typically serve diverse purposes, such as disseminating personal travel narratives and experiences online, nurturing connections with friends and family, engaging socially with kindred spirits, and articulating both favorable and unfavorable travel encounters with fellow sojourners (Pan et al., 2007). Noteworthy instances of blogs adopting this perspective in the realm of dark tourism include "The Bohemian Blog," "Travel Darkling," and "Grief Tourism." Established in 2011 by Darmon Richter, a British freelance writer and photographer, "The Bohemian Blog" functions as an offbeat travel journal devoted to the eerie and extraordinary. Richter utilizes the blog as a platform to showcase his captured photos and written compositions about various dark tourism locales and attractions. These range from the Tombs of Istanbul in Turkey to the Ruins of the Chisinau Ghetto in Moldova and socialist-era monuments in Bulgaria.

Furthermore, he leverages the blog as a conduit to connect with potential collaborators for commercial ventures. In a manner akin to "The Bohemian Blog," "Travel Darkling" stands as a travel blog meticulously curated by freelance travel journalist and editor Anita Isalska, showcasing a distinct fascination with the more esoteric facets of travel. Recent entries showcase visuals and articles spotlighting Soviet-era sculptures in Eastern Europe (December 15, 2014), the Anatomical Theater in Bologna, Italy – an historical site where medical surgeries and autopsies unfolded for centuries, catering to enthusiastic medical students (April 26, 2014) – and various dark destinations in Czech Moravia (February 20, 2014). Isalska also disseminates her contributions to articles on dark sites and attractions, as featured in CNN Travel, *Lonely Planet*, and *TNT* Magazine, through her blog.

On a different note, "Grief Tourism," penned by Sharon Slayton, is devoted to diverse manifestations of what is termed grief tourism. The most recent entry, dated August 14, 2012, delves into plantation slavery and prison confinement in Angola. Unlike certain bloggers, Slayton abstains from uploading images on her blog but includes a link to another travel blog showcasing vacation idea unrelated to dark or grief tourism. This aligns with the observable trend in mobile apps that integrate elements of dark tourism without exclusively concentrating on that particular form of tourism. Schmallegger and Carson (2008) underscore the pivotal role of B2B blogs in the tourism sector, functioning as networking hubs for professionals, academics, businesses, and tourism organizations. These platforms facilitate the exchange of industry insights, technological advancements, research updates, and marketing strategies among experts.

Conversely, B2C and G2C blogs operate as platforms geared toward promoting tourism services, destinations, and elevating customer engagement. Typically overseen by organizational staff, both B2C and G2C blogs strive to disseminate the latest news to readers, providing insider perspectives on the company and details about various interest categories within destinations. These categories encompass attractions, art, events, culinary experiences, and entertainment. The overarching objective of these blogs is to captivate readers while furnishing comprehensive information about the offered products or the distinctive features of a destination.

An exemplary instance of a dark tourism site leveraging a blog for promotional purposes is the National September 11 Memorial and Museum (9/11 Memorial) situated at the World Trade Center in New York. Known as "The Memo Blog" and accessible through a link on the Memorial's website, this blog is crafted by the 9/11 Memorial's staff. It functions as a central hub for updates, discussions, and pertinent information concerning the 9/11 Memorial. Recent entries delve into episodes from the "Our City, Our Story" podcast series, featuring individuals such as Rudy Giuliani and Marilyn Goldberg recounting their experiences during the World Trade Center tragedy on 9/11 (June 3, 2016). Additionally, there's a podcast episode spotlighting the renaming of a Brooklyn school in honor of 9/11 firefighter Capt. Vernon A. Richard (June 2, 2016).

Fundamentally, for travel blogs to attain success and maintain reader engagement, they must consistently deliver updates, provide follow-ups to prior entries, and generate fresh content on a regular basis. Nevertheless, this ideal scenario is not universally realized, and numerous travel blogs, including those centered on dark tourism, grapple with irregular updates or complete abandonment. Presently, instances of dark tourism blogs that have not received recent updates include "Travel Darkling" and "Grief Tourism." The most recent post on "Travel Darkling" dates back to a year ago (December 15, 2014), and "Grief Tourism" has not seen an update in three years (August 14, 2012).

Detractors frequently highlight a significant challenge associated with blogs and user-generated content (UGC) platforms: their susceptibility to the presence of "fake" content, primarily attributed to the option for users to submit comments anonymously or using false identities (Cox et al., 2009; Schmallegger & Carson, 2008). Platforms such as TripAdvisor, for instance, have encountered instances where negative comments are posted by competitors or their proxies, while positive comments may be artificially generated by employees or owners. Despite the generally straightforward nature of detecting fake posts, the ongoing potential for manipulation remains a concern within the blogosphere and UGC sites.

Connectivity via Wi-Fi

The connected experiences through social media and mobile usage offer substantial opportunities for dark tourism businesses. Yet, leveraging this trend effectively requires the development of novel digital strategies. Ensuring robust and high-quality Wi-Fi provision becomes crucial to the effectiveness and relevance of these strategies. Despite the long-standing recognition of this need, the implementation of proper, high-quality Wi-Fi services at event venues, tourism attractions, and accommodations, especially in the United Kingdom, remains inconsistent. While some operators have made strides in this regard, many have yet to meet the expectations of tourists in the second decade of the 21st century.

Therefore, to cultivate a genuine sense of connection that aligns with the evolving expectations of tourists, it is imperative to establish "smart" facilities. These encompass smart hotels, smart restaurants, and smart visiting attractions, all equipped with consistently high-quality, free, and readily available Wi-Fi. Fulfilling this requirement not only enhances user experience but also brings about heightened benefits, leveraging cost-effective marketing and public relations through social media on users' mobile devices. This is particularly crucial for (dark) tourism locations that are equipped with readily available Wi-Fi. In this context, customized apps tailored to individual preferences and behaviors, featuring interactive elements, play an increasingly pivotal role. The integration of beacons has become commonplace, utilizing Bluetooth Low Energy (BLE 4.0) to communicate with beacon-enabled devices in users' smartphones or tablets, providing information as users navigate through these locations.

This technology, originally prominent in retail, is now expanding its influence into events and event management. An illustrative example took place at the 2015 US PGA Golf tournament, where live online streaming introduced inventive interactive features for mobile devices. This included offering comprehensive laser-generated data for every shot on the golf course. Beacons were seamlessly incorporated into the scoreboards carried by volunteers, consistently updating and transmitting data captured by remote drones. Subsequently, this information was integrated into the tournament app accessible to spectators on their smartphones. Such a configuration significantly elevated the spectator experience, delivering real-time, detailed information through their mobile devices throughout the event.

Diverse approaches akin to these could be applied in dark tourism settings to dynamically and interactively offer visitors interpretive details, ultimately enriching their overall visit. Acknowledging the significance of tailoring communication strategies to the digital realm, emphasized by Weber (2009, p. 3), mastering social web marketing becomes imperative. Though proponents like Weber have stressed the importance of social and digital tech in marketing overall, the focus on tourism, especially dark tourism, goes beyond mere initial marketing aspects. Utilizing digital capacities to enhance the visitor's on-site experience at attractions or destinations is increasingly essential. This requires offering robust Wi-Fi and incorporating technological frameworks like the beacons mentioned earlier (Bolan, 2015).

Conclusions

The realm of digital technology is in a constant state of evolution. While the widespread adoption of tablets and smartphones, along with how tourists engage with social media using these devices, is rapidly expanding, the next major disruptor, as of the current writing, poised to have a substantial impact is "wearable tech." Wearable tech refers to technological devices worn by consumers (Chaffey, 2015), increasingly possessing capabilities comparable to those of smartphones or tablets. Initially, developments in wearable technology were centered on health and fitness gadgets, such as Nike's Fuelband. However, recent innovations, including smartwatches (like the Apple Watch or Samsung Galaxy Gear band) and iterations of the Google Glass concept, have proven significantly influential. These devices encapsulate smartphone capabilities in a more immersive, portable, and user-accessible format. As the demand for innovative mobile devices grows with the rise of wearable tech, new opportunities and potential applications will emerge for the tourism sector.

Within the domain of dark tourism, Stone (2006, p. 146) emphasizes the crucial need to grasp the motivations propelling what he terms as "dark tourists." This understanding is particularly vital across diverse social, cultural, and geographical settings. Beyond endorsing this perspective, scholars advocate for its extension, intricately connecting it to the examination of dark tourists in the digital sphere. This integration introduces a new layer to comprehending these tourists and imparts fresh significance and experiences to both visitors and those overseeing dark tourism sites. If this aspect is thoroughly explored and effectively applied, it empowers us to leverage digital tools to illuminate the realm of dark tourism, enhancing the experiences of all parties involved.

To sum up, this chapter has adopted a threefold approach, delving into the digital realms of social media, mobile technology, and blogging, which include online forums. In this context, the pivotal importance of Wi-Fi connectivity has been underscored as essential to bolster the functionality of these digital aspects, both in travel and at tourist destinations. These observations hold significant implications for the dark tourism sector.

Effectively leveraging the discussed digital elements can yield substantial benefits and has become crucial for achievement in the travel industry sector (Baggio & Chiappa, 2014; Tussyadiah & Zach, 2012). Stakeholders managing dark tourism sites must actively adapt to meet the digital expectations of contemporary "dark tourists." Captivating their interest through social media, mobile interfaces, and tailored apps (while catering to essentials like Wi-Fi connectivity) is crucial for attaining success and fostering sustainable growth in the future. Moreover, future research could expand by delving into further investigations utilizing both qualitative and quantitative methods, exploring the viewpoints, attitudes, and firsthand encounters of dark tourists regarding their use of digital platforms, a subject covered in this chapter. An area of specific interest would involve exploring the influencing and significance of social media in shaping this niche tourism segment. Ultimately, forthcoming research initiatives should aim to uncover the ways engagement in these digital social realms impacts diverse aspects, spanning from decision-making processes to the overall journey and shared mutual exchanges inside the domains of "gloomy tourism."

References

Alexander, J. (2002). On the social construction of moral universals: The 'Holocaust' from war crime to trauma drama. *European Journal of Social Theory*, *5*, 5–85.

Arnold, D. (2005). Virtual tourism: A niche in cultural heritage. In M. Novelli (Ed.), *Niche tourism: Contemporary issues, trends, and cases* (pp. 223–232). Elsevier.

Ashworth, G. J. (2002). Holocaust tourism: The experience of Kraków-Kazimierz. *International Research in Geographical & Environmental Education*, *11*(4), 363–367.

Auschwitz-Birkenau State Museum. (2015a). A virtual tour of Auschwitz/Birkenau by alan and Krysia Jacobs. http://remember.org/auschwitz/. Accessed 05 June 2016.

Auschwitz-Birkenau State Museum. (2015b). *Auschwitz report 2015*. Oswiecim: Auschwitz-Birkenau State Museum.

Bolan, P. (2015). A perspective on the near future: Mobilizing events and social media. In I. Yeoman, R. Robertson, U. McMahon-Beattie, K. A. Smith, & E. Backer (Eds.), *The future of events and festivals* (pp. 200–209). Routledge.

Baggio, R., & Chiappa, G. D. (2014). Real and virtual relationships in tourism digital ecosystems. *Information Technology & Tourism*, *14*(1), 3–19.

Causevic, S., & Lynch, P. (2007, April 11). The significance of dark tourism in the process of tourism development after a long-term political conflict: An issue of Northern Ireland. In Unpublished paper presented at: *The ASA Conference 2007: Thinking through tourism*. London Metropolitan University.

Chaffey, D. (2015). *Digital business and e-commerce management: Strategy, implementation and practice* (6th ed.). Pearson.

Cowan, D. (2005). *Cyberhenge: Modern pagans on the internet*. Routledge.

Cox, C., Burgess, S., Sellitto, C., & Buultjens, J. (2009). The role of user-generated content in tourists' travel planning behaviour. *Journal of Hospitality Marketing & Management*, *18*, 743–764.

Dark Tourism Worldwide. (2016). Description. https://www.face-book.com/groups/60258462039/?ref=br_tf. Accessed 04 June 2016.

Dawson, L. (2005). The mediation of religious experience in cyberspace. In M. Højsgaard & M. Warburg (Eds.), *Religion and cyberspace* (pp. 15–37). Routledge.

Dickinson, J. E., Ghali, K., Cherrett, T., Speed, C., Davies, N., & Norgate, S. (2014). Tourism and the smartphone app: Capabilities, emerging practice and scope in the travel domain. *Current Issues in Tourism*, *17*(1), 84–101.

Durkin, K. F. (2003). Death, dying and the dead in popular culture. In C. D. Bryant & D. L. Peck (Eds.), *The handbook of death and dying* (pp. 43–49). Sage.

Ellis, S. R. (2016). What are virtual environments? http://hsi.arc.nasa.gov/publications/ellis_what_ve.pdf. Accessed 04 June 2016.

Flectcher, A., & Lee, M. J. (2012). Digital heritage: Current social media uses and evaluations in American museums. *Museum Management and Curatorship*, *27*(5), 505–521.

Florida Center for Instructional Technology. (2009). Virtual reality movies. http://fcit.usf.edu/holocaust/resource/vr.htm. Accessed 05 June 2016.

Garde-Hansen, J. (2011). *Media and memory*. Edinburgh University Press.

Gruber, R. E. (2012). *iPad app tells story of Oswiecim, home of Auschwitz*. Jta – Jewish Telegraphic Agency. http://www.jta.org/2012/07/04/news-opin-ion/world/ipad-app-tells-story-of-oswiecim-home-of-auschwitz. Accessed 18 December 2015.

Harris, D. (2002). The kitschification of Sept. 11. http://www.salon.com/2002/01/26/kitsch_2/. Accessed 04 June 2016.

History. (2016). Deconstructing history: Alcatraz. http://www.history.com/topics/alcatraz. Accessed 05 June 2016.

Hodalska, J. (2015, April 23–24). Smiling Holocaust selfies: Dark tourism and digital narcissism. Unpublished paper presented at: *The international conference on the Holocaust and the contemporary world, Krakow, Poland.*

Ivanov, S. H., & Webster, C. (2013). Tourism's contribution to economic growth: A global analysis for the first decade of the millennium. *Tourism Economics*, *19*(3), 477–508.

Kaelber, L. (2006a). Place and pilgrimage: Real and imagined. In W. H. Swatos (Ed.), *On the road to being there: Studies in pilgrimage and tourism* (pp. 277–295). Brill.

Kaelber, R. (2006b). Paradigms of travel: From religious pilgrimage to postmodern tourism. In D. Timothy & D. H. Olsen (Eds.), *Tourism, religion, and spiritual journeys* (pp. 49–63). Routledge.

Kaelber, L. (2007). A memorial as virtual traumascape: Darkest tourism in 3D and cyber-space to the gas chambers of Auschwitz. *E-Review of Tourism Research (eRTR)*, *5*(2), 24–33.

Levine, A. (2014). Selfie in the age of digital recursion. *Invisible Culture: An Electronic Journal for Visual Culture*, *20*. http://ivc.lib.rochester.edu/the-selfie-in-the-age-of-digital-recursion/. Accessed 04 June 2016.

Miles, W. F. S. (2002). Auschwitz: Museum interpretation and darker tourism. *Annals of Tourism Research*, *29*(4), 1175–1178.

Munar, A. M., & Jacobsen, J. K. S. (2014). Motivations for sharing tourism experiences through social media. *Tourism Management*, *43*, 46–54.

Nair, M. (2011, March/April). Understanding and measuring the value of social media. *Journal of Corporate Accounting & Finance*, *22*(3), 45–51.

Ntunda, J. (2012, December). *Investigating the challenges of promoting dark tourism in Rwanda*. Bachelor thesis, Rwanda Tourism University College.

Nusair, K. K., Bilgihan, A., & Okumus, F. (2013). The role of online social network travel websites in creating social interaction for gen Y travelers. *International Journal of Tourism Research*, *15*, 458–472.

Pan, B., MacLaurin, T., & Crotts, J. (2007). Travel blogs and their implications for destination marketing. *Journal of Travel Research*, *46*(1), 35–45.

Ritzer, G. (2005). *Enchanting a disenchanted world: Revolutionizing the means of consumption* (2nd ed.). Sage.

Schmallegger, D., & Carson, D. (2008). Blogs in tourism: Changing approaches to information exchange. *Journal of Vacation Marketing*, *14*(2), 99–110.

Seaton, A. V. (1996). From thanatopsis to thanatourism: Guided by the dark. *Journal of International Heritage Studies*, *2*(2), 234–244.

Seaton, A. V. (1999). War and thanatourism: Waterloo 1815–1914. *Annals of Tourism Research*, *26*(1), 130–158.

Simone-Charteris, M. T., & Boyd, S. W. (2010). Northern Ireland re-emerges from the ashes: The contribution of political tourism towards a more visited and peaceful environment. In O. Moufakkir & I. Kelly (Eds.), *Tourism, progress and peace* (pp. 179–198). CABI.

Sontag, S. (1977). *On photography*. Anchor Books Doubleday.

Stelzner, M. A. (2015). *Social media marketing industry report: How marketers are using social media to grow their businesses*, Social Media Examiner.

Stone, P. R. (2006). A dark tourism spectrum: Towards a typology of death and macabre related tourist sites, attractions and exhibitions. *Tourism*, *54*(2), 145–160.

Stone, P. R., & Sharpley, R. (2008). Consuming dark tourism: A thanatological perspective. *Annals of Tourism Research*, *35*(2), 574–595.

The Digital Panopticon. (2014). Digital dark tourism: New directions in the heritage of crime and punishment? http://www.digitalpanopticon.org/?tag=dark-tourism. Accessed 05 June 2016.

Tourism Review. (2016). Dark tourism: Honest interest or rubbernecking? http://www.tourism-review.com/dark-tourism-growing-in-popularity-news5007. Accessed 04 June 2016.

Trotta, J. (2006). Grief tourism. http://www.grief-tourism.com

Tussyadiah, I. P., & Zach, F. J. (2012). The role of geo-based technology in place experiences. *Annals of Tourism Research*, *39*(2), 780–800.

Virtuar. (2016). *Alcatraz*. http://www.virtuar.com/alcatraz/. Accessed 05 June 2016.

Wang, Y., Yu, Q., & Fesenmaier, D. R. (2002). Defining the virtual tourist community: Implications for tourism marketing. *Tourism Management*, *23*, 407–417.

Weber, L. (2009). *Marketing to the social web: How digital customer communities build your business* (2nd ed.). Wiley.

White, L. (2009). Facebook, friends and photos: A snapshot into social networking for generating travel ideas. In *Tourism informatics: Visual travel recommender systems, social communities and user Interface design* (pp. 115–129). IGI Global.

WP Virtuoso. (2013). How many blogs are on the Internet? http://www.wpvirtuoso.com/how-many-blogs-are-on-the-internet/. Accessed 20 December 2015.

Chapter 20

Unveiling Shadows: Prospects for Dark Tourism Development in India

*Ravi Kant Modi and Ruchi Garg**

Nirwan University, India

Abstract

Tourism often conjures notions of relaxation, joy, and pleasure. The tourism industry caters to providing leisure experiences to people, whose preferences have evolved significantly over time. With a shift toward alternative tourism rather than traditional mass travel, the emergence of Niche Tourism becomes pivotal. While India has made strides in niche tourism, one particular facet remains unexplored: Dark Tourism. This form of tourism has gained immense popularity globally in the 21st century. As tourist preferences evolve and the concept of special interest tourism takes root, policymakers in the tourism sector worldwide are compelled to include this segment in their strategies. Several nations, previously impacted by conflict or lacking in tourism infrastructure, have embraced dark tourism to gain a competitive edge in attracting visitors. India's tourism industry has yet to delve into offering dark tourism as a product. A successful product hinges on customers' perceptions of its ability to meet their expectations. Without awareness among potential customers, even the most exceptional product can falter. This chapter aims to explore the potential of dark tourism in India by gauging the public perception of this form of tourism. It will also address potential challenges in developing dark tourism in India and propose recommendations to foster the growth of this tourism segment.

Keywords: Dark tourism; special interest tourism; multiplier; cultural heritage; development

*Corresponding author.

Dark Tourism, 297–310

Published under exclusive licence by Emerald Publishing Limited
doi:10.1108/978-1-83797-336-120241020

Introduction: Concept and Sites

As the daylight dims, a peculiar energy takes hold at Bhangarh Fort. Scores of young men and women feel a rush of excitement as chilling tales of the fort's history unfold through the words of locals. Meanwhile, in Amritsar, Punjab, roughly 750 km away, another group of visitors explores the infamous JallianwalaBagh. The guide narrates the haunting story of how the brutal forces of the British Empire massacred hundreds of innocent Indians gathered there to celebrate a festival. Further southeast, some 2,800 km from Amritsar, lies the Andaman and Nicobar Islands, home to Cellular Jail – a stark reminder of the political prison constructed by the British to torment Indian freedom fighters. This site stands as one of the most prominent tourist destinations on the island. In recent times, the sorrowful chapters of history have transformed into tourist attractions. While the locations mentioned may not align with our conventional notion of tourism, they are rapidly gaining popularity as destinations. These places draw individuals from diverse age groups, each with their own reasons for visiting. The underlying purpose of these visits is often to experience the resilience encapsulated in faith, hope, and the belief that adversities can be overcome through perseverance. This form of tourism is commonly referred to as "Dark Tourism" (Sarkar et al., 2022). The phrase "Dark Tourism" was introduced by Lennon and Foley in 1996, academics from the Department of Hospitality, Tourism & Leisure Management at Glasgow Caledonian University. It refers to travel to locations historically linked with death, tragedy, or events such as natural disasters, acts of violence, or crimes against humanity. Despite being a relatively recent concept, the growing number of tourists engaging in this form of travel indicates promising trends for the market.

In the past decade, India has made remarkable progress in niche tourism sectors that were once considered specialized. Segments like medical tourism, previously niche, have now become strongholds within the country's tourism sector (Kumar & Shekhar, 2020). However, there remain certain segments immensely popular globally that are yet to be explored in India. Dark tourism stands as one of these untouched sectors. The development of dark tourism could prove highly beneficial for India's tourism industry. This endeavor wouldn't just entice foreign tourists but would also deter domestic travelers from seeking similar experiences abroad. The influx of tourists could facilitate the economic growth of these destinations, alleviating the government's burden in fostering such development. Moreover, in states constrained by geographical limitations that impede their tourism potential, the advent of dark tourism could unlock the economic benefits typically associated with the tourism industry (Sharma, 2022). To cultivate dark tourism effectively, meticulous planning and systematic implementation are imperative. First, one must find out the sites which are serving for the cause worldwide. Second, similar sites should be identified within India and their history should be carefully explored. Third, a study should be conducted to assess tourist's attitude toward dark tourism. Fourth, strategies should be formulated on the basis of results (Jain et al., 2020).

Places Renowned for Dark Tourism in India

India is rich in a tapestry of civilizations, kingdoms, and empires, rendering its historical sites perennially popular among tourists (Dhatrak, 2020). However, amid these historical landmarks lie destinations intertwined with narratives of death, tragedy, and enigmatic fates. These places often struggle to attract visitors due to the stigma associated with their past tragedies (Shekhar & Valeri, 2022). Some locations qualify as dark tourism destinations due to recent human-inflicted events like terrorist attacks or natural calamities such as earthquakes. In India, numerous locations have been identified, either currently serving or with the potential to serve the purpose of dark tourism. These destinations include:

- Bhangarh Fort, Rajasthan – These crumbling remnants were once an awe-inspiring palace. While historical confirmation is lacking, local folklore spins tales of the entire village being wiped out overnight due to a curse, leaving the fort haunted. Its mysterious aura has drawn tourists' attention over the past decade (Das & Chakraborty, 2021).
- Kuldhara, Rajasthan – Kuldhara was once home to a Brahmin community, now reduced to ruins. Various accounts detail the village's decline, but local legends have significantly contributed to its publicity, attracting tourists. Legend has it that the residents cursed this place, preventing its reestablishment despite attempts. In 2015, the Rajasthan government took an active decision to develop the village as a tourist attraction. This initiative is a public–private partnership with Jindal Steel Works (Tandon). The proposed plan encompasses visitor amenities like a cafe, a lounge, a space for folk dance performances, overnight accommodation in cottages, and shops.
- The Residency, Lucknow, Uttar Pradesh – Symbolizing India's fight for independence, the Residency stands as a testament to the famous Siege of Lucknow. These ruins serve as a poignant reminder of the sacrifices made by the valiant souls who fought for freedom. While it isn't extensively developed as a primary tourist spot, it often attracts local and domestic tourists on word-of-mouth recommendations. Adjacent to the ruins, a cemetery within a ruined church houses the graves of 2,000 individuals, including men, women, and children. An evening light and sound show vividly narrates the history of the Residency.
- JallianwalaBagh, Amritsar – When discussing dark tourism in India, JallianwalaBagh inevitably takes precedence. Its bloodstained walls, marked with bullets, and the somber atmosphere vividly narrate the tragic events of April 13, 1919. On this fateful day, individuals had gathered for a peaceful protest against the arrest and deportation of two national leaders, Satyapal and SifuddinKitchlew (Naik & Botre, 2019). Unbeknownst to them, they would face a harrowing struggle for survival, ultimately meeting a tragic end.

  Upon General Dyer's orders, British army troops abruptly opened fire on the unarmed protestors, sealing off entry and exit points, trapping the people within. This horrific incident transformed JallianwalaBagh into a memorial commemorating those who sacrificed their lives on that day.

- The Cellular Jail, Andaman & Nicobar – This prison stands as a haunting reminder of our dark past when the British Empire ruthlessly governed the nation, subjecting thousands of innocent individuals to torture while they fought for freedom. Under British rule, those imprisoned here faced the dreaded punishment of Kala Pani, where escape from this isolated island was deemed impossible due to stringent security measures.

  The jail gained notoriety for incarcerating Indian freedom fighters, among them Veer Savarkar, who endured brutal torture and often met their demise within its confines.
- Bhopal Disaster, Madhya Pradesh – Bhopal stands as a site marked by one of the world's most catastrophic industrial disasters, often studied in esteemed institutions and universities worldwide. The tragic gas leak occurred in 1984 at the Union Carbide plant, exposing over five lakh people to toxic methyl isocyanate. The lingering aftermath of this incident still affects the region today. The notoriety of this place stems from the infamous leakage of methyl isocyanate from the Union Carbide plant.
- Taj Palace Hotel, Mumbai – (2008 Mumbai Terrorist Attack) – The Taj Palace Hotel, a jewel in India's crown, embodies beauty, luxury, grace, and grandeur all in one magnificent establishment. Renowned for its upscale amenities and accommodations, this hotel is a pride of Mumbai. However, in 2008, this esteemed hotel became witness to a heinous act against humanity when terrorist organizations like Lashkar-e-Taiba orchestrated an attack, targeting this hotel along with other locations in Mumbai. These series of terrorist assaults in Mumbai sent shockwaves across the nation.
- Wagah Border Crossing Points and Ceremonies, Punjab – The border dividing India and Pakistan hosts a nightly spectacle: a ceremony where troops from both nations parade and showcase their prowess (Stone, 2010). The Beating Retreat Ceremony has emerged as a significant tourist draw in India, captivating the attention of both foreign and domestic visitors in substantial numbers.
- Dumas Beach, Surat, Gujarat – This beach is distinct for its black sand and has been associated with numerous reported paranormal activities. Stories circulate about people walking along the beach at night who have reportedly vanished without a trace. It is often considered a mysterious and eerie location, often referred to as a dark spot in India.

Challenges Encountered by Dark Tourism in India

- The primary challenge facing dark tourism in India is the lack of adequate and widespread publicity.
- As dark tourism is relatively new, there is limited awareness among people about these tourist sites, resulting in inadequacies in services provided by travel agencies.

- Accessibility to these dark destinations is hindered due to poor connectivity, making them less accessible to tourists.
- Insufficient local support hampers the government's efforts in promoting dark tourism and ensuring the satisfaction of visitors.
- Limited funds for the expansion of the tourism industry pose a significant obstacle to the development of dark tourism.
- Maintenance issues plague dark tourism sites, causing their deterioration over time. This deterioration diminishes the authenticity and appeal of the sites, rendering them less appealing and less acceptable to visitors (Baloch et al., 2023).

Literature Review

The availability of literature on dark tourism remains limited due to the comparative lack of emphasis on this segment in many countries, unlike religious, medical, cultural, or leisure tourism. Literature focusing on India is notably sparse. Walter (2009) delved into the psychological and sociological dimensions of dark tourism, advocating for reconciling different ideologies to effectively promote this form of tourism. The study also highlighted the role of media in its promotion and the emotional aspect driving human interest, emphasizing the need to develop dark tourism as a product. Niemelä (2010) explored the motivations driving tourists to visit dark attractions, identifying key characteristics of dark tourists and comparing them with heritage tourists. The study investigated tourist experiences in dark tourism, adapting Maslow's model of motivation to uncover the needs of dark tourists. Bittner (2011) acknowledged the historical existence of dark tourism sites, yet noted the scarcity in demand for this product. The research aimed to determine whether dark tourism aligns more with the purpose of religious or heritage tourism. Additionally, it highlighted the economic benefits of promoting such tourism in Croatia and shed light on methods to develop and promote dark tourism products, emphasizing the necessity for further quantitative research. Verma (2013) analyzed the initiatives undertaken by authorities concerning dark tourism in India, assessing the potential for developing identified sites into tourism products. The study underscored the importance of understanding the psychology of Indian tourists before marketing dark tourism. Podoshen (2013) examined the motivations driving people toward dark tourism. The research investigated various factors influencing tourists' interest in dark tourism and established connections between different art forms and dark tourism experiences. Robinson (2015) delved into the motivations and activities of tourists across various dark tourism sites. The study concluded that the primary reasons for visiting such sites were education and curiosity. Moreover, it compared the behavior patterns of tourists at dark sites with those at regular tourist destinations, revealing significant differences in collecting artifacts and souvenirs. The study also scrutinized the challenges in defining the term "dark tourism." Jeffrey et al. (2015) examined the escalating interest in dystopian dark tourism. Focusing on three specific destinations where tourists engage with death and dystopia, the researchers presented a model of dystopian dark tourism that integrates dark esthetics, simulation, emotional contagion, and the contemporary global fascination with utopia and dystopia. The study highlighted the surge in this form of tourism due to rising societal insecurities about death, violence, and cultural production. Fonseca et al. (2016) investigated typologies within dark tourism, encompassing war/battlefield tourism, disaster tourism, ghost tourism, cemetery tourism, and more. The study explored human beings' conflicting attitudes toward death, potentially impacting the acceptance of dark tourism as a tourism product. It also emphasized the emotional and psychological impacts on individuals visiting these sites. Dar and Reshi (2016) explored the feasibility of developing dark tourism in the state of Jammu and Kashmir. The study found that dark tourism made negligible contributions to the state's GDP. It advocated for effective marketing by aligning with tourists' needs and tailoring the product accordingly. The study also touched upon ethical issues involved in marketing such places. Singh et al. (2016) assessed the potential of dark tourism in India and identified challenges hindering its development. The study concluded that a lack of awareness was a major obstacle in exploiting this sector effectively. Additionally, it noted the government's reluctance, citing its own reasons for not prioritizing this unconventional form of tourism. Yan et al. (2016) centered on the relationship between motivation and experience, highlighting the regulatory role of emotional reactions in dark tourism. The study unveiled that tourists' emotional responses significantly influence their experiences more than cognitive elements do. Curious visitors tend to engage cognitively by seeking knowledge about incidents or related issues. Light (2017) conducted a comprehensive literature review on dark tourism and thanatourism, concluding that distinguishing dark tourism from heritage tourism isn't straightforward. The study traced the evolution of dark tourism, exploring its future prospects and challenges. It scrutinized the motives driving visitors to dark sites and underscored the imperative of comprehending dark tourists better to fully develop dark tourism products. Lennon (2017) scrutinized the alignment between theoretical concepts and the practical application of dark tourism principles. The study highlighted ethical dilemmas concerning the promotion of dark tourism, specifically thanatourism. It emphasized the necessity to alter the prevailing attitudes surrounding dark tourism destinations, urging a shift in perception.

Objectives of the Study

- Investigate the perceptions of tourists and the general public regarding dark tourism in India.
- Identify prominent sites with potential for development as dark tourism destinations in India.
- Propose strategies and recommendations conducive to the advancement of dark tourism in the country.

Research Methodology

It involves the systematic process of identifying, collecting, summarizing, processing, and analyzing data to address the research problem. The methodology section of this chapter is divided into two parts: data collection and data analysis.

Data Collection: For this study, data has been gathered from both primary and secondary sources. Secondary sources encompass reports from the Ministry of Tourism, Government of India, reports from the World Travel and Tourism Council, UNWTO reports, journals and online articles, electronic media, and pertinent literature. Primary data was acquired through a survey conducted among 100 tourists across various age groups.

Data Analysis: The data gathered from interviews underwent analysis using SPSS and MS Excel tools were employed to conduct necessary tests and further analyze the data.

Analysis and Findings

After carrying out the analysis, the research yielded results pertaining to different facets of dark tourism. These findings align closely with the objectives outlined in the research paper. The results obtained address various questions posed during the study.

Understanding of Dark Tourism

Responses regarding familiarity with the term "dark tourism" were documented using a five-point Likert scale. The scale ranged from 0, indicating no familiarity, to 5, denoting a comprehensive understanding of the term.

Analyzing the data presented in Table 20.1, it's apparent that a significant portion of the surveyed individuals lack familiarity with dark tourism. Dark tourism refers to visiting places associated with tragedy, death, or suffering, like

Table 20.1. Understanding of Dark Tourism.

Response	Number of Responses
0	42
1	12
2	14
3	19
4	10
5	03
Total	***100***

Source: Drawn from primary data.

war zones or disaster sites, for educational or commemorative purposes. The table suggests that approximately 68% of the respondents demonstrated minimal or no understanding of this concept.

Places Visited for Dark Tourism

Responses regarding the specific destinations visited by tourists for dark tourism purposes were collected as follows:

Table 20.2. Places Visited for Dark Tourism.

Name of the Place	Type of Place	No of Responses
Bhangarh Quila	Haunted fort	08
Cellular Jail	Former prison	03
Jallianwalabagh	War memorial	06
Purana Quila	Fort	01
Humayun Tomb	Graveyard	01
Residency	War memorial/Battlefield site	07
Nepal/Bhuj	A place where any natural calamity had occurred	01
Others (Hijlidetention camp, Taj Mahal, forts)	A place of holocaust, Museum related with war and death	11

Source: Drawn from primary data.

When examining Table 20.2, we found that renowned destinations such as Bhangarh Fort in Rajasthan, Cellular Jail in Andaman & Nicobar Island, JallianwalaBagh in Punjab, and Residency in Lucknow emerged as commonly visited sites among the respondents. These places were initially identified through diverse secondary sources. Additionally, many respondents mentioned other forts and tombs. Hijli Detention Camp, Golden Temple, and PuranaQila were among the additional places referenced by the respondents.

How Did You Know About the Place?

Awareness about a destination can stem from diverse sources. In this digital age, accessing information is relatively effortless. However, individuals also rely on recommendations from acquaintances to navigate uncertainty during visits. Respondents shared the following insights regarding how they learned about destinations:

Table 20.3. Source of Information.

Particulars	No of Responses
Friends or family	28
Social media	10
Television or movie	5
Academic survey	5
Traveling agencies	1

Source: Drawn from primary data.

Upon scrutinizing the data presented in Table 20.3, suggestions from friends or family persist as the primary source of information regarding these destinations. Social media stands as the second most influential platform in enlightening individuals about such places. Surprisingly, travel agencies did not rank among the top three sources of information. This could be attributed to many Indians independently organizing their trips without engaging a dedicated travel agent. Alternative sources such as television shows or movies and academic surveys also contribute to the dissemination of information about these destinations.

How Long Did You Stay at the Destination?

The duration of stays at these locations holds significant importance for policy-makers. The economic contribution made by tourists is evident. Tourism not only influences the economic growth of an area through job creation and infrastructure development but also serves as a livelihood for locals due to tourist spending. According to various online travel agencies and previous studies, a budget traveler spends approximately Rs 2,000 per day in India. Multiplying this by various tourism multipliers – such as spending multipliers or the employment multiplier of tourism – can illustrate the potential economic impact resulting from the development of dark tourism.

According to Table 20.4, the responses indicate that, on average, visitors spend less than 3 days at dark tourism sites. This presents a challenge for policymakers aiming to extend the duration of stays to a level that can maximize the benefits derived from the tourism industry.

What Was the Purpose Behind Visiting Such a Place?

Table 20.5 shows that the survey underscores the evolving mindset of Indian tourists. Previously, tourism was largely associated with leisure and enjoyment. However, contemporary travelers display a shift toward curiosity and a quest for adventure. Many individuals visiting these sites seek adventure or harbor a curiosity to explore such places. Educational purposes also drive some to travel,

Table 20.4. Average Duration of Stay.

Duration of Stay	No of Responses
Less than 3 days	31
3–7 Days	12
7–14 Days	03
Above 14 Days	01

Source: Drawn from primary data.

Table 20.5. Purpose Behind Visiting.

Particulars	No of Responses
Adventure	15
Curiosity	14
Education	10
Remembrance	01
Artifacts	00
Empathy	00
Impulsive decision	08

Source: Drawn from primary data.

delving deeper into the location and the lifestyles of the locals. Moreover, several researchers frequent these areas for their studies. An intriguing discovery was that 16.7% of respondents made impromptu decisions to visit these places. This behavior diverges from the typical approach of Indian tourists who traditionally plan their visits months in advance. Consequently, this evolving behavioral pattern among Indian tourists holds immense potential for unconventional tourism prospects.

Do You Believe Social Media and Films Contribute to the Promotion of Dark Tourism?

Looking closely Table 20.6, we have come to the conclusion that an astounding 84% of the respondents acknowledged the pivotal role played by social media in promoting dark tourism. This aligns with the observation that a substantial portion of our respondents gathered information primarily from social media platforms. Many individuals actively follow various pages on social media dedicated to these destinations, shaping their travel plans based on

Table 20.6. Impact of Social Media and Films.

Impact	No of Responses
1	06
2	09
3	31
4	31
5	21

Source: Drawn from primary data.

recommendations obtained there. With not just the younger generation but also middle-aged and older individuals becoming tech-savvy, they are increasingly investing more time on social media. Hence, social media emerges as a potent channel for promoting dark tourism. For instance, the film "Schindler's List," lauded as one of Hollywood's finest, depicts the Holocaust of Jews by the Nazis. The movie shed light on the harrowing conditions faced by prisoners, consequently fostering dark tourism in Germany.

Do You Believe Dark Tourism Holds Potential in India?

The success of a product hinges on the favorable perception of its prospective customers. All efforts by policymakers may prove futile unless people perceive the product as having potential.

Table 20.7, titled "Dark Tourism Potential," shows responses regarding interest in dark tourism. Among the participants, 57 expressed interest, 6 declined, and 37 were uncertain. This indicates mixed levels of interest and awareness about dark tourism among the surveyed population.

What Hinders the Growth of Dark Tourism in India?

Upon analyzing the responses using software and conducting a word frequency analysis, it was revealed that approximately 39% of respondents identified a lack of knowledge as the primary factor impeding the growth of dark tourism in India.

Table 20.7. Dark Tourism Potential.

Particulars	No of Responses
Yes	57
No	06
Can't say	37

Source: Drawn from primary data.

This deficit in knowledge arises from insufficient research on the subject and a general lack of interest exhibited by stakeholders in cultivating it as a comprehensive tourism product. Other significant reasons cited included the government's inadequate promotion, which significantly contributes to the overall lack of awareness among people, and the psychological disposition of Indians to avoid places associated with death. This reluctance to visit such sites stems from a natural human inclination to evade experiences related to pain and mortality. Moreover, the insufficient development of infrastructure at these locations poses a challenge by limiting tourist inflow. The absence of adequate infrastructure may also be a result of low demand (Korstanje, 2016); thus, an increase in demand could potentially stimulate the development and maintenance of infrastructure at these sites.

Conclusion and Suggestions

In summary, dark tourism remains largely unfamiliar among Indian tourists, primarily due to the authorities' lack of initiative in formulating promotional strategies for this sector. India boasts destinations and infrastructural capabilities conducive to supporting dark tourism. Moreover, there's acknowledgment that this form of tourism holds potential for success, a crucial aspect for policymakers given the importance of tourists' optimistic perspectives. What's imperative is a long-term vision or strategic plan to integrate dark tourism as an emerging niche trend in the country. Here are several suggestions that authorities may contemplate for the development of dark tourism:

- Extending Average Length of Stay
  Paying attention to extending the average duration of tourists' stays is essential. Agencies should prioritize efforts to prolong tourists' visits. Longer stays correspond to increased spending, contributing significantly to the economic development of the region.
- Destination Incubation
  Developing new destinations specifically tailored for dark tourism is crucial. Identifying locations that have either deteriorated over time or have faced recent calamities and transforming them into purpose-built sites for tourists is imperative. Policymakers need to identify raw, unexplored destinations with potential for dark tourism. Evaluating these sites based on the dark tourism index and assessing their commercial viability should precede the formulation of comprehensive plans for their development and promotion as prime dark tourism destinations.
- Preparing Destinations for Market Readiness
  To ensure these destinations thrive comparably to other niche tourism sectors, essential amenities and facilities must be provided. Developing adequate infrastructure while preserving the emotional essence of the place is crucial. Training locals knowledgeable about the site to serve as guides not only enhances the visitor experience but also addresses unemployment issues.

Furthermore, as these places are branded as dark tourism destinations, markets should showcase the incidents or tragedies that occurred there. Artifacts, images, and paintings should evoke the emotions associated with the historical trauma, appealing to tourists' sentiments.

- Promoting Dark Tourism Locally and Domestically
  Given the emotional connection inherent in dark tourism, authorities can actively promote these destinations among communities and locals affected by past tragedies. This strategy ensures a steady flow of visitors, facilitating infrastructural development at the destination. It also helps resolve the typical challenge faced by underdeveloped destinations concerning the balance between tourist influx and infrastructure. Educational institutions in the area should integrate visits to such sites into their curricula, focusing more on inviting domestic tourists than foreign visitors to these destinations.
- Strategic Destination Branding
  The moral and ethical quandary surrounding the promotion of death remains a significant barrier in dark tourism's advancement. Effectively addressing how to ethically market such sensitive narratives has deterred authorities from actively including this tourism facet in key promotional strategies. An effective solution lies in leveraging destination branding. Initially, crafting a distinctive image for the destination followed by strategic communication of this image to the intended audience is crucial. Advertisements should be thought-provoking, appealing to emotions without causing trauma or exploiting the audience emotionally.

Apart from these, respondents also proposed measures they believe would aid in the development of dark tourism in India. Given its novelty, few states have actively promoted this form of tourism. Some states could gain a competitive edge by pioneering dark tourism, identifying relevant sites, establishing basic infrastructure, and implementing robust marketing campaigns. Government collaboration with educational institutions to disseminate information about dark tourism sites and their historical significance is essential. Moreover, fostering a sense of nationalism and patriotism through dark tourism experiences, for instance, sites like Wagha border or Taj Hotel, can be explored. Digital innovations, particularly leveraging social media, should be employed for destination advertising. Developing documentaries and short films could effectively disseminate information and inspire interest in visiting these historically significant places.

References

Baloch, A. G., de Almeida Leite, E. M., Ribeiro, H. N. R., Alves, S. R. P., & Camacho, E. (2023). Dark tourism in India: Significance and challenges. In *Economic and social development: Book of proceedings* (pp. 312–318). 98th International Scientific Conference on Economic and Social Development – Porto.

Bittner, M. (2011). Historical dark tourism sites and their demand. *Journal of Heritage Tourism*, *6*(2), 123–135.

Dar, A. A., & Reshi, N. A. (2016). Feasibility of dark tourism in Jammu and Kashmir. *International Journal of Tourism Sciences*, *16*(1), 15–28.

Das, J., & Chakraborty, S. (2021). Scope of dark tourism as a revival strategy for the industry - A study with special reference to Rajasthan. *Business Studies*, *XLII*(1 & 2).

Dhatrak, S. P. (2020). Dark tourism sites in India: A review. *Humanities*, *8*(2), 53–57.

Fonseca, A., Seabra, C., & Silva, C. (2016). Typologies within dark tourism. *Journal of Tourism and Cultural Change*, *14*(1), 27–41.

Jain, D., Madireddy, M., & Rao, R. N. (2020). Potential assessment of dark tourism in India. *International Journal of Management*, *11*(8), 1284–1296.

Jeffrey, J., Smith, R., & Williams, K. (2015). Dystopian dark tourism: Engagement with death and dystopia. *Tourism Geographies*, *17*(3), 331–348.

Korstanje, M. (2016). England and the culture of achievement: The roots of dark tourism. In M. Korstanje (Ed.), *Gazing at death: Dark tourism as an emergent horizon of research* (1st ed.). Nova Science Publishers.

Kumar, S., & Shekhar, A. (2020). Technology and innovation: Changing concept of rural tourism – A systematic review. *Open Geosciences*, *12*(1), 737–752.

Lennon, J. J. (2017). Theoretical concepts and ethical dilemmas in dark tourism. *Journal of Tourism Ethics*, *9*(2), 112–128.

Light, D. (2017). Review on dark tourism and thanatourism. *Tourism Management*, *61*, 275–291.

Naik, S. N., & Botre, P. B. (2019). Dark tourism concept, features and significance in India. *IJRAR*, *6*(2).

Niemelä, T. (2010). Motivations of dark tourists: A comparative study with heritage tourists. *Tourism Management*, *31*(6), 949–957.

Podoshen, J. S. (2013). Motivations and influences in dark tourism. *Annals of Tourism Research*, *42*(1), 318–341.

Robinson, P. (2015). Motivations and behaviors of dark tourists. *Tourism Analysis*, *20*(5), 573–584.

Sarkar, A., Chakraborty, P., & Valeri, M. (2022). People's perception on dark tourism: A quantitative exploration. *Current Issues in Tourism*, *25*(13), 2042–2047.

Sharma, N. (2022). Acknowledging the shades of grey: The past, present and future of dark tourism in India. In *Indian tourism: Diaspora perspectives* (pp. 125–142). Emerald Publishing Limited.

Shekhar, S., & Valeri, M. (2022). Evolving themes in dark tourism research: A review study. *Tourism: An International Interdisciplinary Journal*, *70*(4), 624–641.

Singh, R., Sharma, V., & Gupta, P. (2016). Potential and challenges of dark tourism in India. *Tourism Management Perspectives*, *20*, 41–49.

Stone, P. (2010). *Death, dying and dark tourism in contemporary society: A theoretical and empirical analysis*. Doctoral dissertation, University of Central Lancashire.

Tandon, M. (2017). The inception of "dark tourism" in India: A new prescience (with special reference of Rajasthan). In M. Didwania, R. Sharma, P. Kautish, & S. K. Rai (Eds.), *New paradigms in management, economics & commerce* (p. 63). Global Alliance Publishers & Distributors.

Verma, R. (2013). Dark tourism initiatives and development in India. *International Journal of Tourism Research*, *15*(4), 354–367.

Walter, T. (2009). Psychological and sociological dimensions of dark tourism. *Journal of Tourism Studies*, *21*(3), 45–58.

Yan, C., Zhang, H., & Liu, H. (2016). Emotional reactions and experiences in dark tourism. *Journal of Destination Marketing & Management*, *5*(4), 294–305.

Chapter 21

Uncovering Psychological Shadows: Delving Into the Soul of Dark Tourism in India Through Narrative Inquiry

Neharshi Srivastava

Amity University Rajasthan, India

Abstract

The tourism business is seeing a major boom worldwide, especially in India, where there has been a noticeable rise in visitors to destinations related to dark tourism. This subset of tourism in general involves travel to locations associated with death, tragedy, or melancholy. The purpose of this study is to investigate the psychological facets of human nature that contribute to the rising popularity of dark tourism. In particular, it aims to clarify the behavioral goals, cues, and incentives that lead tourists to investigate sites with tumultuous histories. This study chronicles visitors' experiences using a narrative inquiry technique, looking at how their emotions change before, during, and after participating in dark tourism. Interviews with people who have taken part in these excursions provide primary data. Findings imply that people frequently engage in dark tourism because of curiosity, a desire for thrills, emotional triggers, or memories of previous experiences. This study indicates that there are dynamic changes in the feelings of tourists as they travel, ranging from mildly curious at first to finally accepting bad energy and feeling satisfied after the trip is over. This study advances our understanding of the psychological and emotional factors that lead people to engage in gloomy tourism, which involves visiting sites associated with past tragedies and ills. The study's conclusion, which shed light on the changing tourist tastes and motives, have both academic and practical consequences for tourism management, marketing tactics, and the hospitality industry as a whole.

Keywords: Dark tourism; behavior intention; motivation; triggers; emotions and feelings

Dark Tourism, 311–333

Published under exclusive licence by Emerald Publishing Limited
doi:10.1108/978-1-83797-336-120241021

Introduction

"In June 2023, a mysterious mission was launched into the Atlantic Ocean, a risky and audacious journey to discover the eerie remnants of the past. A Canadian research submarine named Titan sank into the RMS Titanic's watery grave as it set out to explore the incomprehensible depths. The endeavor, led by OceanGate Expeditions, sought to reveal the mysteries hidden beneath the renowned ship, the Titanic, which perished in 1912. The expedition gained notice for its daring exploration of the eerie remnants of history. The Titan, which OceanGate had custom-built, tragically disappeared into the ocean when the expedition was exploring the Titanic. The marine community was shocked by the US Coast Guard's massive search and rescue operation that followed this unannounced absence. The project took on an unexpected depth that is frequently linked with 'dark tourism' due to its unexpected turn of events, which elevated it beyond a simple exploratory trip. The narrative highlights the attraction of historical artifacts and the unanticipated hazards connected with deep-sea missions through a combination of audacious exploration, mystery, and an unexpected turn of events."

This emphasizes the importance of comprehending the etymology of the term tour, which is derived from the Latin word tornus, meaning "a tool for making a circle." Tourism is defined as traveling from one's usual place of residence to another for the express purpose of leisure and enjoyment with the intention of returning, for a minimum of 24 hours and a maximum of six months. Tourism is a social, cultural, and economic phenomena that includes people traveling for pleasure or business to countries or areas that are different from their usual surroundings, according to the UNWTO. They engage in tourism-related activities, some of which include paying for trip. Traveling can evoke a range of emotions depending on the destination. While some places can improve one's quality of life, make them happier, and make them feel good about themselves, other places can cause emotional upheaval, sadness, and heavy sentiments.

Every one of these journeys has some sort of effect on the guest. It is possible to enhance one's quality of life by visiting happy and joyful areas. These visits also have a positive impact on the person, increase happiness, and encourage social connection, relaxation, and wellness. This kind of healthy and beneficial travel improves the traveler's overall welfare by lowering stress, improving mental and physical health, building relationships, promoting personal development and self-discovery, and promoting relaxation. This gives rise to the idea of tourism wellness, which is concerned with visitors' social interactions, cultural enrichment, and mental, physical, and emotional well-being. There's also the other side of tourism that provide some negative experiences to tourists as the sites of dark tourism impact the tourists emotional and psychological well-being, provides unique learning, pave way to reflection on history and their own thoughts, may evoke feelings of contemplation and empathy, adding onto a mix of positive and negative impacts on the tourists (Lewis et al., 2021). This is beauty and mystery of dark tourism sites. As there is a growing demand for new, non-commercial tourism destinations in a world where travel is becoming more and more popular among all age groups, that's where the dark tourism niche industry finds its appeal for people seeking out adventures far removed from mainstream safe and sound travel. According to Seaton (1996), "travel to a location wholly, or partially, motivated by the desire for actual or symbolic encounters with death" is what is meant to be understood when one refers to "dark tourism" or "thana-tourism." The ancient Greek term "Thanatos" implies "death" when translated into English. The definition of "Thanatos," according to Gerard Corsane, is "the personification of death, more specifically violent death." Dark tourism is the academic name we give to sites that commemorate and remember disasters and atrocities (Fes, 2022).

"The phenomenon which encompasses the presentation and consumption (by visitors) of real and commodified death and disaster sites" is how Foley and Lennon defined it in 2002. According to Merriam-Webster (n.d.), the definition of "dark" or "darkness" is the absence of light, the presence of evil, devastation, and sadness. According to Seaton, there exist five potential categories of nighttime travel activities: to see death enacted in public; to see death enacted again; visiting the locations of individual or collective fatalities; to view images that symbolize death; and visiting memorials or locations of incarceration.

There have been various sites for the interests of tourists in the world that attract them and make them curious. To mention a few in the world are Concentration Camp Auschwitz, Poland; Chernobyl, Ukraine; Volcano Creeks in Pompeii, Italy; Hiroshima Peace Memorial, Japan; and more such sites. There have been some sites for the same in India as well that has a great tourist footfall across the globe, to mention a few are Bhangarh Fort, Rajasthan; Jomblang Cave, Goa; Dumas Beach, Surat; Roopkund Lake, Uttarakhand; Cellular Jail, Andaman and Nicobar Island, and more.

This instills a curiosity of knowing why there are so many tourists who fall at the destination of dark tourism. So, tourists visit these places as a result of educational or historical interests, to enhance cultural understanding, for emotional connection, for remembrance, to get into personal reflection or contemplation, for morbid curiosity, and widely as a result of heard stories, and social media influences (Molderez & Perera, 2022). All this has resulted in a paradigm shift in the footfall of the tourists from serene destinations to the destinations of dark tourism. The shift is visible because tourists have started to explore authentic or meaningful experiences, undergoing high curiosity for haunted places, interest in history, and the influence of pop culture or media (Sampson, 2019). Various other reasons for this shift are changes in travel motivations from pleasure to thrill, psychological fascinations of the tourists, and cultural enrichment and experiences (Çakar, 2020).

These types of tours impact the tourists in varied ways ranging from the experiences on a continuum from positive to negative depending upon the factors which are tourists who visit dark tourism sites are capable of developing positive emotions, strong geopolitical disposition, and empathy (Chang, 2017). The ones who visit these sites are curious, interested, and intrigued by dark experiences with paranormal activity, resulting in travel choices made for themselves based on personal beliefs and preferences, with minimal outside influence from others. However, it's important to note that dark tourism is a controversial form of tourism that raises ethical concerns (Lewis et al., 2021). While some view it as a way to pay respect to the victims of tragedy and learn from history, others see it as exploitative and disrespectful (Çakar, 2020).

Therefore, it's essential to approach dark tourism with sensitivity and respect for the victims and their families. Adding onto this there could be some negative impacts for the where tourists who visit dark tourism sites may experience rumination on sadness, self-hatred, hostility, and psychological vulnerability (Magano et al., 2022). In addition, inappropriate behavior such as photographing people in moments of sorrow or treating them as museum exhibits can also be a disadvantage of dark tourism (Friedman, 2023). The most common criticism of dark tourism is that it exploits human suffering. Operators can exploit these sites to make money or simply to provide entertainment, which disrespects the victims of the event. This type of behavior may be unethical and can cause psychological harm to both the tourists and the local communities. It's important to approach dark tourism with sensitivity and respect for the victims and their families. Tourists should be mindful of the ethical implications of visiting such sites and avoid tourism sites being run purely for profit rather than to educate.

This well defines the association of dark tourism sites or destinations with the psychological attributes. All these things provide a well connection of these sites with the behavioral intention, motivations, feelings, thoughts, emotions, and other psychological attributes.

Significance of the Study

In the mystical realms of Rajasthan, nestled within the Alwar region like a silent guardian of secrets, there existed the enigmatic Bhangarh Fort. This wasn't just a relic of history; it was a time-worn canvas painted with eerie folklore and tales that whispered of curses and unexplainable occurrences. As if the very air around it carried echoes of ancient legends, the fort became a mysterious tapestry woven with threads of enigma.

Bhangarh Fort, standing proudly amid the landscape, was no mere architectural marvel; it was a convergence of history and the supernatural. Its ancient walls seemed to breathe with the weight of centuries, and within its confines lay a portal to a world where reality and the paranormal danced hand in hand. Legends, like ethereal mist, clung to the stones, creating an aura of mystery that drew in those curious souls enthralled by the whispers of the unknown.

This extraordinary fortress transcended its historical significance to become a beacon for the adventurous and the thrill-seekers. Despite the spine-tingling tales that wrapped around it like a cloak, Bhangarh Fort proudly showcased not just a chilling connection to curses and paranormal activity but also the resilient spirit of history and architectural brilliance.

In the heart of Rajasthan, Bhangarh Fort beckoned, inviting all who dared to step into its haunted embrace. It stood as a testament to the entwining of history and the supernatural, a notorious dark tourism destination that cast a mesmerizing spell on adventurers, drawing them from the far reaches of the globe. This fascinated me to a great extent and developed a strong interest in these historical events and locations. This curiosity within me encouraged the need to investigate the reasons behind many visitors to these destinations. This gave me an idea to research on this topic where I could draw a line connecting two disparate sets to determine the psychological characteristics that play a part in dark tourism. This has also made me enthusiastic to explore the motives behind the tours at dark tourism destinations. Tour on these destinations are not merely for the pleasure or quality time but it also dwell into the depths of history, curiosity, and various aspects of tourists psyche. So, as tourists even after having negative experiences and emotional fluctuations they still want to continue it and explore more. Therefore, I wanted to concentrate on the large research gap and want to work in a direction to cater my curiosity.

> Exploring places touched by tragedy goes beyond a mere stroll through history; it unfolds as a voyage into the intricate realms of the human psyche, where curiosity, empathy, and the quest for understanding gracefully converge.

Review of Literature

International Scenario

There have been various places of dark tourism that interest the tourists. Various research has been done in a way that focuses on studying the profile, practices, motivations, and well-being of tourists who carry out dark tourism. Studies have shown that women and participants with higher levels of education are more knowledgeable about dark tourism; individuals who are aware of dark tourism have visited more Holocaust museums, places of natural and human tragedy, concentration camps, and prisons; they also exhibit greater curiosity, a need to learn and comprehend, and a need to see macabre things (Magano et al., 2022). The majority of research also shows that darker behaviors are linked to ruminating on feelings of melancholy, self-loathing, animosity, and psychological vulnerability. The study has also indicated that participants who were aware of dark tourism beforehand did not exhibit higher levels of well-being (Chang, 2014). It's interesting to note that individuals who visit graveyards had higher scores for antagonism and tourist well-being than participants who do not. The majority of tourists find the history or destination fascinating since it teaches them about the past through the important strategic role that the location had during the specific event (Hohenhaus, 2019). Travelers are attracted to and intrigued by mysterious encounters with the paranormal; as a result, they make their own travel decisions based mostly on their preferences and beliefs, with little external influence, adding onto media does play a major role in depicting push and pull theory (Gaya, 2013). Dark experience was found to have the greatest influence on attitudes and subjective norms among the dark tourism constructs examined (Lewis et al., 2021). There are a variety of reasons why people travel, including education, ethics, politics, philosophy, agony, and amazement (O'Brien, 2017; Reid, 2021). These tourists visit dark tourism sites for reasons that vary from learning about historical events to voyeuristic fascination with the macabre. The engagement with philosophy would range from deep reflection on the human condition to simple thrill-seeking. In dark tourism, there is an inherent moral dimension, most notably ethical concerns surrounding voyeurism and the timing of visits to disaster areas. A balance needs to be struck between educational benefits and respect for the victims (Benedetto, 2018).

Indian Scenario

There have been various researches in the Indian context but they're only limited to the definition of the dark tourism and the information of sites. But when there's a need to assess the psychological attributes related to dark tourism, the review is very limited. In this line some researchers have studied the changing perception of travelers for Indian tourism on dark tourism. In this line they found that the existence, acceptance, and promotion of dark locations in India are the main concerns and they shed light on how travelers' perceptions are evolving, showing a striking change from regular vacationing to traveling to explore the allure of tragedy and death (Soni & Hussain, 2018). Many of the researches and articles have stated a lot of famous destinations of dark tourism that depict the above stories; which are, Bhangarh Fort in Rajasthan; Kuldhara in Rajasthan; Jallianwala Bagh in Amritsar; Dumas Beach in Gujarat; Three Kings Church in Goa; Skeleton Lake in Uttarakhand; Shaniwar Wada in Pune; Cellular Jail in the Andaman and Nicobar Islands; and more (Gupta, 2023; Sumanth & Narendar, 2022). It has also been found that the behavior, affective, and cognitive components are positively and significantly correlated. According to a study, place attachment completely influences the relationship between motivation and satisfaction which in combination impacts the dark tourism visit (Dandotiya & Aggarwal, 2022).

The reviews have found that there are some studies on dark tourism and some of the psychological attributes but there's no such good research or quality research available in the Indian context. Most of the researches that are there in the Indian scenario mostly cover the description of dark tourism along with the destinations, accessibility, and more.

Methodology

The authors emphasized that their proposition is made for a quantitative model of research. As a result, some changes have been made to the research that is being presented. Qualitative narrative inquiry methods have taken the role of statistical approaches in this study since the primary objective was to investigate unknown occurrences. Therefore, rather than testable hypotheses, research questions, and a qualitative analytic method have been used.

Research Objectives

- Explore the emotional, mental, and social impacts of dark tourism.
- Uncover tourists' dark fascination through narrative inquiry.

Interview Questions

- To find out the motivators for dark tourism.
- To figure out the behavior intentions among the tourists of dark tourism.
- To find out the emotion and mood fluctuations among the dark tourists.

Sample and Data Collection

A semi-structured interview was administered to 10 tourists who visited various dark sites. The tourists were within the age range of 18–40 years and explored various destinations of dark tourism within India from Bhangarh Fort, Leh-Ladhak, Kuldhara, Cellular Jail, Putul Bari, South Park Cemetery, and more. They've visited these places in the last 5–7 years. The interview aimed to uncover motivators for dark tourism, behavioral intentions, and emotional/mood fluctuations among dark tourists.

Narrative Inquiry Process

Employed narrative inquiry on interview transcripts to identify narratives, themes, and sub-themes. Commendable narratives and themes were those revealing crucial nuances about the impact of dark tourism on emotions, mental states, and social interactions.

Limitations of the Methodology

Small Sample Size: The study acknowledged a limitation due to a small sample size of 10 tourists, potentially limiting the generalizability of findings.

Selection Bias: The participants were chosen from different tourist sites, introducing selection bias and potentially influencing the homogeneity of experiences.

In-Depth Literature Review: A thorough review of existing literature was conducted to enhance understanding and contextualize findings. Special attention was given to past studies elucidating the dark fascination of tourists in various contexts.

On the basis of in-depth narrative analysis, the researcher came up with four themes: Feelings, Impact, Triggers, and Motive. These themes highlight the experience of tourists toward dark tourism.

Data of Tourists

Tourist-1

Gender: Female Age: 25 years

A tourist went to Bhangarh five or six years ago with friends. She went there out of curiosity, and since she studied history, she was delighted to see the palace. Her goal was to investigate any paranormal activity that might present. According to her account, she experienced significant emotional shifts both before and after the tour. After the tour is over, the curiosity has given way to fear.

Themes	Sub-themes	Narratives
Feelings	Excited, happiness, weird, and fear	"All of us got scared while we encountered with a lot of bats there." "I felt weird there." "I was very happy about the tour as I'll spend good time with friends." "I ran to go there because I had seen a lot of documentaries."
Impact	Negative	"I started feeling scared."
Triggers	Friends, documentaries, and history	"Being a student of history these places fascinates me." "I have watched a lot of documentaries." "I went with my friends."
Motive	It was to explore the paranormal activity	"Wanted to explore the place as well." "I wanted to see whether paranormal activities exists or not."

Tourist-2

Gender: Male Age: 38 years

A few years ago, tourists traveled to Leh-Ladhak and saw the Rang Mahal in the Nahargarh Fort as well. He traveled to the locations due to the stories he had heard, which excited him. Even after many tourists visited the locations, he felt empty despite not feeling anything. Upon finishing the tour, he experienced a feeling of accomplishment. He felt a sense of accomplishment and exhilaration throughout the entire tour.

Themes	Sub-themes	Narratives
Feelings	Emptiness, excited, and sense of achievement	"I like to visit to the places like this" "So many people used to come to that place every day and even after that there was a lot of emptiness at that place."
Impact	Positive	"After doing the tour, I felt a feeling of achievement that I had completed the tour."
Triggers	Cases, stories, and travel	"I went to travel, so when we came across this place on the way, we visited that too." "I've studied and gone through a lot of cases about various

(Continued)

Themes	Sub-themes	Narratives
		haunted places." "I've heard a lot of stories"
Motive	It was to visit the places as a result of stories	"I had heard about an old case." "The place came on a way and we went to visit the place."

Tourist-3

Gender: Female Age: 21 years

Last year, tourists came to Bhangarh Fort. Along with having fun with her cousins, she went to the palace to investigate and determine whether or not these things actually exist. At first, she was intrigued by the location, but she was also afraid. Following the tour, she began to believe in the existence of negative energy and began to accept the unpleasant experiences from the past, which she had before denied. Overall, the excitement and discovery gave way to acceptance of the energies and fear.

Themes	Sub-themes	Narratives
Feelings	Excitement, fear, unsafe, and frightened	"We all got frightened when the voice intensified and the number of voices increased and we stalled to ran away." "On hearing the sound of the male, we'll felt the place to be unsafe." "I was excited and worried that something might happen to someone."
Impact	Negative	"We were scared."
Triggers	Fun	"We all went there stubbornly so that we could enjoy."
Motive	It was for fun, excitement, and wanted to explore	"We went there for fun and to enjoy the trip." "The aim was to explore whether these negative energies exist or not."

Tourist-4

Gender: Male Age: 20 years

Five years ago, tourists went to Bhangarh Fort. He went there to check for the presence of bad energy. He was intrigued by the tales and had a strong understanding of the area's past, which encouraged him to explore. In addition to being terrified, he was interested in the tour. After finishing the tour and returning home without incident, he felt satisfied.

Themes	Sub-themes	Narratives
Feelings	Fear, curiosity, and satisfaction	"I also have this thought to see whether there is negative energy or not." "Afraid of what will happen or that something will go wrong" "Once I'm done with the tour it provides a sense of satisfaction that I'll be able to complete that without any mishappening."
Impact	Positive	"I always feel satisfied on the completion of tour. After coming there I get peace"
Triggers	YouTube videos, stories from people	"A lot of videos on the experiences of the people who have visited the place are there." "I've heard a lot about the stories and mishappenings of the place from people."
Motive	It was for excitement and exploration	"I wanted to see if all this is there or not" "The thought of visiting there gave me thrill and excitement."

Tourist-5

Gender: Female Age: 24 years

Tourist visited Jallianwala Bagh, Putual Bari, and South Park Cemetery few years back. She visited these places as a result of curiosity that got entangled from stories by people, magazines, and adding to which visiting the historic places is close to her heart. Her journey of the tours has been full of curiosity, heaviness, and emotionally draining. Revisiting the memories of these tours still hampers her.

Themes	Sub-themes	Narratives
Feelings	Heaviness, fear. Tearful, emotionally draining	"All this affected me a lot emotionally." "I got reconnected while listening to all incidents that made me teary." "I don't want to encounter witch."
Impact	Negative	"I felt heaviness in those places, even after coming home."

(Continued)

Themes	Sub-themes	Narratives
Triggers	Stories on magazines, cartoon shows, news articles, and heard from people	"I have read many magazines and articles." "I've also seen the place in the cartoons." "I have heard from many people about that place."
Motive	It was to experience the feelings	"I wanted to experience the feeling at that place, as I have heard from so many people. I wanted to know that something like this happens/exists there or not."

Tourist-6

Gender: Male Age: 27 years

Tourist visited Qutub Shahi Tomb last year. He was excited for the visit but after the tour he felt worried and some physical issues. He got there just for spending some good time with colleague but ended up getting worried even on hearing about the place.

Themes	Sub-themes	Narratives
Feelings	Excitement, worried, and fear	"When I went to the place I was excited." "The thought of the place makes me worried." "Whenever I heard of this place, I gets scared."
Impact	Negative	"I got headaches and recurrent thoughts."
Triggers	Chill	"The place was close to my home, so I got there to chill with my colleague."
Motive	It was for fun	"I just went there for fun and to chill."

Tourist-7

Gender: Male Age: 23 years

Tourist visited Kuldhara few years back. He visited there with his family just for spending some leisure time with them and got fascinated by the stories of the place. He felt scared and worried during the tour and even after the tour at times the thought of the place gave him different vibes. This tour gave him negative vibes and he experienced a shift in his feelings from excitement to emptiness.

Themes	Sub-themes	Narratives
Feelings	Feared, weird, empty, and alone	"I felt emptiness and feeling of being alone after listening to the stories and seeing that place." "I felt scared being at that place." "The place was bit different."
Impact	Negative	"Sometimes when I think about the visit, it gives me chilly vibes."
Triggers	Stories from people	"We had heard many stories from the cab driver on the way, about what the king had done, and about the curse."
Motive	It was just to enjoy with family and also to explore about the place.	"I went there with my family so that we could roam around and enjoy." "We went there to figure out and explore the place."

Tourist-8

Gender: Female Age: 21 Years

Tourist visited Kuldhara few years back. She was excited about the place on hearing the stories from people and she also poses an interest in exploring historical places. She do felt nervous and fearful while exploring the place as a result of the stories she had heard but at the end she didn't find it much haunted.

Themes	Sub-themes	Narratives
Feelings	Excitement. Nervousness and fearful	"We were excited to go there because we had heard about that place." "There was nervousness because the place was haunted."
Impact	Neutral	"We didn't felt much."
Triggers	Stories by people	"Hotel manager and staff have told us about the story of the place."
Motive	It was to explore the place.	"We wanted to see what kind of place it is."

Tourist-9

Gender: Female Age: 22 years

Tourist visited Cellular Jail this year. She visited there with family to have fun. She was curious about the history and wanted to explore and know the place. After the completion of the tour she felt proud and patriotic about the history and the place.

Themes	Sub-themes	Narratives
Feelings	Curious, proud, patriotism, happy	"It made me proud while going through the place." "It instills the thought within me like let's do something for the country." "I felt happy on visiting there." "I was curious to know the place."
Impact	Positive	"I liked the place and would like to visit the place again."
Triggers	Heard from people, read in history	"I've heard a lot about it from the people." "Have also read about this in the history."
Motive	It was to explore the place and see what's there	"I was curious to get into the place and explore it."

Tourist-10

Gender: Male Age: 25 years

Tourist visited Shaniwar Wada this year. He went there to explore as he has heard a lot about the place but didn't felt that it was haunted. According to him it was hyped a lot by the people as there was nothing haunted and specific.

Themes	Sub-themes	Narratives
Feelings	Excited, disappointed	"It was heard a lot that the place is haunted, so we visited there with a lot of excitement." "I felt quite disappointed because there was nothing like that, there was just a garden."
Impact	Neutral	"Nothing was there on the place."
Triggers	Stories by people	"I had heard a lot from people that the voices of people are there which is unnatural."
Motive	It was to explore	"I have heard a lot about that place that it's haunted. So that is why we went there to explore the destination."

Major themes, Sub-themes, and Narratives

Theme: Feelings

Sub-theme	Narratives
Excited (happy/curiosity)	"I ran to go there because I had seen a lot of documentaries." *(Tourist 1)* "I like to visit to the places like this" *(Tourist 2)* "I was excited and worried that something might happen to someone." *(Tourist 3)* "When I went to the place I was excited." *(Tourist 6)* "We were excited to go there because we had heard about that place." *(Tourist 8)* "It was heard a lot that the place is haunted, so we visited there with a lot of excitement." *(Tourist 10)* "I was very happy about the tour as I'll spend good time with friends." *(Tourist 1)* "I felt happy on visiting there." *(Tourist 9)* "I also have this thought to see whether there is negative energy or not." *(Tourist 4)* "I was curious to know the place." *(Tourist 9)*
Vulnerable (weird/unsafe/ worried/disappointed)	"I felt weird there." *(Tourist 1)* "The place was bit different." *(Tourist 7)* "On hearing the sound of the male, we'll felt the place to be unsafe." *(Tourist 3)* "The thought of the place makes me worried." *(Tourist 6)* "I felt quite disappointed because there was nothing like that, there was just a garden." *(Tourist 10)*
Fear (frightened/nervousness)	"All of us got scared while we encountered with a lot of bats there." *(Tourist 1)* "I was quite scared." *(Tourist 3)* "Afraid of what will happen or that something will go wrong" *(Tourist 4)* "I don't want to encounter witch." *(Tourist 5)* "Whenever I heard of this place, I gets scared." *(Tourist 6)* "I felt scared being at that place." *(Tourist 7)*

(Continued)

Sub-theme	Narratives
	"We felt scary on entering the place." *(Tourist 8)* "We all got frightened when the voice intensified and the number of voices increased and we started to ran away." *(Tourist 3)* "There was nervousness because the place was haunted." *(Tourist 8)*
Sadness (tearful/emptiness/ emotionally draining/heaviness/ alone)	"So many people used to come to that place every day and even after that there was a lot of emptiness at that place." *(Tourist 2)* "I felt emptiness and feeling of being alone after listening to the stories and seeing that place." *(Tourist 7)* "I got reconnected while listening to all incidents that made me teary." *(Tourist 5)* "All this affected me a lot emotionally." *(Tourist 5)* "Feeling was not great when I heard stories from guide." *(Tourist 5)*
Satisfaction (proud/patriotic)	"Once I'm done with the tour it provides a sense of satisfaction that I'll be able to complete that without any mishappening." *(Tourist 4)* "It made me proud while going through the place." *(Tourist 9)* "It instills the thought within me like let's do something for the country." *(Tourist 9)*

Theme: Impact

Sub Theme	Narratives
Positive	"After doing the tour, I felt a feeling of achievement that I had completed the tour." *(Tourist 2)* "I always feel satisfied on the completion of tour. After coming there I get peace" *(Tourist 4)* "I liked the place and would like to visit the place again." *(Tourist 9)*

(Continued)

(Continued)

Sub Theme	Narratives
Negative	"I started feeling scared." *(Tourist 1)* "We were scared." *(Tourist 3)* "I felt heaviness in those places, even after coming home." *(Tourist 5)* "I got headaches and recurrent thoughts." *(Tourist 6)* "Sometimes when I think about the visit, it gives me chilly vibes." *(Tourist 7)*
Neutral	"We didn't felt much." *(Tourist 8)* "Nothing was there on the place." *(Tourist 10)*

Theme: Triggers

Sub-theme	Narratives
Fun (friends/chill/travel)	"I went with my friends." *(Tourist 1)* "We all went there stubbornly so that we could enjoy." *(Tourist 3)* "The place was close to my home, so I got there to chill with my colleague." *(Tourist 6)* "I went to travel, so when we came across this place on the way, we visited that too." *(Tourist 2)*
Source of information (documentaries/history/cases/ YouTube videos)	"I have watched a lot of documentaries." *(Tourist 1)* "Being a student of history these places fascinates me." *(Tourist 1)* "Have also read about this in the history." *(Tourist 9)* "I've studied and gone through a lot of cases about various haunted places." *(Tourist 2)* "A lot of videos on the experiences of the people who have visited the place are there." *(Tourist 4)*
Stories (by people/news articles/ magazines/cartoons)	"I've heard a lot about the stories and mishappenings of the place from people." *(Tourist 4)*

(Continued)

Sub-theme	Narratives
	"I have heard from many people about that place." *(Tourist 5)* "We had heard many stories from the cab driver on the way, about what the king had done, and about the curse." *(Tourist 7)* "Hotel manager and staff have told us about the story of the place." *(Tourist 8)* "I've heard a lot about it from the people." *(Tourist 9)* "I had heard a lot from people that the voices of people are there which is unnatural." *(Tourist 10)* "I have read many magazines and articles." *(Tourist 5)* "I've also seen the place in the cartoons." *(Tourist 5)* "I've heard a lot of stories" *(Tourist 2)*

Theme: Motives

Sub-theme	Narratives
Explore paranormal activity (experience)	"I wanted to see whether paranormal activities exists or not." *(Tourist 1)* "I had heard about an old case." *(Tourist 2)* "The aim was to explore whether these negative energies exist or not." *(Tourist 3)* "I wanted to experience the feeling at that place, as I have heard from so many people. I wanted to know that something like this happens/exists there or not." *(Tourist 5)*
Explore (visit) the place	"Wanted to explore the place as well." *(Tourist 1)* "I wanted to see if all this is there or not" *(Tourist 4)*

(Continued)

(Continued)

Sub-theme	Narratives
	"We went there to figure out and explore the place." *(Tourist 7)* "We wanted to see what kind of place it is." *(Tourist 8)* "I was curious to get into the place and explore it." *(Tourist 9)* "I have heard a lot about that place that it's haunted. So that is why we went there to explore the destination." *(Tourist 10)* "The place came on a way and we went to visit the place." *(Tourist 2)*
Fun (excitement/enjoy)	"We went there for fun and to enjoy the trip." *(Tourist 3)* "I just went there for fun and to chill." *(Tourist 6)* "I was very excited." *(Tourist 3)* "The thought of visiting there gave me thrill and excitement." *(Tourist 4)* "I went there with my family so that we could roam around and enjoy." *(Tourist 7)*

Discussions

This chapter explores the psychological attributes that contribute to tourist's footfall in dark tourism destinations. Research has been carried forward by taking up some interviews using narrative inquiry. The themes from the narratives were extracted after which sub-themes were also drawn out as it's known that dark tourism is a form of tourism that means visiting places associated with some tragedy, macabre, or negative events. This type of tour poses a lot of emotional changes and affects the tourists to an extent. After going through these things and experiencing some negative energy at times, tourists continue to explore these places. In this various areas were explored that are feelings, impact, triggers, and motives regarding dark tourism. These areas were explored and it was found that tourists experience a range of feelings.

Feelings and Narratives Associated With Dark Tourism

Feelings are an amazing, self-contained experience. As these are evaluative, subjective, and unaffected by the ideas, emotions, or images that evoke them. The destinations for dark tourism create a lot of emotional shifts, extending from joy to sorrow, susceptibility to excitement, anxiety to contentment, and more. Through the interviews, it has been found that several tourists go through the feeling of excitement ("I ran to go there because I had seen a lot of documentaries."*(Tourist 1)*; "I like to visit to the places like this". *(Tourist 2)* "I was excited and worried that something might happen to someone." *(Tourist 3)*; "When I went to the place I was excited." *(Tourist 6)*), including a number of emotions encompassing happiness ("I was very happy about the tour as I'll spend good time with friends." *(Tourist 1)*; "I felt happy on visiting there." *(Tourist 9)*), and curiosity ("I was curious to know the place." *(Tourist 9)*). The feelings are not only related to the positive side of the tour but the feeling of vulnerability also persists as some travelers experience a range of emotions like feeling weird ("I felt weird there." *(Tourist 1)*; "The place was bit different." *(Tourist 7)*), unsafe ("On hearing the sound of the male, we'll felt the place to be unsafe." *(Tourist 3)*), worried ("The thought of the place makes me worried." *(Tourist 6)*), or disappointed ("I felt quite disappointed because there was nothing like that, there was just a garden." *(Tourist 10)*) at varied levels and phase of the tour. With the gradual elevation of the tour tourists do undergo the experiences of fear ("I don't want to encounter witch." *(Tourist 5)*; "Whenever I heard of this place, I get scared." *(Tourist 6)*) ranging from frightened ("All of us got scared while we encountered with a lot of bats there." *(Tourist 1)* "I was quite scared." *(Tourist 3)*) to nervousness ("There was nervousness because the place was haunted." *(Tourist 8)*). This type of feeling persists and at times tourists also undergo a variety of sad emotions depicting emptiness ("So many people used to come to that place every day and even after that there was a lot of emptiness at that place." *(Tourist 2)*), emotional draining ("All this affected me a lot emotionally." *(Tourist 5)*), heaviness ("Feeling was not great when I heard stories from guide." *(Tourist 5)*), alone feeling ("I felt emptiness and feeling of being alone after listening to the stories and seeing that place." *(Tourist 7)*), and teary state ("I got reconnected while listening to all incidents that made me teary." *(Tourist 5)*). Contrary to this quite a few tourists also encounter the feeling of satisfaction ("Once I'm done with the tour it provides a sense of satisfaction that I'll be able to complete that without any mishappening." *(Tourist 4)*) with a sense of patriotism ("It instills the thought within me like let's do something for the country." *(Tourist 9)*) and proudest achievement ("It made me proud while going through the place." *(Tourist 9)*) depending upon the destination.

The feelings of the tourists can vary to a great extent depending upon the situation. As stated it can range from overly positive to negative. The experience of these destinations can result in unpleasant emotional feelings including anxiety, despair, disgust, rage, and more (Forsdick et al., 2019; Nawijn & Fricke, 2015). The destination doesn't only evoke changes in feelings, it do posses some impact on the tourists.

Impact and Its Narratives Associated With Dark Tourism

The destinations can possess the impact in three basic ways as per the data by the participants. It can be either positive ("After doing the tour, I felt a feeling of achievement that I had completed the tour." *(Tourist 2);* "I always feel satisfied on the completion of tour. After coming there I get peace" *(Tourist 4);* "I liked the place and would like to visit the place again." *(Tourist 9)*), negative ("I started feeling scared." *(Tourist 1);* "We were scared." *(Tourist 3);* "I felt heaviness in those places, even after coming home." *(Tourist 5);* "I got headaches and recurrent thoughts." *(Tourist 6);* "Sometimes when I think about the visit, it gives me chilly vibes." *(Tourist 7)*), or neutral ("We didn't felt much." *(Tourist 8);* "Nothing was there on the place." *(Tourist 10)*). The impact can be a result of certain changes in the experience one has undergone. A positive impact can enhance the tourist's outlook toward the destinations and can enhance their experience whereas the negative experience's intensity can moderately affect the other person's outlook and belief about the places whereas the neutral impact can make the tourists disinterested in the destinations at times (Grebenar, 2018). The impacts are associated with the triggers that contribute to instilling the motivation for the tour.

Triggers and Their Narratives Associated With Dark Tourism

The triggers for dark tourism could range to a great extent from cartoons to magazines, from heard stories to having fun with loved ones. Major triggers for the motivation of dark tourism as per the tourists are fun ("We all went there stubbornly so that we could enjoy." *(Tourist 3);* "The place was close to my home, so I got there to chill with my colleague." *(Tourist 6)*) which comprises spending good time with friends or family ("I went with my friends." *(Tourist 1)*) and by traveling ("I went to travel, so when we came across this place on the way, we visited that too." *(Tourist 2)*). One of the other sources through which tourists get triggered was the sources of information provided by articles or documentaries ("I have watched a lot of documentaries." *(Tourist 1)*), YouTube videos ("A lot of videos on the experiences of the people who have visited the place are there." *(Tourist 4)*), Study of history and case studies ("Being a student of history these places fascinates me." *(Tourist 1);* "Have also read about this in the history." *(Tourist 9)*; "I've studied and gone through a lot of cases about various haunted places." *(Tourist 2)*). Adding on to these stories by people ("I've heard a lot about the stories and mishappenings of the place from people." *(Tourist 4);* "I have heard from many people about that place." *(Tourist 5)*), news or magazines ("I have read many magazines and articles." *(Tourist 5)*), and from cartoons ("I've also seen the place in the cartoons." *(Tourist 5)*) at times triggers the tourists and people to visit or explore the various places. The triggers can motivate the tourists to initiate dark tourism.

Motivation and Its Narratives Associated With Dark Tourism

Motivation serves as the driving force behind human actions, initiating, guiding, and maintaining goal-directed behaviors. It possesses the potential to ignite a journey into dark tourism, influenced by the triggers individuals encounter. The major motives of the dark tourism tourists are entangled as an exploration of the paranormal activities ("The aim was to explore whether these negative energies exist or not." *(Tourist 3);* "I wanted to see whether paranormal activities exists or not." *(Tourist 1)*), at times either with the purpose of experiencing the feeling for the same ("I wanted to experience the feeling at that place, as I have heard from so many people. I wanted to know that something like this happens/exists there or not." *(Tourist 5)*). Narratives have also directed that the motive can be to explore the place ("Wanted to explore the place as well." *(Tourist 1);* "I wanted to see if all this is there or not" *(Tourist 4);* "We went there to figure out and explore the place." *(Tourist 7);* "We wanted to see what kind of place it is." *(Tourist 8);* "I was curious to get into the place and explore it." *(Tourist 9);* "I have heard a lot about that place that it's haunted. So that is why we went there to explore the destination." *(Tourist 10)*), and gain fun ("I just went there for fun and to chill." *(Tourist 6)*) at the destination encompassing the excitement of the tourist ("I was very excited." *(Tourist 3);* "The thought of visiting there gave me thrill and excitement." *(Tourist 4)*) and to enjoy ("I went there with my family so that we could roam around and enjoy." *(Tourist 7)*). These motives provide an insight that one could go to extents to fulfill their drives that are inclined with their triggers.

Each of these aspects collectively suggests that tourists harbor a heightened level of excitement toward dark tourism destinations. They perceive these sites as thrilling and view them as genuine destinations worthy of further exploration. This enthusiasm underscores a palpable interest in delving deeper into the unique experiences offered by dark tourism locations, reflecting a desire to engage with the history and narratives surrounding such destinations.

Conclusion

The research aimed at figuring out the narratives that contribute in assessing the feelings, impact, triggers, and motives of the tourists who have gone for the dark tourism destination. The interviews have been taken up by the people for collecting the narratives from which the themes have been extracted. The themes have suggested that there's a wide range of feelings that the individuals go through that do posses a lot of impact on the tourists during and after the tour. Narratives have also suggested that there could be some triggers associated for the tour which can contribute in instilling some motives among the tourists for the tours. The research delved into the emotional, mental, and social dimensions of dark tourism through a narrative inquiry approach. Despite methodological limitations, the study sought to capture significant patterns in tourists' responses. An extensive literature review complemented the findings, contributing to a comprehensive understanding of the dark fascination experienced by tourists visiting different dark sites.

Implications/Suggestions

The study can be used at various places.

- At the institutional platforms for spreading information.
- Students can take benefit of the research to get an insight into the topic.

- Generic population can get to know about the association between two vibrant fields.
- It can provide a basis for further exploration in this area.

Limitations

- The study has identifiable limitations that could benefit from further exploration in subsequent research endeavors.
- There is potential to enhance the study by expanding the sample size, allowing for a more comprehensive analysis.
- Future research efforts could be advanced by incorporating additional variables, thereby enriching the depth and breadth of the investigation.
- Furthermore, there is an opportunity to elevate the research by introducing quantitative methods, providing a more structured and measurable approach to the inquiry.

References

Benedetto, M. (2018). *What is dark tourism?* University of Applied Sciences and Arts.

Çakar, K. (2020). Investigation of the motivations and experiences of tourists visiting the Gallipoli Peninsula as a dark tourism destination. *European Journal of Tourism Research*, *24*, 2405. https://doi.org/10.54055/ejtr.v24i.407

Chang, T.-Y. (2014). Dark tourism. The effects of motivation and environmental attitudes on the benefits of experience. *Revista Internacional de Sociología*, *72*(2), 69–86. https://doi.org/10.3989/ris.2013.08.06

Chang, L. H. (2017). Tourists' perception of dark tourism and its impact on their emotional experience and geopolitical knowledge: A comparative study of local and non-local tourist. *Journal of Tourism Research & Hospitality*, *6*(3). https://doi.org/10.4172/2324-8807.1000169

Dandotiya, R., & Aggarwal, A. (2022). Study on various attributes that affects visit to a dark tourist destination. *ECS Transactions*, *107*(1), 3749–3756. https://doi.org/10.1149/10701.3749ecst

Fes, N. (2022, April 7). *World's top dark tourism destinations: TR*. World's Top Dark Tourism Destinations | TR. https://www.tourism-review.com/best-dark-tourism-destinations-in-the-world-news10862

Forsdick, C., Kinsley, Z., & Wa lchester, K. (2019). *Keywords for travel writing studies: A critical glossary*. Anthem Press. ISBN 1783089245.

Friedman, Z. (2023, February 3). *Dark tourism ethics and criticisms*. Where the Road Forks. https://wheretheroadforks.com/dark-tourism-ethics-and-criticisms/

Gaya, E. (2013). *Understanding motivation of visitors at dark tourism sites: A case study of August 7th Memorial Park, Kenya (thesis)*. https://core.ac.uk/download/pdf/38095097.pdf

Grebenar, A. (2018). *The commodification of dark tourism: Conceptualising the visitor experience (thesis)*. https://core.ac.uk/download/pdf/159754327.pdf

Gupta, A. (2023). *Into the shadows: Dark tourism explained and 7 Indian must-visits*. Outlook Traveller. https://www.outlooktraveller.com/experiences/places-of-interest/into-the-shadows-dark-tourism-explained-and-7-indian-must-visits

Hohenhaus, S. (2019). *What is "dark tourism"?* What is dark tourism - Dark Tourism - the guide to dark travel destinations around the world. https://www.dark-tourism.com/index.php/18-main-menus/mainmenussubpages/599-what-is-dark-tourism

Lewis, H., Schrier, T., & Xu, S. (2021). Dark tourism: Motivations and visit intentions of tourists. *International Hospitality Review*, *36*(1), 107–123. https://doi.org/10.1108/ihr-01-2021-0004

Magano, J., Fraiz-Brea, J. A., & Leite, Â. (2022). Dark tourists: Profile, practices, motivations and wellbeing. *International Journal of Environmental Research and Public Health*, *19*(19), 12100. https://doi.org/10.3390/ijerph191912100

Merriam-Webster. (n.d.). *Darkness definition & meaning*. Merriam-Webster. https://www.merriam-webster.com/dictionary/darkness

Molderez, I., & Perera, K. (2022). Inclusive tourism, a cultural paradigm shift in approaching tourism development. *Contributions to Management Science*, 279–295. https://doi.org/10.1007/978-3-031-14121-8_15

Nawijn, J., & Fricke, M.-C. (2015). Visitor emotions and behavioral intentions: The case of concentration camp memorial Neuengamme. *International Journal of Tourism Research*, *17*(3), 221–228. https://doi.org/10.1002/jtr.1977

O'Brien, G. (2017, May 15). Spike Island: The Irish "Alcatraz" and the growth of dark tourism. *The Conversation*. https://theconversation.com/spike-island-the-irish-alcatraz-and-the-growth-of-dark-tourism-85289

Reid, R. (2021, May 3). *Is "dark tourism" OK? Travel*. https://www.nationalgeographic.com/travel/article/is-dark-tourism-ok-chernobyl-pripyat-disaster-sites

Sampson, H. (2019, November 13). Dark tourism, explained: Why visitors flock to sites of tragedy. *Washington Post*. https://www.washingtonpost.com/graphics/2019/travel/dark-tourism-explainer/

Seaton, A. V. (1996). Guided by the dark: From thanatopsis to thanatourism. *International Journal of Heritage Studies*, *2*(4), 234–244. https://doi.org/10.1080/13527259608722178

Soni, G., & Hussain, S. (2018). Dark tourism: Changing perception of travelers for Indian tourism. *OmniScience: A Multi-disciplinary Journal*, *8*(3), 1–11.

Sumanth, M. K., & Narendar, K. V. S. (2022). Dark tourism: The attractive dark side of tourism. *International Journal of Creative Research Thoughts*, *10*(4), 266–272.

UNWTO. (n.d.). *World tourism organization*. UNWTO. https://www.unwto.org/glossary-tourism-terms

Printed and bound by CPI Group (UK) Ltd, Croydon, CR0 4YY

13/09/2026

14961027-0003